Advancing Maths for AQA
CORE MATHS 3

Sam Boardman, Tony Clough and David Evans

Series editors
**Sam Boardman   Ted Graham
David Pearson   Roger Williamson**

TOWN CENT
DEANE ROA
BOLTON
BL3 5BG

D0184937

11 / 22

Core Maths 3

heinemann.co.uk
✓ Free online support
✓ Useful weblinks
✓ 24 hour online ordering

01865 888058

Heinemann
Inspiring generations

11/22

Advancing Maths for AQA
CORE MATHS 4

Sam Boardman and Tony Clough

Series editors
Sam Boardman   Ted Graham
David Pearson   Roger Williamson

Core Maths 4

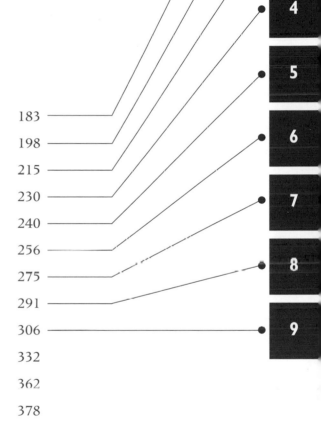

heinemann.co.uk
✓ Free online support
✓ Useful weblinks
✓ 24 hour online ordering

01865 888058

Heinemann
Inspiring generations

Heinemann is an imprint of Pearson Education Limited,
a company incorporated in England and Wales, having
its registered office at Edinburgh Gate, Harlow, Essex, CM20 2JE.
Registered company number: 872828

Heinemann is a registered trademark of
Pearson Education Limited

© Sam Boardman, Tony Clough and David Evans 2005
Complete work © Pearson Educational Limited 2005

First published 2005

10
10 9 8

British Library Cataloguing in Publication Data is available
from the British Library on request.

ISBN: 978 0 435513 31 3

Copyright notice
All rights reserved. No part of this publication may be reproduced in any form
or by any means (including photocopying or storing it in any medium by
electronic means and whether or not transiently or incidentally to some other
use of this publication) without the written permission of the copyright owner,
except in accordance with the provisions of the Copyright, Designs and Patents
Act 1988 or under the terms of a licence issued by the Copyright Licensing
Agency, 90 Tottenham Court Road, London W1T 4LP. Applications for the
copyright owner's written permission should be addressed to the publisher.

Edited by Alex Sharpe, Standard Eight Limited
Typeset and illustrated by Tech-Set Limited, Gateshead, Tyne & Wear.
Original illustrations © Pearson Education Limited, 2004
Cover design by Miller, Craig and Cocking Ltd
Printed and bound in China (CTPS/08)

The answers have been provided by the authors and are not the responsibility
of the examining board.

Every effort has been made to contact copyright holders of material reproduced
in this book. Any omissions will be rectified in subsequent printings if notice is
given to the publishers.

About this book

This book is one in a series of textbooks designed to provide you with exceptional preparation for AQA's 2004 Mathematics Specification. The series authors are all senior members of the examining team and have prepared the textbooks specifically to support you in studying this course.

Finding your way around

The following are there to help you find your way around when you are studying and revising:

- **edge marks** (shown on the front page) – these help you to get to the right chapter quickly;
- **contents list** – this identifies the individual sections dealing with key syllabus concepts so that you can go straight to the areas that you are looking for;
- **index** – a number in bold type indicates where to find the main entry for that topic.

Key points

Key points are not only summarised at the end of each chapter but are also boxed and highlighted within the text like this:

> When a mapping is *one-one* or *many-one* it is called a function. It is usually represented by a single letter such as f, g, or h, etc.

Exercises and exam questions

Worked examples and carefully graded questions familiarise you with the syllabus and bring you up to exam standard. Each book contains:

- Worked examples and Worked exam questions to show you how to tackle typical questions; Examiner's tips will also provide guidance;
- Graded exercises, gradually increasing in difficulty up to exam-level questions, which are marked by an [A];
- Test-yourself sections for each chapter so that you can check your understanding of the key aspects of that chapter and identify any sections that you should review;
- Answers to the questions are included at the end of the book.

Contents C3

4 The number e and calculus

5 Further differentiation and the chain rule

6 Differentiation using the product rule and the quotient rule

7 Numerical solution of equations and iterative methods

Contents C4

CHAPTER 1
C3: Functions

Learning objectives

After studying this chapter you should:
- be familiar with the terms one-one and many-one mappings
- understand the terms domain and range for a mapping
- understand the term function
- be able to find the range of a function
- be able to form composite functions
- understand the condition for an inverse function to exist.

1.1 Notation

In the first chapter of C1 you were introduced to function notation.

It is rather like having a machine into which numbers are fed, and for each value input, the machine determines the output value.

The function f which squares the number input and then adds 3 to the result can be represented by

$$f(x) = x^2 + 3$$

so that $f(1) = 1 + 3 = 4$ and $f(-4) = 16 + 3 = 19$.

The expression $f(x)$ is sometimes called the **image of** x.

The letter f is frequently used to represent a function, since it is the first letter of the word 'function', but it is quite in order to use any other letter instead. However, when you have two different functions, it is usual to call the first one f and the second one g, and so on.

> An alternative notation for f, often used in university texts, is $f : x \mapsto x^2 + 3$.
> However this notation will **not** be used in the AQA examinations.

Worked example 1.1

The functions f and g are defined for all real values of x and are such that $f(x) = x^2 - 4$ and $g(x) = 4x + 1$.

(a) Find $f(-3)$ and $g(0.3)$.

(b) Find the two values of x for which $f(x) = g(x)$.

Solution

(a) $f(-3) = (-3)^2 - 4 = 9 - 4 = 5$

$g(0.3) = (4 \times 0.3) + 1 = 1.2 + 1 = 2.2$

(b) Since $f(x) = g(x)$ you can write $x^2 - 4 = 4x + 1$.

Therefore $x^2 - 4x - 5 = 0$.

Factorising gives $(x - 5)(x + 1) = 0$, so that $x = 5$ or $x = -1$.

Hence the two values of x for which $f(x) = g(x)$ are $x = 5$ and $x = -1$.

EXERCISE 1A

1 The function f is defined for all real values of x by $f(x) = 2x^3$. Find the values of:

   (a) $f(-1)$,   (b) $f(3)$.

2 The function g is defined for all real values of x by $g(x) = (2x)^3$. Find the values of:

   (a) $g(-1)$,   (b) $g(3)$.

3 Given that $s(x) = 3 + 2\sin x$, find:

   (a) $s(0°)$,   (b) $s(90°)$,   (c) $s(30°)$,   (d) $s(270°)$.

4 Given that $t(x) = 4 - \tan x$, find:

   (a) $t(0°)$,   (b) $t(45°)$,   (c) $t(180°)$,   (d) $t(135°)$.

5 The functions f and g are defined for all real values of x and are such that $f(x) = 3x - 5$ and $g(x) = 4x + 1$.

   (a) Find $f(-1)$ and $g(2)$.

   (b) Find the value of x for which $f(x) = g(x)$.

6 The functions f and g are defined for all real values of x and are such that $f(x) = 2x^2 - 1$ and $g(x) = 5x + 2$.

   (a) Find $f(-3)$ and $g(-5)$.

   (b) Find the two values of x for which $f(x) = g(x)$.

7 Given that $f(x) = x^3 - 3x^2 + 2x - 1$, evaluate:

   (a) $f(0)$,   (b) $f(1)$,   (c) $f(2)$,

   (d) $f(-1)$,   (e) $f(-2)$.

8 Given that $g(x) = (x + 3)^{-3}$, evaluate:

   (a) $g(0)$,   (b) $g(1)$,   (c) $g(2)$,

   (d) $g(-1)$,   (e) $g(-2)$,   (f) $g(-4)$,

   (g) $g(-2.99)$,   (h) $g(b)$,   (i) $g(a - 3)$.

9 Given that $f(x) = \sqrt{(2x + 5)}$, find the exact values of:

   (a) $f(0)$,   (b) $f(1)$,   (c) $f(2)$,

   (d) $f(-1)$,   (e) $f(-2)$.

1.2 Mapping diagram

Instead of finding a single value of f(*x*), imagine that each number in the set {−2, 0, 1, 3, 4} is input in turn to a function machine. The corresponding output values could be represented as a **mapping diagram** as shown in the diagram.

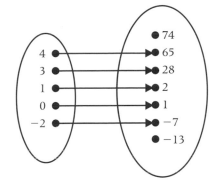

Can you recognise what the mapping is actually doing? Each number is cubed and then one is added to the result.

It doesn't matter that no elements are mapped onto **74** and −13.

Because each element of the first set is mapped to exactly one element of the second set, we say the mapping is **one-one**.

The set of input values is called the **domain**. So for this mapping the domain is the set {−2, 0, 1, 3, 4}.

If you map from the domain using arrows, the set of values where the arrows map onto is called the **range**.

Here, the range is the set {−7, 1, 2, 28, 65}.

Consider a second mapping diagram as shown.

The larger set on the right that contains the range is called the codomain so that, in this example, the codomain is the set {−13,−7, 1, 2, 28, 65, 74}. The term codomain will not be used in examination questions.

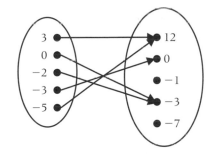

Notice that both of the numbers 3 and −5 are mapped onto 12. Also the numbers 0 and −2 are mapped onto −3.

This time the domain is the set {−5, −3, −2, 0, 3} and the range is the set {−3, 0, 12}.

In this case more than one element of the domain maps onto the same element in the range. The mapping is **many-one**.

When a mapping is *one-one* or *many-one* it is called a **function**. It is usually represented by a single letter such as f, g, or h, etc.

The set of numbers for which a function is defined is called the **domain**.

A mapping such as illustrated below is one-many and cannot represent a function.

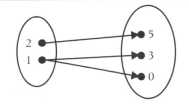

A function f consists of two things:
- a defining **rule** such as $f(x) = x^2 + 3$;
- its **domain**.

The set of values the function can take for a given domain is called the **range**.

Worked example 1.2

The function h has domain $\{-2, -1, 0, 3, 7\}$ and is defined by $h(x) = (x - 3)^2 + 2$. Find the range of h.

Solution

$h(-2) = (-2 - 3)^2 + 2 = 25 + 2 = 27$
$h(-1) = (-1 - 3)^2 + 2 = 16 + 2 = 18$
$h(0) = (0 - 3)^2 + 2 = 9 + 2 = 11$
$h(3) = (3 - 3)^2 + 2 = 0 + 2 = 2$
$h(7) = (7 - 3)^2 + 2 = 16 + 2 = 18$

The range of h is $\{2, 11, 18, 27\}$

> Because h(-1) and h(7) give the same value, we only write the value 18 once in the range.

1.3 Functions with continuous intervals as domains

It is more common for a function to have an interval of values as its domain rather than the domain consisting of just a finite set of values such as $\{-2, 0, 1, 3\}$.

Suppose the function f is defined for the domain $-2 \leqslant x \leqslant 3$ by $f(x) = 3x + 2$.

The graph of $y = 3x + 2$ is a straight line. The section of the line you are restricted to is where $-2 \leqslant x \leqslant 3$, since this is the domain of the function f.

Since $f(-2) = -6 + 2 = -4$ and $f(3) = 9 + 2 = 11$, the only part of the line to be considered is the section between the two points with coordinates $(-2, -4)$ and $(3, 11)$.

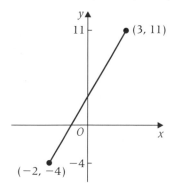

> The graph of $y = f(x)$ is shown opposite.

> It is a good idea to put 'blobs' at the end points to remind you that these are values the function can actually take.

The possible values that y can take are therefore $-4 \leqslant y \leqslant 11$. This defines the range of the function.

The range of f can be written as $-4 \leqslant y \leqslant 11$ or $-4 \leqslant f(x) \leqslant 11$.

> When the domain of f is a continuous interval, the range can be found by considering the graph of $y = f(x)$. The range consists of the possible values that y can take. The range of f is written as an inequality involving $f(x)$.

Worked example 1.3

The function g has domain $-1 \leqslant x \leqslant 2$ and is defined by $g(x) = x^2 - 1$.

(a) Sketch the graph of $y = g(x)$.

(b) Find the range of g.

Solution

(a) The quadratic graph $y = x^2 - 1$ has a minimum point at $(0, -1)$. It is useful to evaluate the function at the end points of the domain. Here $g(-1) = 1 - 1 = 0$ and $g(2) = 4 - 1 - 3$.

The section of the parabola required is sketched below and the end points are indicated by small blobs.

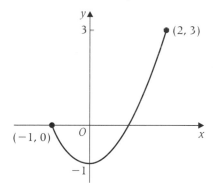

> The graph of $y = g(x)$ is shown opposite.

(b) You need to consider more than the two end points when considering the range of values that y can take. Notice that the graph comes down as low as $y = -1$.
The greatest value y can take is 3, as given by the right-hand extremity of the graph.
Hence the range of g is given by $-1 \leqslant g(x) \leqslant 3$.

EXERCISE 1B

1 The function f has domain $\{-2, -1, 0, 1, 2, 3\}$ and is defined by $f(x) = x^2 + 2$. Find the range of f.

2 The function g has domain $\{-2, -1, 0, 1, 2\}$ and is defined by $g(x) = (2x - 1)^3 - 7$. Find the range of g.

3 The function h is defined for $-1 \leqslant x \leqslant 3$ by $h(x) = 3x + 1$.
Find the range of h.

4 The function f is defined for $-2 \leqslant x \leqslant 1$ by $f(x) = x^2 + 1$.
(a) Sketch the graph of $y = f(x)$.
(b) Find the range of f.

5 The function g is defined for $-1 \leqslant x \leqslant 4$ by $g(x) = 10 - x^2$.
(a) Sketch the graph of $y = g(x)$.
(b) Find the range of g.

6 The function f is defined for all real values of x by
$f(x) = (1 - 2x)(1 + 2x)$.
(a) (i) Find the coordinates of the points where the graph
of $y = f(x)$ cuts the coordinate axes.
(ii) Sketch the graph of $y = f(x)$.
(b) State the range of f. [A]

1.4 Further examples involving domain and range

Sometimes the domain is defined in such a way as to exclude the end points of the interval. For instance, the domain may be of the form $x > 3$, or perhaps something like $-2 < x < 1$.

Worked example 1.4

The function g with domain $-1 \leqslant x < 2$ is defined by
$g(x) = x^3 - 4$.

Sketch the graph of $y = g(x)$ and find the range of g.

Solution

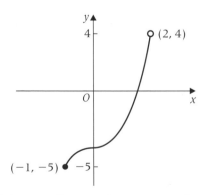

You need to find the smallest and greatest values of $y = g(x)$ from the graph.

$g(-1) = (-1)^3 - 4 = -1 - 4 = -5$
$g(2) = 2^3 - 4 = 8 - 4 = 4$ (even though 2 is not in the domain)

You can use a graphics calculator or recognise that the basic graph of $y = x^3$ has been translated through $\begin{bmatrix} 0 \\ -4 \end{bmatrix}$.

It is important to sketch the graph only for the domain indicated.

There is a difference between the graph of $y = x^3 - 4$, which exists for all real values of x, and the graph of $y = g(x)$ which is sketched here.

Since the function is not defined when $x = 2$ it is a good idea to represent this in some special way. This is usually done by drawing a small circle to remind you that the point $(2, 4)$ is not actually included in the graph.

The graph shows that y can take all values between -5 and 4. The function can take the value -5 but not the value 4. Hence, the range of g is given by

$$-5 \leqslant g(x) < 4.$$

Worked example 1.5

The function f is defined by $f(x) = 4 - x^2, \quad x > 2$. Sketch the graph of $y = f(x)$ and find the range of f.

Domains may be defined in this way in the C3 examination. Notice that the word domain is not actually mentioned.

Solution

It is likely that your first attempt to obtain a sketch will produce something like the one opposite, particularly if you are using a graphics calculator.
However, you need to restrict the set of values so that only the part of the graph for $x > 2$ is drawn.

The graph of $y = f(x)$ for $x > 2$ is shown below.

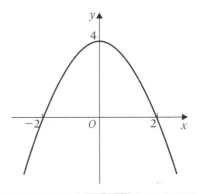

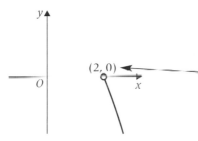

Although the value $x = 2$ is not actually in the domain you need to find $f(2) = 4 - 2^2 = 0$.

As x gets larger than 2 you can see that $f(x)$ decreases continuously from the value 0.

The range is therefore given by $f(x) < 0$.

EXERCISE 1C

1 The function g with domain $-4 < x \leqslant 1$ is defined by $g(x) = 7 - x$. Sketch the graph of $y = g(x)$ and find the range of g.

You may wish to use a graphics calculator to help you with your sketches.

2 The function f with domain $1 < x < 3$ is defined by $f(x) = 1 + x^2$. Sketch the graph of $y = f(x)$ and find the range of f.

3 The function q with domain $-1 \leqslant x < 2$ is defined by $q(x) = 1 - x^3$. Sketch the graph of $y = q(x)$ and find the range of q.

4 The function g is defined by $g(x) = 3 - 5x$, $x < 1$. Sketch the graph of $y = g(x)$ and find the range of g.

5 The function f with domain $-2 \leqslant x < 1$ is defined by $f(x) = 3 + x^5$. Sketch the graph of $y = f(x)$ and find the range of f.

6 The function h with domain $-2 < x \leqslant 3$ is defined by $h(x) = 2 - x^3$. Sketch the graph of $y = h(x)$ and find the range of h.

7 The function f with domain $x \geqslant 1$ is defined by $f(x) = 3 + \dfrac{1}{x}$. Sketch the graph of $y = f(x)$ and find the range of f.

8 The function g with domain $x > 3$ is defined by $g(x) = \dfrac{6}{x}$. Sketch the graph of $y = g(x)$ and find the range of g.

9 The function f with domain $0° \leqslant x \leqslant 180°$ is defined by $f(x) = 2 \sin x$. Sketch the graph of $y = f(x)$ and find the range of f.

10 The function h with domain $0 < x < \pi$ is defined by $h(x) = \cos\left(x + \dfrac{\pi}{3}\right)$. Sketch the graph of $y = h(x)$ and find the range of h.

1.5 Greatest possible domain

*x* 'belongs to' $\mathbb{R}$.

A function is sometimes defined for all real values of x. We say the domain is the set of real numbers, $\mathbb{R}$, or $x \in \mathbb{R}$. When this is the case, the domain is sometimes omitted and implicitly understood to be the set of real numbers.

For instance $f(x) = x^2 + 3x - 2$ (*with no mention of a domain*) implies that x can take all real values.

Sometimes restrictions on the domain are necessary.

If you try to find f(0) on your calculator you will get an error message.

For instance $f(x) = 3 + \dfrac{1}{x}$ cannot be defined for $x = 0$.

Its greatest possible domain is therefore $x \in \mathbb{R}$, $x \neq 0$.

Since we cannot find square roots of negative quantities, the function g, where $g(x) = \sqrt{x - 3}$, does not exist for $x < 3$.

The greatest possible domain for g is $x \geqslant 3$.

EXERCISE 1D

Determine the greatest possible domain for each of the following functions, f.

1 $f(x) = \dfrac{1}{(x - 1)}$

2 $f(x) = \sqrt{3 + x}$

3 $f(x) = \dfrac{1+x}{2-x}$

4 $f(x) = \dfrac{3-x}{(1+x)^2}$

5 $f(x) = \dfrac{4+x}{1+x^2}$

6 $f(x) = \dfrac{1}{\sqrt{x-2}}$

7 $f(x) = \dfrac{(x+4)}{(x-1)(x-2)}$

8 $f(x) = \dfrac{3}{x\sqrt{4-x}}$

1.6 Graphs that represent functions

Consider the two graphs opposite.

In graph (a), for each value of x you can draw a vertical line and see that it gives a unique y value. Any horizontal line for a particular value of y also corresponds to a unique value of x. Graph (a) represents a **one-one** function.

Repeating the procedure for graph (b). Any vertical line gives one value of y and so the graph represents a function. This time, however, some of the horizontal lines pass through more than one point on the curve and indicate that more than one value of x maps onto a particular value of y.

The function represented by graph (b) is **many-one**.

1.7 Graphs that do not represent functions

Contrast the graphs shown here with those in the previous section.

For certain values of x you can draw a vertical line and see that it does not correspond to a unique value of y. These graphs do **not** represent functions.

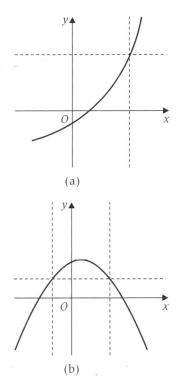

(a)

(b)

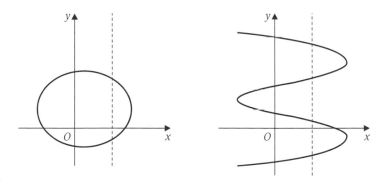

For certain values of x there is more than one value of y. The corresponding mapping diagram would be one-many and cannot represent the function.

EXERCISE 1E

For each of the following graphs, state whether it represents a function or not. For the functions, identify them as one-one or many-one.

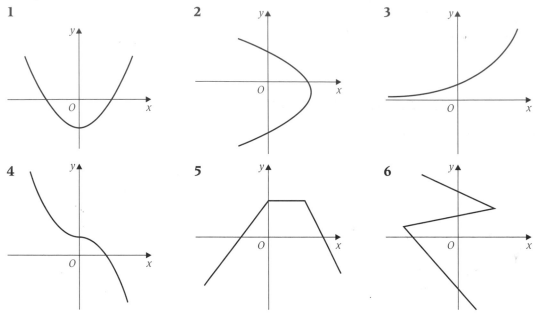

1.8 Composite functions

The term composition is used when one operation is performed after another operation. For instance:

$$x \xrightarrow{\quad} \boxed{\text{Add } 3} \xrightarrow{\quad x + 3 \quad} \boxed{\text{Multiply by 5}} \xrightarrow{\quad 5(x + 3)}$$

This function can be written as $h(x) = 5(x + 3)$.

Sometimes you have two given functions such as f and g and need to perform one function after another.

Suppose $f(x) = x^2$ and $g(x) = 2 + 3x$, $x \in \mathbb{R}$.

What is $f[g(x)]$?

$$g(x) = 2 + 3x$$
$$f[g(x)] = f(2 + 3x) = (2+3x)^2.$$

> You could try with a number
> $g(2) = 2 + 6 = 8$
> $f[g(2)] = f(8) = 8^2 = 64$

The expression $f[g(x)]$ is usually written without the extra brackets as $fg(x)$ and fg is said to be a **composite function**.

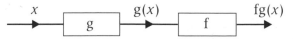

$$x \xrightarrow{\quad} \boxed{g} \xrightarrow{\quad g(x) \quad} \boxed{f} \xrightarrow{\quad fg(x)}$$

> Although we write $fg(x)$ the function g operates first on x because it is closest to x.

The function gf can be found in a similar way.

$$f(x) = x^2$$

so that $gf(x) = g[f(x)] = g(x^2) = 2 + 3(x^2) = 2 + 3x^2$

The composite function gf is such that

$$gf(x) = 2 + 3x^2$$

Worked example 1.6

The functions f and g are defined for all real values of x by
$f(x) = x^3 + 5$ and $g(x) = 3 - 2x$. The composite functions fg and
gf are such that $p = fg$ and $q = gf$. Find $p(x)$ and $q(x)$.

Solution

$g(x) = 3 - 2x$

$\Rightarrow f[g(x)] = f[3 - 2x] = (3 - 2x)^3 + 5$

The composite function $fg = p$.

Hence $p(x) = (3 - 2x)^3 + 5$.

$f(x) = x^3 + 5$
$g(f(x)) = g(x^3 + 5) = 3 - 2(x^3 + 5)$

Since $gf = q$,

$q(x) = 3 - 2(x^3 + 5)$

> Note that fg is not the same as gf
> Although we could multiply out
> the brackets, it is best to leave
> the functions in this more
> compact form.

EXERCISE 1F

Assume that the domain of the functions in this exercise is the
set of real numbers.

1 Find an expression for $fg(x)$ for each of these functions:
   (a) $f(x) = x - 1$ and $g(x) = 5 - 2x$,
   (b) $f(x) = x^2 - 3$ and $g(x) = 2 + x$,
   (c) $f(x) = x^3 + 1$ and $g(x) = 3x - 1$,
   (d) $f(x) = x^4 - 2$ and $g(x) = (x + 1)^2$.

2 For each pair of functions in question **1**, find an expression
   for $gf(x)$.

3 Given that $f(x) = x^2 - 2x + 1$ and $g(x) = 1 - 3x$, find:
   (a) $fg(x)$,
   (b) $gf(x)$, simplifying your answers.

4 Given that $f(x) = 2x - 3$, find **(a)** $ff(2)$,   **(b)** $ff(a)$.
   Solve the equation $ff(a) = a$.

5 Given that $f(x) = kx - 2$ and $g(x) = 4 - 3x$, find in terms of k
   and x:
   (a) $fg(x)$,
   (b) $gf(x)$.
   State the value of k for which $fg(x) = gf(x)$.

6 Given that $f(x) = x^2$ and $g(x) = 5 + x$, find $fg(x)$ and $gf(x)$.
   Show that there is a single value of x for which $fg(x) = gf(x)$
   and find this value of x.

7 The functions f and g are defined with their respective domains by

$$f(x) = \frac{3}{2x - 1} \quad x \in \mathbb{R}, \quad x \neq \tfrac{1}{2}$$
$$g(x) = x^2 + 1 \quad x \in \mathbb{R}.$$

(a) Find the range of g.

(b) The domain of the composite function fg is $\mathbb{R}$. Find fg(x) and state the range of fg.

1.9 Domains of composite functions

In the examples of composite functions considered so far the domains have been the real numbers, but sometimes the domains need more careful consideration.

Consider the composite function fg

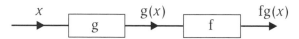

The whole of the range of g must be included in the domain of f, otherwise the domain of g needs restricting.

Worked example 1.7

Given the functions f and g such that

$$f(x) = \sqrt{x + 1}, \; x > 0, \; \text{and} \; g(x) = 5 - x, \; x \in \mathbb{R},$$

find the maximum possible domain of fg.

Solution

It is necessary to solve the inequality g$(x) > 0$ so that the range of g consists only of positive numbers.

$$5 - x > 0 \implies x < 5.$$

The maximum domain of fg is therefore $x < 5$.

> Initially it might seem that any real number can be part of the domain of fg since the domain of g is $\mathbb{R}$.
>
> However g$(7) = -2$, for example, and this is not acceptable to be fed into f (see the diagram above).

1.10 Inverse functions

The function f defined for all real values of x by f$(x) = 3x - 4$ can be thought of as a sequence of operations.

If you reverse the operations and the flow,

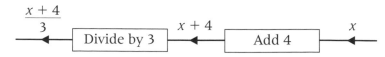

The new function, g say, can be written as $g(x) = \dfrac{x+4}{3}$.

In general $fg(x) = f\left(\dfrac{x+4}{3}\right) = 3 \times \left(\dfrac{x+4}{3}\right) - 4 = x$

Also $gf(x) = g(3x - 4) = \dfrac{(3x - 4) + 4}{3} = \dfrac{3x}{x} = x.$

A function g such that $fg(x) = x$ and $gf(x) = x$ is said to be the **inverse function** of f and is denoted by f^{-1}.

In this case, $f^{-1}(x) = \dfrac{x+4}{3}$.

> Notice that $f(3) = 5$ and $g(5) = 3$.
> Similarly $f(-1) = -7$ and
> $g(-7) = -1$, etc.

> This is purely a symbol and should not be thought of as a reciprocal.

A reverse flow diagram can be used to find an inverse function when x occurs only once in $f(x)$. You consider how $f(x)$ has been constructed as a sequence of simple operations and set up a flow diagram. Then you reverse each operation and reverse the direction of the flow to find $f^{-1}(x)$.

Worked example 1.8

Find: **(a)** $f^{-1}(x)$ and **(b)** $g^{-1}(x)$, where

$f(x) = x^3 + 5, \; x \in \mathbb{R}$ and

$g(x) = \dfrac{x+1}{x-2}, \; x \in \mathbb{R}, \; x \neq 2.$

> The reverse flow diagram method can be used to find inverse functions in simple cases when x occurs only once in $f(x)$. This worked example gives a more general method for finding inverse functions.

Solution

(a) A flow diagram approach gives

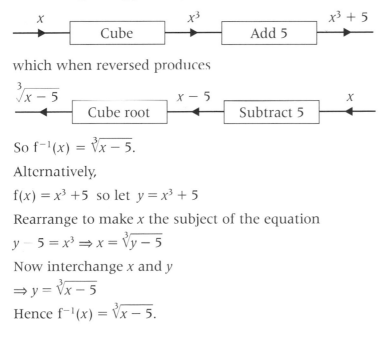

which when reversed produces

So $f^{-1}(x) = \sqrt[3]{x - 5}$.

Alternatively,

$f(x) = x^3 + 5$ so let $y = x^3 + 5$

Rearrange to make x the subject of the equation

$y - 5 = x^3 \Rightarrow x = \sqrt[3]{y - 5}$

Now interchange x and y

$\Rightarrow y = \sqrt[3]{x - 5}$

Hence $f^{-1}(x) = \sqrt[3]{x - 5}$.

> The forward flow diagram has the boxes 'cube' then 'add 5'.
>
> The reverse flow diagram would be 'subtract 5' then 'take the cube root' – giving the same answer for the inverse function.

(b) Since x occurs more than once in the expression for g(x), a flow diagram cannot be used.

$$g(x) = \frac{x+1}{x-2} \text{ so let } y = \frac{x+1}{x-2}$$

$\Rightarrow (x-2)y = x+1$ Multiply up by ($x-2$).

$\Rightarrow xy - 2y = x + 1$

$\Rightarrow xy - x = 2y + 1$ Collect all the terms involving x onto one side.

$\Rightarrow x(y-1) = 2y + 1$

$\Rightarrow x = \dfrac{2y+1}{(y-1)}$ Make x the subject of the formula.

Now interchange x and y

$\Rightarrow y = \dfrac{2x+1}{(x-1)}$ Essentially we are interchanging the domain and range to produce an inverse function.

Hence

$$g^{-1}(x) = \frac{2x+1}{(x-1)}$$ The domain for g^{-1} is $x \in \mathbb{R}$, $x \neq 1$.

You can check that the answer is correct by choosing a value of x from the domain. For instance, when $x = 3$,

$g(x) = \dfrac{x+1}{x-2}$ gives $g(3) = \dfrac{3+1}{3-2} = 4$. Using the answer for

$g^{-1}(x)$, $g^{-1}(4) = \dfrac{8+1}{4-1} = \dfrac{9}{3} = 3$, which suggests the answer

is correct.

The inverse of f can be found by the following procedure:
- Write $y = f(x)$.
- Rearrange the equation to make x the new subject.
- Interchange x and y (equivalent to reflecting in $y = x$).
- The new expression for y is equal to $f^{-1}(x)$.

1.11 Condition for an inverse function to exist

In order for f^{-1} to exist, the function f must be one-one.

You can easily draw the graph of f^{-1} when it exists.

The graph of f^{-1} is obtained from the graph of $y = f(x)$ by reflection in the line $y = x$ provided you have equal scales on the x- and y-axes.

This is because the domain and range are interchanged when you perform an inverse mapping. If a function were many-one, the inverse mapping would be one-many and, as you have seen in Section 1.7, this could not be a function.

Worked example 1.9

The function f is defined by

$$f(x) = (x - 2)^3, \ x \geqslant 0$$

and is sketched with equal scales on the axes.

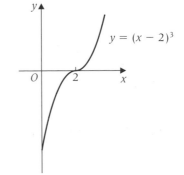

(a) Find the range of f.

(b) State why the inverse function f^{-1} exists.

(c) Find $f^{-1}(x)$ and sketch the graph of f^{-1}.

(d) State the domain and range of f^{-1}.

Solution

(a) $f(0) = -8$ and the graph shows that the values of $f(x)$ increase as x increases.

$$\Rightarrow \text{Range is } f(x) \geqslant -8.$$

(b) For each value of $f(x)$, there is a unique value of x. The function f is one-one.

(c) $f(x) = (x - 2)^3$ so let $y = (x - 2)^3$

Rearranging to make x the new subject.

$$\Rightarrow \sqrt[3]{y} = (x - 2) \Rightarrow x = 2 + \sqrt[3]{y}$$

Interchanging x and y.

$$y = 2 + \sqrt[3]{x} \Rightarrow f^{-1}(x) = 2 + \sqrt[3]{x}$$

> Graph of f^{-1} obtained by reflection of graph of f in line $y = x$.

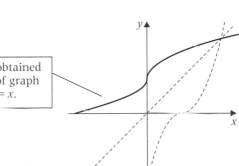

(d) The domain of f^{-1} is $x \geqslant -8$ (since the range of f is $f(x) \geqslant -8$).
The range of f^{-1} is $f^{-1}(x) \geqslant 0$ (since the domain of f is $x \geqslant 0$).

1.12 Self-inverse functions

Suppose you take the reciprocal of 5. You get $\frac{1}{5} = 0.2$. Taking the reciprocal of 0.2 gives $\frac{1}{0.2} = 5$. Doing the operation twice brings you back to the number you started with. This is true for every non-zero number you try to take the reciprocal of. 'Taking the reciprocal' is an example of a self-inverse operation. The function f defined for all non-zero values of x by $f(x) = \frac{1}{x}$ is called a **self-inverse function**.

Consider the function g given by $g(x) = 2 - x, x \in \mathbb{R}$.
If the inverse of g is g^{-1}, then $g \, g^{-1}(x) = x$ and $g^{-1} g(x) = x$.

But $gg(x) = g(2 - x) = 2 - (2 - x) = 2 - 2 + x = x$.

This proves that the inverse of g is itself g.

Hence g is a self-inverse function.

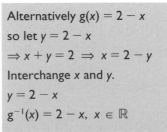

Alternatively $g(x) = 2 - x$
so let $y = 2 - x$
$\Rightarrow x + y = 2 \Rightarrow x = 2 - y$
Interchange x and y.
$y = 2 - x$
$g^{-1}(x) = 2 - x, \ x \in \mathbb{R}$

Other examples of self-inverse operations are 'divide into 6' and 'subtract from 7'. Hence, when $f(x) = \dfrac{6}{x}$, you can write $f^{-1}(x) = \dfrac{6}{x}$ also. Similarly, when $g(x) = 7 - x$, then $g^{-1}(x) = 7 - x$ as well.

You need to remember this if you are using the flow diagram method to find the inverse function when some operations are self-inverse.

Worked example 1.10

The function f is defined for all non-zero values of x by $f(x) = 3 - \dfrac{2}{x}$. Use a reverse flow diagram to find $f^{-1}(x)$.

Solution

The function f can be thought of as a sequence of operations.

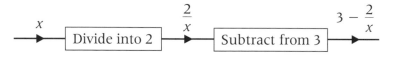

If you reverse the operations and the flow,

$$\xleftarrow{\;\frac{2}{3-x}\;} \boxed{\text{Divide into 2}} \xleftarrow{\;3-x\;} \boxed{\text{Subtract from 3}} \xleftarrow{\;x\;}$$

The inverse function f^{-1} is given by $f^{-1}(x) = \dfrac{2}{3 - x}$ and is defined for all real values of x not equal to 3.

EXERCISE 1G

1 Each of the following functions, f, has domain $\mathbb{R}$. Find $f^{-1}(x)$ by means of a reverse flow diagram.

(a) $f(x) = 5x + 7$
(b) $f(x) = (x + 2)^3$
(c) $f(x) = \dfrac{(2x + 1)}{6}$
(d) $f(x) = (2x - 1)^{\frac{1}{3}}$
(e) $f(x) = \dfrac{(-2)x + 3}{5}$
(f) $f(x) = \dfrac{7 - 3x}{4}$

2 For each of the functions in question **1**, sketch the graphs of $y = f(x)$ and $y = f^{-1}(x)$.

3 The function f has domain $x > 5$ and is defined by
$$f(x) = \dfrac{3}{4 - x}.$$
(a) Sketch the graph of $y = f(x)$.
(b) Find the range of f.
(c) The inverse of f is f^{-1}. Find $f^{-1}(x)$.

4 The function f has domain $x \geq 4$ and is defined by

$$f(x) = (x - 3)^2 + 1.$$

  (a) **(i)**   Find the value of $f(4)$ and sketch the graph of $y = f(x)$.

       (ii)   Hence find the range of f.

  (b) Explain why the equation $f(x) = 1$ has no solution.

  (c) The inverse function of f is f^{-1}. Find $f^{-1}(x)$.   [A]

5 For each of the following:

  (i)   find the range of the function,

  (ii) find the inverse function, stating its domain,

  (iii) state the range of the inverse function.

  (a) $f(x) = (3x - 1)^3, \ x \geq -1,$ **(b)** $g(x) = 1 - \dfrac{2}{x}, \ x \geq 1,$

  (c) $h(x) = (2x + 3)^5, \ x > 0,$ **(d)** $q(x) = 1 + \dfrac{5}{x}, \ x < -5,$

  (e) $r(x) = \dfrac{4}{3 - x}, \ x \geq 5.$

6 For each of the following functions f and g:

  (i)   find the range of the function,

  (ii) find the inverse function, stating its domain,

  (iii) state the range of the inverse function.

  (a) $f(x) = \dfrac{2x + 5}{x - 3}, \quad x \in \mathbb{R}, x \neq 3,$

  (b) $g(x) = \dfrac{5x - 4}{2x + 1}, \quad x \in \mathbb{R}, x \neq -\dfrac{1}{2}.$

7 **(a)** Sketch the graph of the function f, where

$$f(x) = x^2 + 3, \ x \in \mathbb{R}.$$

     Explain why f does not have an inverse function.

  (b) Sketch the graph of the function g given by

$$g(x) = x^2 + 3, \ x \geq 1.$$

     Explain why g has an inverse and find $g^{-1}(x)$.

     State the domain and range of g^{-1}.

8 The function h has domain $x < 0$ and is defined by:

$$h(x) = x^2 - 3$$

  (a) Sketch the graph of $y = h(x)$ and explain why h has an inverse.

  (b) Find $h^{-1}(x)$ and state the domain and range of h^{-1}.

9 Determine whether any of the functions f, g and h are self-inverse functions.

  (a) $f(x) = \dfrac{2x}{x - 2}, \quad x \in \mathbb{R}, x \neq 2,$

  (b) $g(x) = \dfrac{3x + 4}{2x + 1}, \quad x \in \mathbb{R}, x \neq -\dfrac{1}{2},$

  (c) $h(x) = \dfrac{3x - 5}{x - 3}, \quad x \in \mathbb{R}, x \neq 3.$

10 The function f is defined by

$$f(x) = 1 - \frac{2}{x}, x \geqslant 2.$$

(a) Sketch the graph of $y = f(x)$ and state the range of f.

(b) Explain why the inverse function f^{-1} exists and state its domain. Find an expression for $f^{-1}(x)$. [A]

11 The function f has domain $x \geqslant 2$ and is defined by

$$f(x) = \frac{2x - 3}{x}.$$

(a) Find $f(2)$ and $f(100)$.

(b) Determine the range of f.

(c) The inverse of f is f^{-1}. Find $f^{-1}(x)$.

12 The function f with domain $x \geqslant 2$ is defined by

$$f(x) = \sqrt{x - 2}.$$

(a) Describe geometrically how the graph of $y = \sqrt{x}$, $x \geqslant 0$ is transformed into the graph of $y = f(x)$.

(b) Sketch the graph of $y = f(x)$.

(c) Explain briefly why f has an inverse function, state the domain of f^{-1}, and express $f^{-1}(x)$ in terms of x.

Worked examination question

The function f has domain $0 \leqslant x \leqslant 2$ and is defined by

$$f(x) = x^3 + 1.$$

(a) Find $f(0)$ and $f(2)$.

(b) Sketch the graph of $y = f(x)$.

(c) Find the range of f.

(d) State, with a reason, whether the inverse function, f^{-1}, exists.

(e) Find $ff(x)$, giving your answer in the form $x^9 + ax^6 + bx^3 + c$.

Solution

(a) $f(0) = 0 + 1 = 1$
$f(2) = 8 + 1 = 9$

(b)

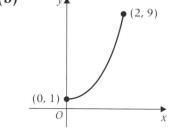

> The sketch is not intended to be an accurate plot and so you should not get too worried about the relative positions of the two endpoints of the graph.

(c) The lowest point on the curve is (0, 1). The range, therefore, is

$$1 \leqslant f(x) \leqslant 9.$$

(d) Since the function is one-one, the inverse function does exist.

(e) $ff(x) = f(x^3 + 1) = (x^3 + 1)^3 + 1 = x^9 + 3x^6 + 3x^3 + 1 + 1$

Hence, $ff(x) = x^9 + 3x^6 + 3x^3 + 2$.

MIXED EXERCISE

1 The function f has domain $-1 \leqslant x \leqslant 2$ and is defined by $f(x) = x^2 + 5$.

(a) Find $f(-1)$ and $f(2)$.

(b) Sketch the graph of $y = f(x)$.

(c) Find the range of f.

(d) State, with a reason, whether the inverse function, f^{-1}, exists.

(e) Find $ff(x)$, giving your answer in the form $x^4 + px^2 + q$. [A]

2 The functions f and g are defined with their respective domains by

$$f(x) = \frac{6}{2x - 1}, x \in \mathbb{R}, x \neq \frac{1}{2}$$

$$g(x) = x^2 + 2, x \in \mathbb{R}.$$

(a) Find the range of g.

(b) The composite function fg is defined for all real values of x. Find $fg(x)$, giving your answer in the simplest form.

(c) The inverse of f is f^{-1}. Find an expression for $f^{-1}(x)$.

(d) The graph of $y = f(x)$ and the graph of $y = f^{-1}(x)$ intersect at two points. Find the coordinates of the two points.

3 The functions f and g are defined with their respective domains by

$$f(x) = \frac{4}{3 + x}, x > 0$$

$$g(x) = 9 - 2x^2, x \in \mathbb{R}.$$

(a) Find $fg(x)$, giving your answer in its simplest form.

(b) **(i)** Sketch the graph of $y = g(x)$.
(ii) Find the range of g.

(c) **(i)** Solve the equation $g(x) = 1$.
(ii) Explain why the function g does not have an inverse.

(d) The inverse of f is f^{-1}.
(i) Find $f^{-1}(x)$.
(ii) Solve the equation $f^{-1}(x) = f(x)$. [A]

4 The function $y = f(x)$ with domain $\{x : x \geqslant 0\}$ is defined by

$$f(x) = \frac{8}{x + 2}.$$

(a) Sketch the graph of $y = f(x)$ and state the range of f.

(b) Find $f^{-1}(x)$, where f^{-1} denotes the inverse of f.

(c) Calculate the value of x for which $f(x) = f^{-1}(x)$. [A]

5 The functions f and g are defined by

$$f(x) = 3x + 4, \qquad x \in \mathbb{R}$$
$$g(x) = \frac{1}{x^2}, \qquad x \in \mathbb{R}, x \neq 0.$$

Write down, in a similar form:

(a) the composite function fg,

(b) the inverse function f^{-1}. [A]

6 (a) State which of the following graphs, G_1, G_2 or G_3, does **not** represent a function. Give a reason for your answer.

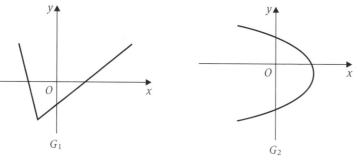

G_1 $\qquad\qquad\qquad$ G_2 $\qquad\qquad\qquad$ G_3

(b) The function f has domain $x \geqslant 2$ and is defined by

$$f(x) = \frac{1}{1 - x} + 5.$$

A sketch of $y = f(x)$ is shown opposite.

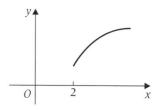

(i) Calculate $f(2)$ and $f(101)$.

(ii) Find the range of f.

(iii) The inverse of f is f^{-1}. Find $f^{-1}(x)$. [A]

7 The functions f and g are defined for all real values of x by

$$f(x) = 5 - 3x$$
$$g(x) = x^3 - 4.$$

(a) Solve the inequality $f(x) < 1$.

(b) The composite function fg is defined for all real values of x. Find $fg(x)$, expressing your answer in the form $p + qx^3$, where the values of p and q are to be found.

(c) The graph of $y = g(x)$ is sketched opposite with equal scales on the x- and y-axes.

Copy the graph of $y = g(x)$ and, on the same axes, sketch the graph of $y = g^{-1}(x)$.

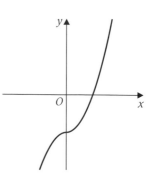

(d) Find an expression for $g^{-1}(x)$.

Key point summary

1 A function is a one-one or a many-one mapping. *p3*

2 The set of numbers for which a function is defined is called the **domain**. *p3*

3 A function f consists of two things: *p4*
- a defining rule such as $f(x) = x^2 + 3$;
- its domain.

4 The set of values the function takes for the given domain is called the **range**. *p4*

5 When the domain of f is a continuous interval, the range can be found by considering the graph of $y = f(x)$. The range consists of the possible values that y can take. The range of f is written as an inequality involving $f(x)$. *p5*

6 The composite function fg means first g then f, since: *p10*

$$fg(x) = f[g(x)].$$

7 A function f has an inverse only when f is one-one. Its graph is obtained by reflecting the graph of $y = f(x)$ in the line $y = x$. *p12*

8 A reverse flow diagram can be used to find an inverse function when x occurs only once in $f(x)$. You consider how $f(x)$ has been constructed as a sequence of simple operations and set up a flow diagram. Then you reverse each operation and reverse the direction of the flow to find $f^{-1}(x)$. *p13*

9 The inverse of f can be found by the following procedure: *p14*
- Write $y = f(x)$.
- Rearrange the equation to make x the new subject.
- Interchange x and y (equivalent to reflecting in $y = x$).
- The new expression for y is equal to $f^{-1}(x)$.

Test yourself	What to review
1 The function f is defined for all real values of x by $$f(x) = x^3 - 3x + 2. \text{ Find:}$$ **(a)** f(0), **(b)** f(1), **(c)** f(-2).	*Section 1.1*
2 Find the range of the function f where f is defined by $$f(x) = x^3 - 1, x \geqslant 3.$$	*Section 1.3*
3 State the maximum possible domain for the function g where $$g(x) = \sqrt{4 - x}.$$	*Section 1.5*
4 The functions f and g are defined by $f(x) = x^2 + 5$ and $g(x) = 7 - x$, and each has domain $\mathbb{R}$. Find an expression for $gf(x)$ in its simplest form.	*Section 1.7*
5 The function f with domain $x < 1$ is defined by $$f(x) = 2x^3 + 7.$$ **(a)** Find the range of f. **(b)** Find the inverse function, f^{-1}, and state its domain.	*Section 1.10*
6 The function g is defined for all real values of x, $x \neq 4$, by $$g(x) = \frac{2x - 3}{4 - x}.$$ Find the inverse function g^{-1} and state its domain.	*Section 1.10*

Test yourself ANSWERS

6 $f^{-1}(x) = \dfrac{4x + 3}{2 + x}$, domain is all real values of x, $x \neq -2$.

5 **(a)** $f(x) < 9$;   **(b)** $f^{-1}(x) = \sqrt[3]{\dfrac{x - 7}{2}}$, $x < 9$.

4 $gf(x) = 2 - x^2$.

3 $x \leqslant 4$.

2 $f(x) \geqslant 26$.

1 **(a)** 2;   **(b)** 0;   **(c)** 0.

CHAPTER 2

C3: Transformations of graphs and the modulus function

Learning objectives

After studying this chapter, you should be able to:
- transform simple graphs to produce other graphs
- understand the effect of composite transformations on equations of curves and describe them geometrically
- understand what is meant by a modulus function
- sketch graphs of functions involving modulus functions
- solve equations and inequalities involving modulus functions.

2.1 Review of simple transformations of graphs

Some simple transformations of graphs were introduced in chapter 5 of C2. The basic results are reviewed below. For instance, the graph of $y = x^2$ can be transformed into the graph of $y = (x - 3)^2 + 4$ by a translation of $\begin{bmatrix} 3 \\ 4 \end{bmatrix}$.

Although you may use a graphics calculator to draw graphs, it is important to see how the graph of one curve can be obtained from the graph of a simpler curve using a sequence of transformations.

> In general, a translation of $\begin{bmatrix} a \\ b \end{bmatrix}$ transforms the graph of
> $y = f(x)$ into the graph of $y = f(x - a) + b$.

You learnt how to find the equations of new curves after a reflection in one of the coordinate axes.
For example, the graph of $y = 2 + x^3$ is sketched opposite.

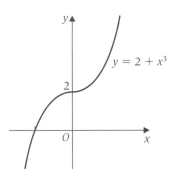

After reflection in the x-axis, the new curve will have equation $y = -2 - x^3$. The general result is given below.

> The graph of $y = f(x)$ is transformed into the graph of $y = -f(x)$ by a reflection in the line $y = 0$ (the x-axis).

When the curve $y = 2 + x^3$ is reflected in the y-axis, the new curve has equation $y = 2 - x^3$.

> The graph of $y = f(x)$ is transformed into the graph of $y = f(-x)$ by a reflection in the line $x = 0$ (the y-axis).

The graph of $y = \sin x$ is transformed into the graph of $y = 3 \sin x$ by a stretch of scale factor 3 in the y-direction.

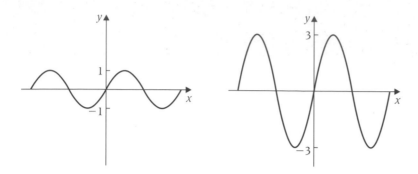

The general result is:

> The graph of $y = f(x)$ is transformed into the graph of $y = d\,f(x)$ by a stretch of scale factor d in the y-direction.

A stretch of scale factor 2 in the x-direction transforms the graph of $y = \sin x$ into the graph with equation $y = \sin\left(\dfrac{x}{2}\right)$.

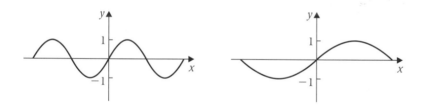

> The graph of $y = f(x)$ is transformed into the graph of $y = f\left(\dfrac{x}{c}\right)$ by a stretch of scale factor c in the x-direction.

Worked example 2.1

Find the equation of the resulting curve when the curve $y = 2 + \tan x$ is transformed by:

(a) a reflection in the y-axis,

(b) a translation of $\begin{bmatrix} \dfrac{\pi}{3} \\ 5 \end{bmatrix}$,

(c) a stretch of scale factor 0.5 in the x-direction.

You are transforming the original curve in each case. Successive transformations will be discussed in the next section.

Solution

(a) The graph of $y = f(x)$ is transformed into the graph of $y = f(-x)$ by a reflection in the y-axis.
Hence the new curve has equation $y = 2 + \tan(-x)$.
However, since $\tan(-x) = -\tan x$, the equation of the new curve can be written as $y = 2 - \tan x$.

(b) Recall that a translation of $\begin{bmatrix} a \\ b \end{bmatrix}$ transforms the graph of $y = f(x)$ into the graph of $y = f(x - a) + b$.

After translation through the vector $\begin{bmatrix} \frac{\pi}{3} \\ 5 \end{bmatrix}$, the curve

$y = 2 + \tan x$ has equation $y = 2 + \tan\left(x - \frac{\pi}{3}\right) + 5$

or $\qquad\qquad\qquad y = 7 + \tan\left(x - \frac{\pi}{3}\right)$.

(c) The graph of $y = f(x)$ is transformed into the graph of $y = f\left(\dfrac{x}{c}\right)$ by a stretch of scale factor c in the x-direction.
Hence $y = 2 + \tan x$ is transformed into $y = 2 + \tan\left(\dfrac{x}{0.5}\right)$
or $y = 2 + \tan 2x$.

Worked example 2.2

Describe geometrically how the first curve is transformed into the second curve in each of the following cases:

(a) $y = x^2$, $\quad y = (x + 3)^2 + 2$

(b) $y = \cos x$, $y = \cos 3x$

(c) $y = 2^x$, $\quad y = 2^{-x}$

Solution

(a) The first curve has been translated through the vector $\begin{bmatrix} -3 \\ 2 \end{bmatrix}$ to give the second curve.

(b) A one-way stretch in the x-direction has taken place. The scale factor is $\frac{1}{3}$.

> Notice that x has been divided by $\frac{1}{3}$ to give $3x$.

(c) The curve has been reflected in the y-axis (or $x = 0$).

EXERCISE 2A

1 Find the equation of the resulting curve after each of the following has been translated through $\begin{bmatrix} 2 \\ -1 \end{bmatrix}$.

(a) $y = x^3$ 　　　　　　　(b) $y = x^2 + 4x + 5$

(c) $y = \tan x$ 　　　　　　(d) $y = 3^x$

2

2 Find the equation of the resulting curve after each of the following has been stretched in the x-direction by scale factor $\frac{1}{2}$.

    (a) $y = x^3$ **(b)** $y = x^2 + 4x + 5$

    (c) $y = \tan x$ **(d)** $y = 3^x$

3 Describe geometrically how the curve $y = 1 + \sin x$ is transformed into the following curves:

    (a) $y = \sin x$ **(b)** $y = 1 - \sin x$

    (c) $y = 5 + \sin (x + 2)$ **(d)** $y = 4 + 4 \sin x$

    (e) $y = -1 - \sin x$ **(f)** $y = 1 + \sin 2x$

4 Describe geometrically how the curve $y = 2^x$ is transformed into the following curves:

    (a) $y = 2^{5x}$ **(b)** $y = 2^{x-3}$ **(c)** $y = 2^{\frac{x}{3}}$

    (d) $y = 2^{x+7}$ **(e)** $y = -2^x$

5 Describe a geometrical transformation which maps the graph of $y = 3^x$ onto:

    (a) $y = 3^{-x}$ **(b)** $y = 3^{x-4}$ **(c)** $y = 3^{\frac{5x}{2}}$

    (d) $y = 2 \times 3^x$ **(e)** $y = 4 + 3^x$ **(f)** $y = 9^x$

2.2 Composite transformations

You can perform each of the transformations described above in sequence so as to produce a composite transformation.

Worked example 2.3

Find the equation of the resulting curve when the following transformations are performed in sequence on the curve with equation $y = x^3$:

(a) a stretch by a factor 2 in the y-direction,

(b) a translation through $\begin{bmatrix} -1 \\ -3 \end{bmatrix}$,

(c) a reflection in the x-axis.

Solution

(a) After a stretch by a factor 2 in the y-direction, the curve $y = x^3$ becomes the curve $y = 2x^3$.

(b) Applying a translation of $\begin{bmatrix} -1 \\ -3 \end{bmatrix}$ to the curve $y = 2x^3$ gives the new curve $y = 2(x + 1)^3 - 3$.

(c) Finally the effect of a reflection in the *x*-axis is to transform
$y = f(x)$ into the curve $y = -f(x)$.

The curve $y = 2(x + 1)^3 - 3$ therefore becomes the curve
with equation $y = 3 - 2(x + 1)^3$.

It is important to perform the transformations in the given
order or you will not obtain the correct final equation.

Worked example 2.4

Describe geometrically a sequence of transformations that
transforms $y = \sin x$ into $y = 2 + 5 \sin 3x$.

Solution

The curve $y = \sin x$ is stretched by scale factor $\frac{1}{3}$ in the *x*-direction
to give the curve $y = \sin 3x$.

Next the curve $y = \sin 3x$ is stretched by scale factor 5 in the
y-direction to give $y = 5 \sin 3x$.

Finally $y = 5 \sin 3x$ is translated through $\begin{bmatrix} 0 \\ 2 \end{bmatrix}$ to give the final
curve with equation $y = 2 + 5 \sin 3x$.

> The two stretches could have
> been done in any order to map
> $y = \sin x$ onto $y = 5 \sin 3x$.

EXERCISE 2B

1 Describe a sequence of transformations that would map the
graph of $y = x^3$ onto the graph of $y = 3(x - 5)^3$.

2 Find the resulting curve when the following transformations
are applied in sequence to the curve $y = \cos x$.

  (a) a reflection in the *x*-axis,

  (b) a translation through $\begin{bmatrix} 0 \\ 2 \end{bmatrix}$,

  (c) a stretch by a scale factor 3 in the *x*-direction.

3 Find the equation of the resulting graph when the graph of
$y = 3^x$ is translated through $\begin{bmatrix} -1 \\ 5 \end{bmatrix}$ then reflected in the *y*-axis.

4 Find the equation of the new curve when the $y = \sin x$ is
transformed by the following sequence of transformations:

  (a) a translation through $\begin{bmatrix} \pi \\ 0 \end{bmatrix}$,

  (b) a reflection in the *x*-axis,

  (c) a stretch by a scale factor 4 in the *y*-direction.

5 Express $2x^2 - 12x + 19$ in the form $2(x - a)^2 + b$. Hence
describe geometrically how the graph of $y = x^2$ can be
transformed into the graph of $y = 2x^2 - 12x + 19$.

6 Describe geometrically how the first curve can be transformed into the second curve by a sequence of transformations:

   (a) $y = x^2$, $y = 4(x - 2)^2$
   (b) $y = x^2$, $y = 4 + 3(x + 1)^2$
   (c) $y = x^3$, $y = (2x - 1)^3$
   (d) $y = x^3$, $y = -(x - 3)^3$
   (e) $y = x^4$, $y = (3x + 5)^4$
   (f) $y = x^5$, $y = 4\left(\dfrac{x}{3} - 2\right)^5$

7 Describe geometrically how the curve $y = 3x^2 - 5$ can be transformed into the curve $y = x^2$ by a sequence of transformations.

8 Describe geometrically how the curve $y = 5\sin(x - 3)$ can be transformed into the curve $y = \sin(x + 1)$ by a sequence of transformations.

9 Describe geometrically how the curve $y = 3 + \cos 2x$ can be transformed into the curve $y = 5\cos x$ by a sequence of transformations.

10 Describe geometrically how the curve $y = 4^{x + 3}$ can be transformed into the curve $y = \dfrac{2^x}{5}$ by a sequence of stretches.

11 (a) Describe the geometrical transformation that transforms the graph of $y = 3x$ into the graph of $3y = x$.

   (b) Find the equations of the new graphs after each of the folowing has been reflected in the line $y = x$:
      (i) $y = 3x + 2$,   (ii) $y = x^2$,   (iii) $(x - 1)^2 + y^2 = 4$.

12 The graph of $y = f(x)$ is reflected in the x-axis and then the y-axis to produce the graph with equation $y = g(x)$.

   (a) Find $g(x)$ in terms of f and x.

   (b) Describe geometrically the single transformation that maps the graph of $y = f(x)$ onto the graph of $y = g(x)$.

13 (a) Find the new equation resulting from reflecting the curve $y = x^2$ in the line $y = 1$.

   (b) Describe a sequence of simple geometrical transformations that will map $y = x^2$ onto your answer to (a).

14 (a) Find the new equation resulting from reflecting the curve $y = 2^x$ in the line $x = 5$.

   (b) Describe a sequence of simple geometrical transformations that will map $y = 2^x$ onto your answer to (a).

15 (a) Describe a sequence of geometrical transformations that will transform the graph of $y = f(x)$ into the graph of $y = 6 - f(x)$.

(b) Describe a single geometrical transformation that will transform the graph of $y = f(x)$ into the graph of $y = 6 - f(x)$.

(c) Describe a single geometrical transformation that will transform the graph of $y = f(x)$ into the graph of $y = 2p - f(x)$.

16 (a) Describe a sequence of geometrical transformations that will transform the graph of $y = f(x)$ into the graph of $y = f(4 - x)$.

(b) Describe a single geometrical transformation that will transform the graph of $y = f(x)$ into the graph of $y = f(4 - x)$.

(c) Describe a single geometrical transformation that will transform the graph of $y = f(x)$ into the graph of $y = f(2q - x)$.

2.3 Modulus function

The modulus function finds the absolute value of a number. Any negative sign in front of a number is disregarded and a positive answer is returned. Consider the function box below.

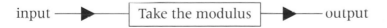

When the value 3 is input then the output is 3, whereas when -7 is input the output is 7. An input of zero gives an output of zero.

The modulus of x is written as $|x|$ and is usually read as 'mod x'.

By taking a set of values of x it is easy to see that the graph of $y = |x|$ would have the appearance of the V shape below.

Often the key on a calculator which finds the modulus is denoted by ABS since it finds the absolute value of a number.

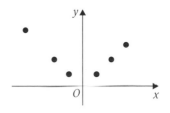

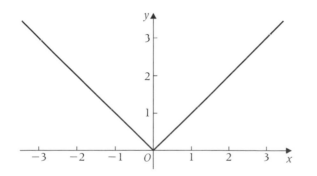

The **modulus function** is actually defined as follows:

$$|x| = \begin{cases} x & \text{when } x \geq 0 \\ -x & \text{when } x < 0 \end{cases}$$

Worked example 2.5

Sketch the graph of $y = |x - 3|$.

Solution

Method 1

When $x \geq 3$, $|x - 3| = x - 3$ since $x - 3 \geq 0$.

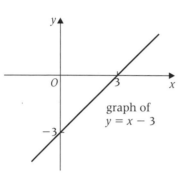

graph of
$y = x - 3$

So the graph for $x \geq 3$ is the same as the graph of $y = x - 3$.

However when $x < 3$, $x - 3$ is negative and so $|x - 3| = -(x - 3)$.

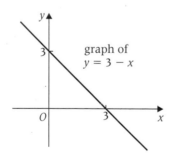

graph of
$y = 3 - x$

This means that when $x < 3$, the graph of $y = |x - 3|$ is the same as the graph of $y = -(x - 3) = 3 - x$.

The graph of $y = |x - 3|$ is therefore as drawn below.

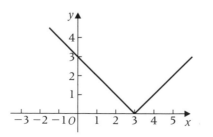

Method 2
An alternative approach is to draw the graph of $y = x - 3$ and then to reflect the section of the graph that lies below the x-axis in the x-axis.

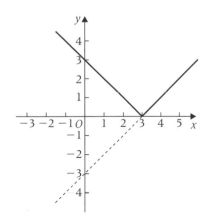

2

Worked example 2.6

Sketch the graphs of

(a) $y = |x^2 - 4|$ for $-3 \leqslant x \leqslant 3$,

(b) $y = |\sin x|$ for $0 \leqslant x \leqslant 2\pi$.

Solution

(a) Draw the graph of $y = x^2 - 4$.

Now reflect in the x-axis all the parts of the graph which lie below the x-axis.

The resulting graph is that of $y = |x^2 - 4|$.

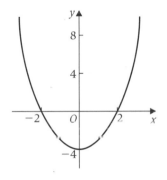

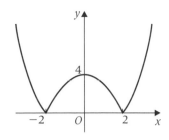

> There are now some 'sharp' corners on the graph called **cusps**. Do not be tempted to smooth these out.

(b) Draw the graph of $y = \sin x$ for $0 \leqslant x \leqslant 2\pi$.

Reflect in the x-axis the portion of the graph where y is negative so as to produce the graph of $y = |\sin x|$.

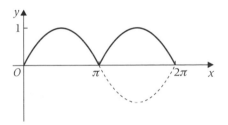

Worked example 2.7

Sketch the graphs of:

(a) $y = 3 - |x|$, **(b)** $y = |x - 1| + |x - 2|$.

Solution

(a) For $x > 0$, the graph is identical to that of $y = 3 - x$.

When $x = 0$, $y = 3$.

For $x < 0$, the graph is identical to that of $y = 3 - (-x) = 3 + x$.

Hence we can sketch the graph of $y = 3 - |x|$.

> Alternatively, the graph of $y = 3 - |x|$ can be obtained from the graph of $y = |x|$ by a reflection in the x-axis followed by a translation of $\begin{bmatrix} 0 \\ 3 \end{bmatrix}$.

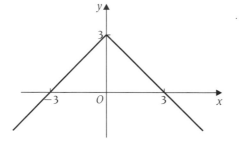

(b) It is necessary to consider three separate intervals in order to sketch $y = |x - 1| + |x - 2|$.

Firstly, when $x < 1$, $|x - 1| = -(x - 1) = -x + 1$ and also $|x - 2| = -(x - 2) = -x + 2$.
Therefore $y = |x - 1| + |x - 2| = -x + 1 + (-x + 2) = 3 - 2x$.
The graph is identical to $y = 3 - 2x$ for $x < 1$.

Next, when $1 < x < 2$, $|x - 1| = x - 1$.
But $|x - 2| = -(x - 2) = -x + 2$.
Therefore $y = |x - 1| + |x - 2| = x - 1 + (-x + 2) = 1$.
The graph is identical to $y = 1$ for $1 < x < 2$.

Finally, when $x > 2$ $|x - 1| = x - 1$ and also $|x - 2| = x - 2$.
Therefore $y = |x - 1| + |x - 2| = x - 1 + x - 2 = 2x - 3$.
The graph is identical to $y = 2x - 3$ for $x > 2$.

The graph of $y = |x - 1| + |x - 2|$ can now be sketched.

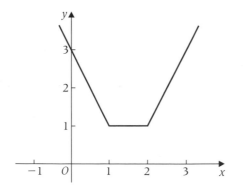

> The end points of these intervals were not included but serve as useful check points.
>
> When $x = 1$, $y = 0 + 1 = 1$.
>
> When $x = 2$, $y = 1 + 0 = 1$.

EXERCISE 2C

Sketch the graph of each of the following, showing the values of any intercepts on the axes.

1 $y = |3x|$ **2** $y = |x + 4|$ **3** $y = |3x - 5|$

4 $y = |5 - 2x|$ **5** $y = |x| - 5$ **6** $y = 4 - |x|$

7 $y = |x^2 - 1|$ **8** $y = |(x - 5)(x + 2)|$

9 $y = |x| + |x - 3|$ **10** $y = |x| - |x - 3|$

11 $y = |(x - 1)(x - 2)(x - 3)|$ **12** $y = |x^2 - 5| + 4$

13 $y = |x| + |x - 1| + |x - 2|$ **14** $y = \sin|x|, -2\pi \leqslant x \leqslant 2\pi$

15 $y = |\cos x|, -2\pi \leqslant x \leqslant 2\pi$ **16** $y = |\tan x|, 0 < x < \pi$

17 $y = |\cos 3x|, -\pi \leqslant x \leqslant \pi$ **18** $y = 1 + |\sin 2x|, -\pi \leqslant x \leqslant \pi$

2.4 Equations involving modulus functions

Often the graphical approach is best since you can see the approximate solutions and how many solutions to expect.

Worked example 2.8

Solve the equation $|3 - 2x| = x - 1$.

Solution

Method 1

Firstly, sketch the graph of $y = |3 - 2x|$. This is the V-shaped graph. The sections have gradients ± 2.

Now add the straight line $y = x - 1$, which has gradient 1. Hence there are two points of intersection.

This means the equation $|3 - 2x| = x - 1$ will have two solutions, and from the graph one of these is less than 1.5 and the other is greater than 1.5.

When $x < \dfrac{3}{2}$, $|3 - 2x| = 3 - 2x$.

One solution is given by $3 - 2x = x - 1$

$$\Rightarrow 3 + 1 = 2x + x \Rightarrow x = \frac{4}{3}.$$

Similarly, when $x > \dfrac{3}{2}$, $|3 - 2x| = -(3 - 2x)$.

The second solution is given by $-(3 - 2x) = x - 1$

$$\Rightarrow -3 + 2x = x - 1 \Rightarrow 2x - x = 3 - 1 \Rightarrow x = 2.$$

The two solutions are $x = 2, \dfrac{4}{3}$.

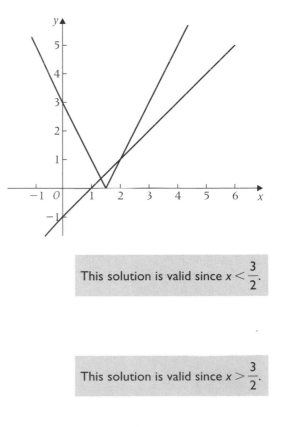

This solution is valid since $x < \dfrac{3}{2}$.

This solution is valid since $x > \dfrac{3}{2}$.

Method 2

Square both sides of the equation $|3 - 2x| = x - 1$.

$(|3 - 2x|)^2 = (x - 1)^2 \Rightarrow 9 - 12x + 4x^2 = x^2 - 2x + 1$.

$\Rightarrow 3x^2 - 10x + 8 = 0 \Rightarrow (3x - 4)(x - 2) = 0$.

The two solutions are $x = 2, \dfrac{4}{3}$.

> It is always a little dangerous when you square both sides of an equation, since you may produce spurious answers. You should therefore check that these solutions satisfy the original equation. In this case the solutions are valid.

Worked example 2.9

Solve the equation $|x - 2| = -3$.

Solution

An approach involving squaring both sides yields

$(|x - 2|)^2 = (-3)^2 \Rightarrow x^2 - 4x + 4 = 9 \Rightarrow x^2 - 4x - 5 = 0$.

$\Rightarrow (x - 5)(x + 1) = 0 \Rightarrow x = 5, x = -1$.

Checking each of these values in the original equation shows that neither of the answers are correct solutions.

When $x = 5$, $|x - 2| = |5 - 2| = 3$.

When $x = -1$, $|x - 2| = |(-1) - 2| = |-3| = 3$.

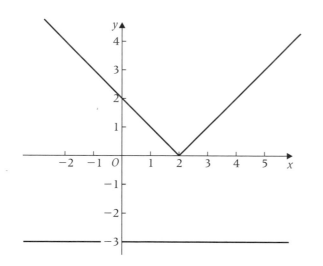

> A sketch can easily reveal when no solutions actually exist since the graphs of $y = |x - 2|$ and $y = -3$ do not intersect.

The graph of $y = |x - 2|$ is never negative and so the equation $|x - 2| = -3$ has no solutions.

Can you spot the flaw in the following argument?

For solutions to $|x - 2| = -3$, solve $x - 2 = -3 \Rightarrow x = 2 - 3 = -1$.

Also solve $-(x - 2) = -3 \Rightarrow -x + 2 = -3 \Rightarrow x = 5$.

Hence the solutions are $x = -1$ and $x = 5$.

Worked example 2.10

Solve the equation $|x^2 - 1| = 6x$.

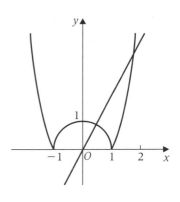

Solution

A sketch showing the graphs of $y = |x^2 - 1|$ and $y = 6x$ enables you to see that there are two solutions, one in the interval $0 < x < 1$ and the other satisfying $x > 1$.

For $0 < x < 1$, you can write $|x^2 - 1| = -(x^2 - 1) = -x^2 + 1$
Hence, solving

$$-x^2 + 1 = 6x \Rightarrow 0 = x^2 + 6x - 1 \Rightarrow x = \frac{-6 \pm \sqrt{36 + 4}}{2}$$

$$\Rightarrow \quad x = -3 \pm \frac{\sqrt{40}}{2} = -3 \pm \sqrt{10}.$$

> Using the quadratic equation formula $x = \dfrac{-b \pm \sqrt{b^2 - 4ac}}{2a}$.

But the only valid solution in the interval $0 < x < 1$ is $x = -3 + \sqrt{10} \approx 0.162\ldots$. You must reject the solution that is negative.

For $x > 1$, you can write $|x^2 - 1| = x^2 - 1$

Hence, solving $x^2 - 1 = 6x \Rightarrow 0 = x^2 - 6x - 1 \Rightarrow x = \frac{6 \pm \sqrt{36 + 4}}{2}$

$\Rightarrow x = 3 \pm \dfrac{\sqrt{40}}{2} = 3 \pm \sqrt{10}$. Once again, rejecting the negative

solution gives the solution $x = 3 + \sqrt{10} \approx 6.162\ldots$.

The two solutions are $\quad x = -3 + \sqrt{10} \approx 0.162\ldots$
$\qquad\qquad\qquad\qquad\quad x = 3 + \sqrt{10} \approx 6.162\ldots$.

EXERCISE 2D

Solve the following equations. In some cases there are no solutions.

1 $|x - 3| = 2$ **2** $|4 - x| = 5$ **3** $|x + 2| = 7$

4 $|x - 1| = x$ **5** $|2 - x| = x + 1$ **6** $|2x + 3| = x - 1$

7 $|2x - 3| = -2$ **8** $|4 - 3x| = x$ **9** $|x^2 + 2| = 3x$

10 $|2x^2 - 3x| = -1$ **11** $|4 - 3x^2| = x$

12 $|x^2 + 2x| = 3x + 2$ **13** $|x^2 + x| + 2 = 0$

14 $|4x^2 - x| = -3$ **15** $|4x^2 - 3| = x$

2.5 Inequalities involving modulus functions

It is advisable to draw a sketch and then to find the critical points when trying to solve inequalities involving modulus functions.

Worked example 2.11

Solve the inequality $|x - 2| < 2x + 3$.

Solution

The graphs $y = |x - 2|$ and $y = 2x + 3$ are drawn opposite and there is a single point of intersection when x is negative.

Since $x < 2$ for this point of intersection, you can write

$$|x - 2| = -(x - 2) = 2 - x.$$

Solving $2 - x = 2x + 3$ gives $-1 = 3x$. Hence $x = -\frac{1}{3}$.

This is the critical point and by looking at the graph, the y-value for the graph of $y = |x - 2|$ is less than the y-value on the graph of $y = 2x + 3$ whenever you are to the right-hand side of this critical point.

The solution is therefore $x > -\frac{1}{3}$.

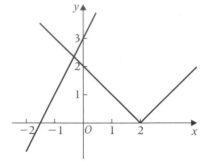

You can check this solution by taking a test value. For example when $x = 0$, $|x - 2| = 2$ and $2x + 3 = 3$. Since $2 < 3$, it gives a check on the solution.

Worked example 2.12

Solve the inequality $|x^2 - 3x| > 2$.

Solution

The graph of $y = x^2 - 3x = x(x - 3)$ is a parabola cutting the x-axis at $(0, 0)$ and $(3, 0)$. The graph of $y = |x^2 - 3x|$ is sketched opposite together with the straight line $y = 2$.

There are four points of intersection and hence four critical points.

Two of these are given by solving
$x^2 - 3x = 2 \Rightarrow x^2 - 3x - 2 = 0$.
This equation must be solved using the formula
$x = \dfrac{-b \pm \sqrt{b^2 - 4ac}}{2a}$. Hence $x = \dfrac{3 \pm \sqrt{9 + 8}}{2}$.
The approximate values of x are $3.56 \ldots$ and $-0.56 \ldots$.

The other two critical points are given by solving
$-(x^2 - 3x) = 2 \Rightarrow x^2 - 3x + 2 = 0 \Rightarrow (x - 1)(x - 2) = 0$.
Therefore $x = 1, x = 2$.

By considering the graph, the intervals satisfying $|x^2 - 3x| > 2$ are
$$x < \frac{3 - \sqrt{17}}{2}, \, 1 < x < 2, \, x > \frac{3 + \sqrt{17}}{2}.$$

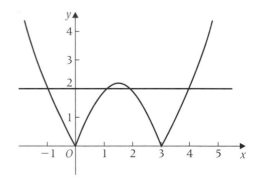

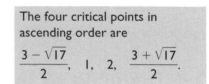

The four critical points in ascending order are
$$\frac{3 - \sqrt{17}}{2}, \quad 1, \quad 2, \quad \frac{3 + \sqrt{17}}{2}.$$

EXERCISE 2E

Solve the following inequalities:

1 $|x - 5| > 2$ **2** $|4 - x| \leqslant 1$ **3** $|x + 3| < 7$

4 $|x - 2| \leqslant x$ **5** $|3 - x| > x + 4$ **6** $|2x + 3| \leqslant x - 2$

7 $|2x - 5| \geqslant 2 - x$ **8** $|2 - 5x| < x$ **9** $|x^2 + 4| \geqslant 5x$

10 $|2x^2 - x| > 1$ **11** $|7 - 3x^2| \leqslant 2x + 1$

12 $|x^2 + 2x| \geqslant 7x + 6$

Worked examination question

The function f is defined for all real values of x by

$$f(x) = |x| + |x - 3|.$$

(a) For values of x such that $x < 0$, show that $f(x) = 3 - 2x$.

(b) Write down expressions for $f(x)$ in a form not involving modulus signs for each of the intervals:

    (i) $x > 3$;   **(ii)** $0 \leqslant x \leqslant 3$.

(c) Sketch the graph of f and write down the equation of its line of symmetry.

(d) State the range of f.

(e) Solve the equation $f(x) = 4$.

(f) Explain whether it is possible to find an inverse of the function f.   [A]

Solution

(a) When $x < 0$, $|x| = -x$ and $|x - 3| = -(x - 3) = 3 - x$.
Hence $f(x) = -x + 3 - x = 3 - 2x$.

(b) **(i)**   When $x > 3$, $|x| = x$ and $|x - 3| = x - 3$.
           Hence $f(x) = x + x - 3 = 2x - 3$.

        (ii)   When $0 \leqslant x \leqslant 3$, $|x| = x$ and $|x - 3| = -(x - 3) = 3 - x$.
           Hence $f(x) = x + 3 - x = 3$.

(c) The graph consists of the three sections defined above and is sketched opposite.

The line of symmetry has equation $x = \dfrac{3}{2}$.

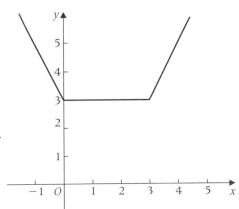

(d) Since the function has least value equal to 3, the range is given by $f(x) \geqslant 3$.

(e) From the graph, the equation $f(x) = 4$ has two solutions.
One of these is when $x < 0 \Rightarrow 3 - 2x = 4 \Rightarrow x = -\frac{1}{2}$.
The other is when $x > 3 \Rightarrow 2x - 3 = 4 \Rightarrow x = 3\frac{1}{2}$.

(f) Since the graph of f is many-one, it does NOT have an inverse.

MIXED EXERCISE

1 The function f is defined for all real values of x by

   $f(x) = |2x - 3| - 1$.

   (a) Sketch the graph of $y = f(x)$. Indicate the coordinates of the points where the graph crosses the coordinate axes.

   (b) State the range of $y = f(x)$.

   (c) Find the values of x for which $f(x) = x$. [A]

2 Sketch the graphs of $y = |2x + 3|$ and $y = |2x - 5|$ on the same axes. Hence solve the inequality $|2x + 3| \geqslant |2x - 5|$.

3 The function f is defined by $f(x) = |x - 3|$, $x \in \mathbb{R}$.

   (a) Sketch the graph of $y = f(x)$.

   (b) Solve the inequality $|x - 3| < \frac{1}{2}x$. [A]

4 The functions f and g are defined for all real values of x by

   $f(x) = x^2 - 10$ and $g(x) = |x - 2|$:

   (a) Show that $ff(x) = x^4 - 20x^2 + 90$.
   Find all the values of x for which $ff(x) = 26$.

   (b) Show that $gf(x) = |x^2 - 12|$. Sketch the graph of $y = gf(x)$. Hence or otherwise, solve the equation $gf(x) = x$. [A]

5 **(a)** Determine the two values of x for which $|2x - 3| = |5 - x|$.

   (b) The function f is defined for all real values of x. The graph of $y = |f(x)|$ is sketched opposite. Sketch two possible graphs of $y = f(x)$ on separate axes. [A]

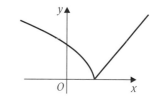

6 **(a)** Sketch the graphs of
   (i) $y = x^2 - 6x + 5$;
   (ii) $y = |x^2 - 6x + 5|$.

   (b) Calculate the four roots of the equation $|x^2 - 6x + 5| = 3$, expressing the irrational solutions in surd form.

   (c) Using this result and the sketch to **(a) (ii)**, or otherwise, solve the inequality

   $|x^2 - 6x + 5| \leqslant 3$. [A]

7 The diagram shows a sketch of the curve with equation

$$y = \sin 2x \text{ for } -\frac{\pi}{2} \leqslant x \leqslant \frac{\pi}{2}.$$

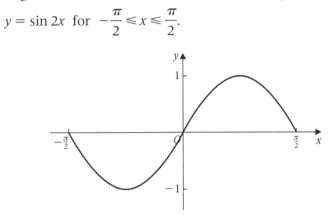

(a) Draw on the same diagram sketches of the graphs with

equations $y = |x|$ and $y = |\sin 2x|$ for $-\frac{\pi}{2} \leqslant x \leqslant \frac{\pi}{2}$.

(b) Hence state the number of times the graph of the curve
with equation $y = |\sin 2x| - |x|$ intersects the x-axis in

the interval $-\frac{\pi}{2} \leqslant x \leqslant \frac{\pi}{2}$.   [A]

8 (a) Sketch the graph of $y = |2x - 4|$. Indicate the
coordinates of the points where the graph meets the
coordinate axes.

(b) (i) The line $y = x$ intersects the graph of $y = |2x - 4|$ at
two points P and Q. Find the x-coordinates of the
points P and Q.

(ii) Hence solve the inequality $|2x - 4| > x$.

(c) The graph of $y = |2x - 4| + k$ touches the line $y = x$ at
only one point. Find the value of the constant k.   [A]

9 A function f is defined for all real values of x by

$$f(x) = 3 - |2x - 1|.$$

(a) (i) Sketch the graph of $y = f(x)$. Indicate the
coordinates of the points where the graph crosses
the coordinate axes.

(ii) Hence show that the equation $f(x) = 4$ has no real
roots.

(b) State the range of f.

(c) By finding the values of x for which $f(x) = x$, solve the
inequality $f(x) < x$.   [A]

Key point summary

1 A translation of $\begin{bmatrix} a \\ b \end{bmatrix}$ transforms the graph of $y = f(x)$ *p23*
   into the graph of $y = f(x - a) + b$.

2 The graph of $y = f(x)$ is transformed into the graph of   *p23*
   $y = -f(x)$ by a reflection in the line $y = 0$ (the x-axis).

3 The graph of $y = f(x)$ is transformed into the graph of   *p23*
   $y = f(-x)$ by a reflection in the line $x = 0$ (the y-axis).

4 The graph of $y = f(x)$ is transformed into the graph of   *p24*
   $y = d\,f(x)$ by a stretch of scale factor d in the y-direction.

5 The graph of $y = f(x)$ is transformed into the graph of   *p24*

   $y = f\left(\dfrac{x}{c}\right)$ by a stretch of scale factor c in the x-direction.

6 The **modulus function** $|x|$ is defined by

   $$|x| = \begin{cases} x & \text{when } x \geqslant 0. \\ -x & \text{when } x < 0. \end{cases}$$ *p30*

Test yourself

Test yourself	What to review
1 Describe geometrically how the curve:	*Section 2.1*
(a) $y = x^5$ is transformed into $y = (x + 1)^5 + 3$,	
(b) $y = \tan x$ is transformed into $y = \tan 4x$.	
2 Describe geometrically a sequence of transformations that transforms $y = x^2$ into $y = 3(x - 1)^2 + 4$.	*Section 2.2*
3 Describe geometrically a sequence of transformations that transforms $y = \sin x$ into $y = 3 + 2\sin 4x$.	*Section 2.2*
4 The function f is defined for all values of x by $$f(x) = |x - 5|.$$	*Section 2.3*
(a) Express $f(x)$ in a form not involving modulus signs when $x < 5$.	
(b) Sketch the graph of $y = f(x)$, indicating any values where the graph meets the axes.	
5 **(a)** Sketch the graph of $y = 4 - x^2$.	*Section 2.3*
(b) Hence sketch the graph of $y = |4 - x^2|$.	
(c) State the number of roots of the equation $|4 - x^2| = 1$.	
6 Solve the equation $|3x - 5| = 2 - x$.	*Section 2.4*
7 Solve the inequality $|2x - 7| < 3 + x$.	*Section 2.5*

Test yourself ANSWERS

1 (a) Translation through $\begin{bmatrix} -1 \\ 3 \end{bmatrix}$;

    (b) Stretch in x-direction with scale factor $\frac{1}{4}$.

2 Translation of $\begin{bmatrix} 1 \\ 0 \end{bmatrix}$, followed by stretch in y-direction with scale factor 3, followed by translation of $\begin{bmatrix} 0 \\ 4 \end{bmatrix}$.

3 Stretch in x-direction with scale factor $\frac{1}{4}$ and stretch in y-direction with scale factor 2, followed by translation by $\begin{bmatrix} 0 \\ 3 \end{bmatrix}$.

4 (a) $f(x) = 5 - x$, when $x < 5$;

    (b)

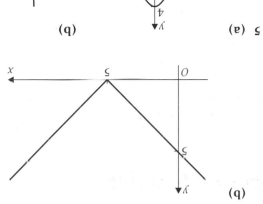

5 (a)   **(b)**

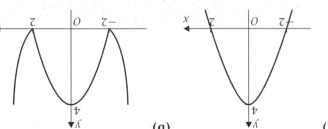

    (c) 4 roots.

6 $x = 1\frac{1}{2},\ 1\frac{3}{4}$.

7 $1\frac{1}{3} < x < 10$.

CHAPTER 3

C3: Inverse trigonometric functions and secant, cosecant and cotangent

Learning objectives

After studying this chapter, you should be able to:
- work with the inverse trigonometric functions $\sin^{-1}$, $\cos^{-1}$ and $\tan^{-1}$ and be able to draw their graphs over appropriate restricted domains
- understand the secant, cosecant and cotangent functions
- sketch the graphs of the secant, cosecant and cotangent functions
- know and use the two identities relating to the squares of the secant, cosecant and cotangent functions to prove other identities and to solve equations.

3.1 Inverse trigonometric functions

In section 1.11 you read that in order for an inverse function f^{-1} to exist, the function f must be one-one. The sine function, defined over the domain of real numbers, is clearly a many-one function as can be seen from its graph.

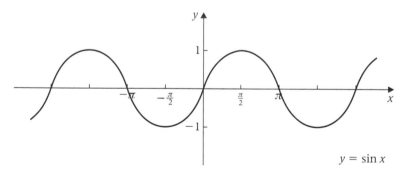

$$y = \sin x$$

So, at first sight, it would seem that the sine function has no inverse. To overcome this problem the domain of the sine function is restricted so that it becomes a one-one function but still takes all real values in the range $-1 \leq \sin x \leq 1$. You can do this by restricting the domain to $-\dfrac{\pi}{2} \leq x \leq \dfrac{\pi}{2}$, where x is in radians.

The inverse function of $\sin x$, $-\dfrac{\pi}{2} \leq x \leq \dfrac{\pi}{2}$, is written as $\sin^{-1} x$.

From the general definition of an inverse function you can deduce that $y = \sin^{-1} x \Leftrightarrow \sin y = x$ and $-\dfrac{\pi}{2} \leq y \leq \dfrac{\pi}{2}$.

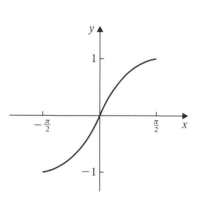

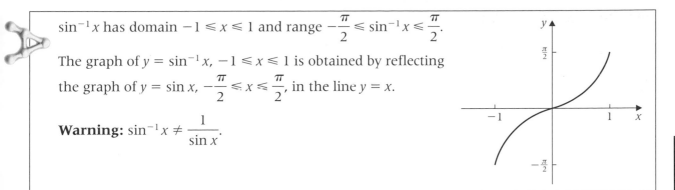

$\sin^{-1} x$ has domain $-1 \leqslant x \leqslant 1$ and range $-\dfrac{\pi}{2} \leqslant \sin^{-1} x \leqslant \dfrac{\pi}{2}$.

The graph of $y = \sin^{-1} x$, $-1 \leqslant x \leqslant 1$ is obtained by reflecting the graph of $y = \sin x$, $-\dfrac{\pi}{2} \leqslant x \leqslant \dfrac{\pi}{2}$, in the line $y = x$.

Warning: $\sin^{-1} x \neq \dfrac{1}{\sin x}$.

3

Worked example 3.1

Find the exact values of:

(a) $\sin^{-1} \dfrac{1}{2}$

(b) $\cos\left(\sin^{-1} \dfrac{1}{3}\right)$

Solution

(a) Since an exact answer is required, using your calculator set in degree mode, $\sin^{-1} \dfrac{1}{2} = 30°$, and $30° = \dfrac{\pi}{180} \times 30 = \dfrac{\pi}{6}$ radians.

So $\sin^{-1} \dfrac{1}{2} = \dfrac{\pi}{6}$, since $\sin \dfrac{\pi}{6} = \dfrac{1}{2}$ and $\dfrac{\pi}{6}$ lies between $-\dfrac{\pi}{2}$ and $\dfrac{\pi}{2}$.

(b) Let $\alpha = \sin^{-1} \dfrac{1}{3} \Rightarrow \sin \alpha = \dfrac{1}{3}$.

Using the identity $\cos^2 \alpha + \sin^2 \alpha = 1$, you have $\cos^2 \alpha = \dfrac{8}{9}$.

Since α has to lie between $-\dfrac{\pi}{2}$ and $\dfrac{\pi}{2}$ you can deduce that $\cos \alpha$ is positive.

$$\Rightarrow \cos(\sin^{-1} \tfrac{1}{3}) = \cos \alpha = \sqrt{\dfrac{8}{9}} = \dfrac{2\sqrt{2}}{3}.$$

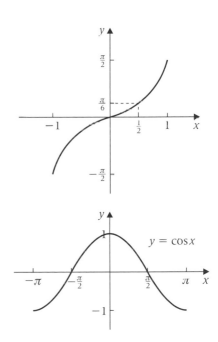

Although the following direct results are true

$\sin^{-1}(\sin x) = x$ for $-\dfrac{\pi}{2} \leqslant x \leqslant \dfrac{\pi}{2}$,

$\sin(\sin^{-1} x) = x$ for $-1 \leqslant x \leqslant 1$.

Composite functions and inverse functions are considered in chapter 1.

you must be careful when x lies outside the given inequalities. The next worked example shows you how to deal with such a case.

Worked example 3.2

Given that $\sin\dfrac{\pi}{3} = \dfrac{\sqrt{3}}{2}$, find the exact value of $\sin^{-1}\!\left(\sin\dfrac{4\pi}{3}\right)$.

Solution

Since $\dfrac{4\pi}{3}$ does not lie between $-\dfrac{\pi}{2}$ and $\dfrac{\pi}{2}$ you cannot use the

direct result. Instead you evaluate

$$\sin\frac{4\pi}{3} = \sin\left(\pi + \frac{\pi}{3}\right) = -\sin\frac{\pi}{3} = -\frac{\sqrt{3}}{2}.$$

Let $\alpha = \sin^{-1}\!\left(\sin\dfrac{4\pi}{3}\right) = \sin^{-1}\!\left(-\dfrac{\sqrt{3}}{2}\right) \Rightarrow \sin\alpha = -\dfrac{\sqrt{3}}{2}.$

The value of α which satisfies $\sin\alpha = -\dfrac{\sqrt{3}}{2}$ and lies between $-\dfrac{\pi}{2}$

and $\dfrac{\pi}{2}$ is $-\dfrac{\pi}{3}$

$\Rightarrow \sin^{-1}\!\left(\sin\dfrac{4\pi}{3}\right) = -\dfrac{\pi}{3}.$

The inverse functions of $\cos x$ and $\tan x$ are dealt with in a similar way.

The restricted domain for $\cos x$ is not the same as that used for $\sin x$.

The inverse function of $\cos x$, $0 \leqslant x \leqslant \pi$, is written as $\cos^{-1} x$.
$y = \cos^{-1} x \Leftrightarrow \cos y = x$ and $0 \leqslant y \leqslant \pi$.

> Restricting the domain of $\cos x$
> to $-\dfrac{\pi}{2} \leqslant x \leqslant \dfrac{\pi}{2}$ would not give
> the negative values in the range 0
> to -1.

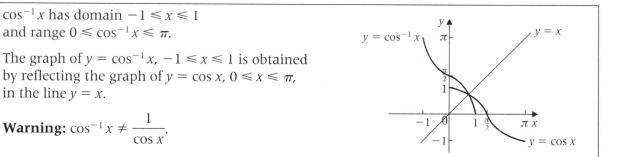

$\cos^{-1} x$ has domain $-1 \leqslant x \leqslant 1$ and range $0 \leqslant \cos^{-1} x \leqslant \pi$.

The graph of $y = \cos^{-1} x$, $-1 \leqslant x \leqslant 1$ is obtained by reflecting the graph of $y = \cos x$, $0 \leqslant x \leqslant \pi$, in the line $y = x$.

Warning: $\cos^{-1} x \neq \dfrac{1}{\cos x}$.

$\cos^{-1}(\cos x) = x$ for $0 \leqslant x \leqslant \pi$,
$\cos(\cos^{-1} x) = x$ for $-1 \leqslant x \leqslant 1$.

The tangent function can be made one-one by restricting its domain to $-\dfrac{\pi}{2} < x < \dfrac{\pi}{2}$.

The inverse function of $\tan x$, $-\dfrac{\pi}{2} < x < \dfrac{\pi}{2}$, is written as $\tan^{-1} x$.

$y = \tan^{-1} x \Leftrightarrow \tan y = x$ and $-\dfrac{\pi}{2} < y < \dfrac{\pi}{2}$.

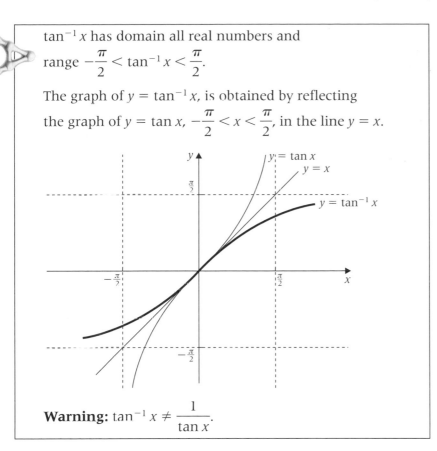

$\tan^{-1} x$ has domain all real numbers and

range $-\dfrac{\pi}{2} < \tan^{-1} x < \dfrac{\pi}{2}$.

The graph of $y = \tan^{-1} x$, is obtained by reflecting

the graph of $y = \tan x$, $-\dfrac{\pi}{2} < x < \dfrac{\pi}{2}$, in the line $y = x$.

Warning: $\tan^{-1} x \neq \dfrac{1}{\tan x}$.

The lines $x = -\dfrac{\pi}{2}$ and $x = \dfrac{\pi}{2}$ are vertical asymptotes of the graph
of $y = \tan x$. When reflected in the line $y = x$ they become
horizontal.

The lines $y = -\dfrac{\pi}{2}$ and $y = \dfrac{\pi}{2}$ are horizontal asymptotes of the
graph of $y = \tan^{-1} x$.

Worked example 3.3

By considering the graphs of $y = 0.5x$ and $y = \cos^{-1} x$ determine
the number of real roots of the equation $2 \cos^{-1} x = x$.

Solution

The equation $2 \cos^{-1} x = x$ can be rearranged into the form
$\cos^{-1} x = 0.5x$.

The number of real roots of the equation correspond to the
number of times the graphs $y = \cos^{-1} x$ and $y = 0.5x$ intersect.

The graphs intersect in just one point so the equation
$2 \cos^{-1} x = x$ has only one real root.

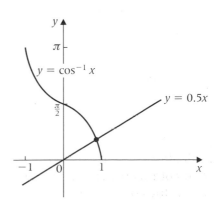

EXERCISE 3A

In this exercise you may assume the results in the table opposite. You do not need to learn them for the exam.

1 Find the exact values of:

    (a) $\cos^{-1}\dfrac{1}{2}$ **(b)** $\cos(\cos^{-1}0)$

    (c) $\cos^{-1}\left(\cos\dfrac{2\pi}{3}\right)$ **(d)** $\cos^{-1}\left(\cos\dfrac{4\pi}{3}\right)$

    (e) $\cos^{-1}\left(\sin\dfrac{2\pi}{3}\right)$ **(f)** $\cos^{-1}\left(\sin\dfrac{4\pi}{3}\right)$

Angle θ in radians	$\sin\theta$	$\cos\theta$	$\tan\theta$
0	0	1	0
$\dfrac{\pi}{6}$	$\dfrac{1}{2}$	$\dfrac{\sqrt{3}}{2}$	$\dfrac{1}{\sqrt{3}}$
$\dfrac{\pi}{4}$	$\dfrac{1}{\sqrt{2}}$	$\dfrac{1}{\sqrt{2}}$	1
$\dfrac{\pi}{3}$	$\dfrac{\sqrt{3}}{2}$	$\dfrac{1}{2}$	$\sqrt{3}$
$\dfrac{\pi}{2}$	1	0	∞
π	0	-1	0

2 Find the exact values of:

    (a) $\tan^{-1}\dfrac{1}{\sqrt{3}}$ **(b)** $\tan[\tan^{-1}(-1)]$

    (c) $\tan^{-1}\left(\tan\dfrac{3\pi}{4}\right)$ **(d)** $\tan\left(\cos^{-1}\dfrac{1}{2}\right)$

    (e) $\tan\left(\cos^{-1}\dfrac{1}{3}\right)$ **(f)** $\sin\left(\tan^{-1}\dfrac{1}{2}\right)$

3 You are given that $h(x) = \sin^{-1}x + \cos^{-1}x - 3\tan^{-1}x$. Find the value of:

    (a) $h(1)$,   **(b)** $h(-1)$.

4 Determine the number of real roots of the equation $2\sin^{-1}x = x$.

5 Show that the equation $\tan^{-1}x = kx$, where k is a constant, has only one real root when $k < 0$ and state its value.

6 For very large positive values of the constant k, state the number of real roots of the equation $k\tan^{-1}x = x$.

3.2 Secant, cosecant and cotangent

So far you have worked with three trigonometric ratios, sine, cosine and tangent. The remaining three trigonometric ratios are secant, cosecant and cotangent. They are written as $\sec x$, $\operatorname{cosec} x$ and $\cot x$, respectively, and are defined as follows:

$$\sec x = \frac{1}{\cos x}$$

$$\operatorname{cosec} x = \frac{1}{\sin x}$$

$$\cot x = \frac{1}{\tan x} = \frac{\cos x}{\sin x}$$

> Secant is the reciprocal of cosine. Do **not** write $\sec x$ as $\cos^{-1}x$, which means something entirely different as you saw in section 3.1. Similarly $\operatorname{cosec} x \neq \sin^{-1}x$ and $\cot x \neq \tan^{-1}x$.

> Used $\tan x = \dfrac{\sin x}{\cos x}$, from C2 section 6.5.

Calculators do not have the function keys for secant, cosecant and cotangent. The next worked example shows you how to find the values of these functions.

Worked example 3.4

(a) Find the values of:
  (i) $\sec 40°$,

  (ii) $\operatorname{cosec} \dfrac{\pi}{5}$,

  (iii) $\cot 250°$.

(b) Given that $\tan \dfrac{\pi}{6} = \dfrac{1}{\sqrt{3}}$, find the exact value of $\cot \dfrac{5\pi}{6}$.

3

Solution

(a) **(i)** $\sec 40° = \dfrac{1}{\cos 40°} = \dfrac{1}{0.766\,04\ldots} = 1.305$ (to 3 d.p.),

  (ii) $\operatorname{cosec} \dfrac{\pi}{5} = \dfrac{1}{\sin \dfrac{\pi}{5}} = \dfrac{1}{\sin 0.628\,318\ldots}$

> Set your calculator in radian mode.

$$= \dfrac{1}{0.587\,785\ldots} = 1.701 \text{ (to 3 d.p.)}.$$

  (iii) $\cot 250° = \dfrac{1}{\tan 250°} = \dfrac{1}{2.747\,47\ldots} = 0.364$ (to 3 d.p.);

(b) $\cot \dfrac{5\pi}{6} = \dfrac{1}{\tan \dfrac{5\pi}{6}} = \dfrac{1}{\tan \left(\pi - \dfrac{\pi}{6} \right)} = \dfrac{1}{-\tan \dfrac{\pi}{6}} = \dfrac{1}{-\dfrac{1}{\sqrt{3}}} = -\sqrt{3}$

> Used $\tan (\pi - \theta) = -\tan \theta$.

The next two worked examples show you how to solve some basic trigonometric equations involving a single secant, cosecant or cotangent function.

Worked example 3.5

Solve these equations for $0° \leqslant x < 360°$. Give your answers to one decimal place.

(a) $\sec x = 1.9$, **(b)** $\operatorname{cosec} x = -1.25$.

Solution

(a) $\sec x = 1.9 \Rightarrow \cos x = \dfrac{1}{1.9} = 0.526\,315\ldots$

Cos x is positive so answers lie in the intervals $0° < x < 90°$ and $270° < x < 360°$. The acute angle $\cos^{-1}(0.526\,315\ldots) = 58.24°$, so $x = 58.24°$ or $x = 360° - 58.24°$

To one decimal place, $x = 58.2°, 301.8°$.

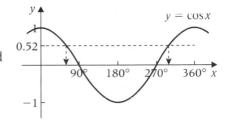

(b) $\csc x = -1.25 \Rightarrow \sin x = -\dfrac{1}{1.25} = -0.8$

Sin x is negative so answers lie in the interval $180° < x < 360°$. The acute angle, $\sin^{-1} 0.8 = 53.13°$, so $x = 180° + 53.13°$ or $x = 360° - 53.13°$.
To one decimal place, $x = 233.1°,\ 306.9°$.

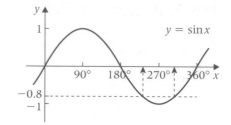

Worked example 3.6

Given that $\tan^{-1} 1 = \dfrac{\pi}{4}$, solve the equation $\cot 2x = -1$ giving your answer, in terms of π, in the interval for $0 \le x \le 2\pi$.

Solution

$\cot 2x = -1 \Rightarrow \tan 2x = -1$.
Since $0 \le x < 2\pi$, you require all values for $2x$ between 0 and 4π.

$\tan^{-1}(-1) = -\dfrac{\pi}{4}$, and tan is periodic with period π,

$\Rightarrow 2x = \pi - \dfrac{\pi}{4}$ or $2\pi - \dfrac{\pi}{4}$ or $3\pi - \dfrac{\pi}{4}$ or $4\pi - \dfrac{\pi}{4}$

$\Rightarrow x = \dfrac{3\pi}{8},\ \dfrac{7\pi}{8},\ \dfrac{11\pi}{8}$ and $\dfrac{15\pi}{8}$.

The next Worked example shows you how to use the definitions of secant, cosecant and cotangent to obtain other trigonometric identities.

Worked example 3.7

Prove the identity $\cot A \sec A \equiv \csc A$.

Solution

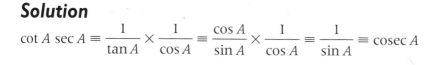

$\cot A \sec A \equiv \dfrac{1}{\tan A} \times \dfrac{1}{\cos A} \equiv \dfrac{\cos A}{\sin A} \times \dfrac{1}{\cos A} \equiv \dfrac{1}{\sin A} \equiv \csc A$

$\boxed{\tan A = \dfrac{\sin A}{\cos A}}$

EXERCISE 3B

1 Find the values of:

(a) $\sec 70°$

(b) $\csc 70°$

(c) $\cot 20°$

(d) $\sec(-70°)$

(e) $\csc 90°$

(f) $4 + \cot 430°$

(g) $\dfrac{1}{1 + \sec 60°}$

(h) $\dfrac{2}{6 + \cot 315°}$

2 Find the values of:

(a) sec 2

(b) cosec 0.7

(c) cot 0.5

(d) sec(−1)

(e) $\operatorname{cosec} \dfrac{\pi}{8}$

(f) $4 + \cot \dfrac{\pi}{8}$

(g) $\dfrac{1}{1 + \sec \dfrac{\pi}{10}}$

(h) $\dfrac{1}{6 + \cot \dfrac{\pi}{5}}$

> The angles are measured in radians in this question.

3 Using the table of results opposite and below, find the exact values of:

(a) sec 60°

(b) cosec 60°

(c) cot 30°

(d) sec(−180°)

(e) cosec 135°

(f) 1 + cot 420°

(g) $\dfrac{1}{\sqrt{3} - \sec 30°}$

(h) $\dfrac{2}{7 + \sqrt{3}\,\cot 150°}$

4 Using the table of results opposite, find the exact values of:

(a) $\operatorname{cosec} \dfrac{\pi}{4}$

(b) $4 + \cot \dfrac{3\pi}{4}$

(c) $\dfrac{1}{1 + \sqrt{3}\,\sec \dfrac{\pi}{6}}$

(d) $\dfrac{2\sqrt{3}}{2 - \cot \dfrac{\pi}{6}}$

Angle θ in degrees	Angle θ in radians	$\sin\theta$	$\cos\theta$	$\tan\theta$
30	$\dfrac{\pi}{6}$	$\dfrac{1}{2}$	$\dfrac{\sqrt{3}}{2}$	$\dfrac{1}{\sqrt{3}}$
45	$\dfrac{\pi}{4}$	$\dfrac{1}{\sqrt{2}}$	$\dfrac{1}{\sqrt{2}}$	1
60	$\dfrac{\pi}{3}$	$\dfrac{\sqrt{3}}{2}$	$\dfrac{1}{2}$	$\sqrt{3}$
180	π	0	−1	0

5 Solve these equations for $0° \leqslant x \leqslant 360°$.
Give your answers to one decimal place.

(a) sec x = 1.8

(b) cosec x = −2.25

(c) cot x = 3

(d) sec x = −1.3

(e) cosec x = 3

(f) cot x = −2.4

(g) 4 sec 2x = −7

(h) 5 cot 2x = −2

6 Solve these equations for $0 \leqslant x \leqslant 2\pi$, giving your answers in radians to three significant figures.

(a) sec x = 2

(b) cosec x = −2

(c) cot 2x = 1

(d) sec 5x = −1

(e) $\sqrt{3}\,\operatorname{cosec} 3x = 2$

(f) $\cot 2x = -\sqrt{3}$

(g) sec 3x = 1

(h) $\sqrt{12}\,\cot 3x = 2$

7 Prove the following identities:

(a) $\tan A \operatorname{cosec} A \equiv \sec A$ **(b)** $\sin A \cot A \equiv \cos A$

(c) $\dfrac{\cot A}{1 + \cot A} \equiv \dfrac{1}{1 + \tan A}$ **(d)** $\tan A \operatorname{cosec} A \equiv \sec A$

(e) $\sec A - \cos A \equiv \tan A \sin A$

(f) $\dfrac{\sin A}{1 - \cos A} \equiv \operatorname{cosec} A + \cot A$

> **Hint for part (f):** Consider
> $(1 - \cos A)(\operatorname{cosec} A + \cot A)$.

3.3 Graphs of sec *x*, cosec *x* and cot *x*

Graph of *y* = sec *x*

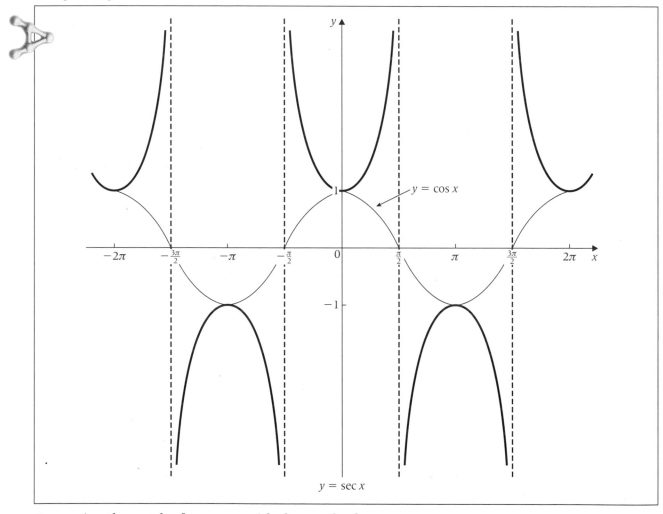

$y = \sec x$

Comparing the graph of $y = \sec x$ with the graph of $y = \cos x$ you can see that $y = \sec x$ has maximum points where $y = \cos x$ has minimum points and has minimum points where $y = \cos x$ has maximum points.

Secant, like cosine, is a periodic function with period 2π and so the graph repeats itself every 2π radians.

The domain of sec x is all real values $\neq (2k + 1)\dfrac{\pi}{2}$ and its range is all real values **except** $-1 < y < 1$.

Graph of y = cosec x

Consider $y = \operatorname{cosec} x = \dfrac{1}{\sin x} = \dfrac{1}{\cos\left(x - \dfrac{\pi}{2}\right)}$

$\Rightarrow \operatorname{cosec} x = \sec\left(x - \dfrac{\pi}{2}\right)$

So a translation of $\begin{bmatrix} \dfrac{\pi}{2} \\ 0 \end{bmatrix}$ transforms the graph of $y = \sec x$ into

the graph of $y = \operatorname{cosec} x$.

So, if $f(x) = \sec x$, then

$f\left(x - \dfrac{\pi}{2}\right) = \operatorname{cosec} x.$

C2, section 5.2.

3

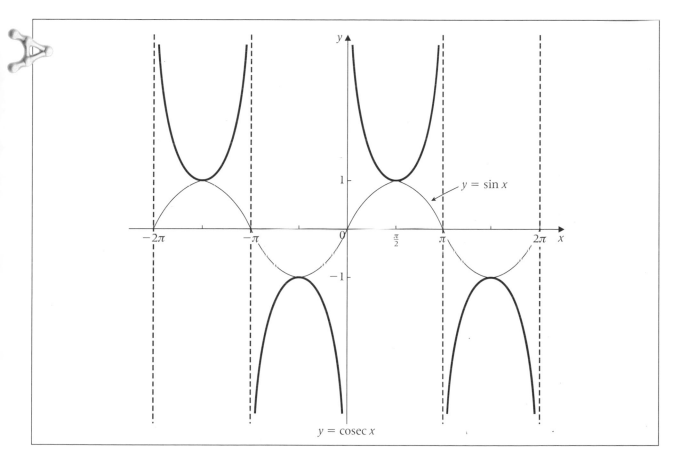

$y = \operatorname{cosec} x$

Comparing the graph of $y = \operatorname{cosec} x$ with the graph of $y = \sin x$
you can see that $y = \operatorname{cosec} x$ has maximum points where $y = \sin x$
has minimum points and has minimum points where $y = \sin x$
has maximum points.
Cosecant, like sine, is a periodic function with period 2π and so
the graph repeats itself every 2π radians.
The domain of $\operatorname{cosec} x$ is all real values $\neq k\pi$ and its range is all
real values **except** $-1 < y < 1$.

Graph of y = cot x

Consider $y = \cot x = \dfrac{\cos x}{\sin x} = \dfrac{\sin\left(\dfrac{\pi}{2} - x\right)}{\cos\left(\dfrac{\pi}{2} - x\right)}$

$\Rightarrow \cot x = \tan\left(\dfrac{\pi}{2} - x\right) = \tan\left[2\left(\dfrac{\pi}{4}\right) - x\right]$

So a reflection in the line $x = \dfrac{\pi}{4}$ transforms the graph of $y = \tan x$ into the graph of $y = \cot x$.

So, if f(x) = tan x, then

$f\left[2\left(\dfrac{\pi}{4}\right) - x\right] = \cot x.$

C3 Ex2B Q16(c).

$y = \tan x$ $y = \cot x$

Cotangent, like tangent, is a periodic function with period π and so the graph repeats itself every π radians.
The domain of cot x is all real values $\neq k\pi$ and its range is all real values.

Worked example 3.8

(a) Determine the transformation that maps $y = \sec x$ onto $y = \sec 2x$.

(b) State the period, in radians, of the graph $y = \sec 2x$.

(c) Sketch the graph of $y = \sec 2x$ for $0 \leqslant x \leqslant \pi$, $x \neq \dfrac{\pi}{4}, \dfrac{3\pi}{4}$.

Solution

(a) Let f(x) = sec x then f(2x) = sec 2x.
The transformation that maps $y = \sec x$ onto $y = \sec 2x$ is a stretch of scale factor $\dfrac{1}{2}$ in the x-direction.

(b) The period of $y = \sec 2x$ is $\dfrac{2\pi}{2} = \pi$ radians.

(c) The graph of $y = \sec 2x$ for $0 \leqslant x \leqslant \pi$, $x \neq \dfrac{\pi}{4}, \dfrac{3\pi}{4}$ is

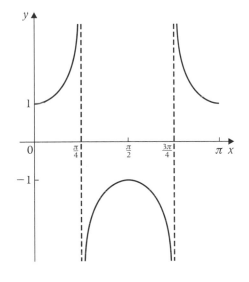

EXERCISE 3C _____

1 (a) Determine the transformation that maps $y = \text{cosec}\, x$ onto $y = \text{cosec}\, 4x$.

   (b) State the period, in radians, of the graph $y = \text{cosec}\, 4x$.

2 (a) Determine the transformation that maps $y = \cot x$ onto $y = \cot \dfrac{x}{4}$.

   (b) State the period, in radians, of the graph $y = \cot \dfrac{x}{4}$.

3 Sketch the graph of $y = 1 + \sec \dfrac{x}{2}$ for $-\pi < x < \pi$.

4 (a) Sketch the graph of $y = 1 + 3\,\text{cosec}\, 2x$ for $-180° < x < 180°$, $x \neq 0°, \pm 90°$.

   (b) Describe a sequence of transformations that maps the curve $y = \sec x$ onto the curve $y = 1 + 3\,\text{cosec}\, 2x$.

3.4 Identities involving the squares of secant, cosecant and cotangent

In C2, section 6.4, you used the identity $\cos^2 \theta + \sin^2 \theta \equiv 1$. In this section you will be shown two similar identities.

If you divide each term in the identity $\cos^2\theta + \sin^2\theta \equiv 1$ by $\cos^2 \theta$ you get

$$\frac{\cos^2 \theta}{\cos^2 \theta} + \frac{\sin^2 \theta}{\cos^2 \theta} \equiv \frac{1}{\cos^2 \theta}$$

$$\Rightarrow 1 + \tan^2 \theta \equiv \sec^2 \theta$$

Similarly, if you divide each term in the identity $\cos^2 \theta + \sin^2 \theta \equiv 1$ by $\sin^2 \theta$, you get

$$\frac{\cos^2 \theta}{\sin^2 \theta} + \frac{\sin^2 \theta}{\sin^2 \theta} \equiv \frac{1}{\sin^2 \theta}$$

$$\Rightarrow \cot^2 \theta + 1 \equiv \operatorname{cosec}^2 \theta$$

These two identities, along with $\cos^2 \theta + \sin^2 \theta \equiv 1$, are frequently used to solve trigonometric equations and to prove other identities. They are **not** given in the examination formulae booklet so you must memorise them and know how to apply them.

Worked example 3.9

Solve the equation $\sec^2 x = 4 + 2 \tan x$, giving all solutions in the interval $0° \leqslant x < 360°$.

Solution

To solve the equation $\sec^2 x = 4 + 2 \tan x$
use the identity $\sec^2 x \equiv 1 + \tan^2 x$ to get $1 + \tan^2 x = 4 + 2 \tan x$.

$\Rightarrow \tan^2 x - 2 \tan x - 3 = 0$

$\Rightarrow (\tan x - 3)(\tan x + 1) = 0$

$\Rightarrow \tan x = 3,\ \tan x = -1.$

> As a starter, try to get a quadratic equation in the non-squared trigonometric term, $\tan x$ here.

For $\tan x = 3$, $\tan^{-1} 3 = 71.57°$.
For $0° \leqslant x < 360°$, $\tan x = 3 \Rightarrow x = 71.57°,\ 180° + 71.57°$
For $\tan x = -1$, $\tan^{-1}(-1) = -45°$.
For $0° \leqslant x < 360°$, $\tan x = -1 \Rightarrow x = 180° - 45°,\ 360° - 45°$

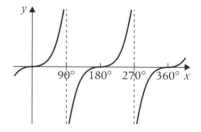

The solutions of the equation $\sec^2 x = 4 + 2 \tan x$, in the interval $0° \leqslant x < 360°$, are $x = 71.6°,\ 135°,\ 251.6°,\ 315°$.

The general strategy when asked to solve trigonometric equations which involve the same angle but with a squared trigonometric function is to try to write the equation as a quadratic equation in the non-squared trigonometric function.

Worked example 3.10

Solve the equation $\operatorname{cosec}^2 x = 5 + 3 \cot x$, giving all solutions in the interval $0 \leqslant x < 2\pi$, to three significant figures.

Solution

To solve the equation $\operatorname{cosec}^2 x = 5 + 3 \cot x$
use the identity $\operatorname{cosec}^2 x \equiv 1 + \cot^2 x$ to get $1 + \cot^2 x = 5 + 3 \cot x$

> Try to write the equation as a quadratic equation in cot x.

$\Rightarrow \cot^2 x - 3 \cot x - 4 = 0$

$\Rightarrow (\cot x - 4)(\cot x + 1) = 0$

$\Rightarrow \cot x = 4, \cot x = -1$

$\Rightarrow \tan x = \dfrac{1}{4}, \tan x = -1$

> Used $\tan x = \dfrac{1}{\cot x}$.

For $\tan x = \dfrac{1}{4}$, $\tan^{-1} \dfrac{1}{4} = 0.245$ rads.

> Set calculator in radian mode.

For $0 \leqslant x < 2\pi$, $\tan x = \dfrac{1}{4} \Rightarrow x = 0.245$ rads, $(\pi + 0.245)$ rads.

For $\tan x = -1$, $\tan^{-1}(-1) = -0.785$ rads.
For $0 \leqslant x < 2\pi$,
$\tan x = -1 \Rightarrow x = (\pi - 0.785)$ rads, $(2\pi - 0.785)$ rads.

The solutions of the equation $\operatorname{cosec}^2 x = 5 + 3 \cot x$, in the interval $0 \leqslant x < 2\pi$, are 0.245^c, 2.36^c, 3.39^c and 5.50^c, to 3 sf.

Worked example 3.11

Prove the identity $\sec^2 A - \operatorname{cosec}^2 A \equiv (\tan A + \cot A)(\tan A - \cot A)$.

Solution

$\sec^2 A - \operatorname{cosec}^2 A \equiv 1 + \tan^2 A - (1 + \cot^2 A)$

$\equiv \tan^2 A - \cot^2 A$

> Difference of two squares.

$\equiv (\tan A + \cot A)(\tan A - \cot A)$

So $\sec^2 A - \operatorname{cosec}^2 A \equiv (\tan A + \cot A)(\tan A - \cot A)$.

EXERCISE 3D

1 Prove the identity
$\tan^2 A - \cot^2 A \equiv (\sec A + \operatorname{cosec} A)(\sec A - \operatorname{cosec} A)$.

2 Prove the identity
$\cot^2 A + \sin^2 A \equiv (\operatorname{cosec} A + \cos A)(\operatorname{cosec} A - \cos A)$.

3 Prove the identity $\operatorname{cosec}^2 A \cos^2 A \equiv \operatorname{cosec}^2 A - 1$.

4 Given that $2\cos^2 x - \sin^2 x = 1$, show that $\cos^2 x = 2\sin^2 x$ and hence find the possible values of $\cot x$.

5 Given that $5\sec^2 x + 3\tan^2 x = 9$, find the possible values of $\sin x$.

6 Given that $x = \sec\theta$ and $y = \tan\theta$, show that $x^2 - y^2 = 1$.

7 Prove the identity

$$\frac{(\sec A - \tan A)(\tan A + \sec A)}{\operatorname{cosec} A - \cot A} \equiv \cot A + \operatorname{cosec} A.$$

8 Prove the identity $(\operatorname{cosec} A + \cot A)^2 \equiv \dfrac{1 + \cos A}{1 - \cos A}$.

9 Eliminate θ from equations $x = \operatorname{cosec}\theta$ and $y = \dfrac{1}{4}\cot\theta$.

10 Eliminate θ from equations $x = 2 + \operatorname{cosec}\theta$ and $y = \dfrac{1}{4}\tan\theta$.

11 Solve the equation $\sec^2 x = 3 + \tan x$, giving all solutions in the interval $0° \leqslant x < 360°$.

12 Solve the equation $3\sec^2 x = 5 + \tan x$, giving all solutions in the interval $-180° \leqslant x < 180°$.

13 Solve the equation $2\operatorname{cosec}^2 x = 1 + 3\cot x$, giving all solutions in the interval $0° \leqslant x < 360°$.

14 Solve the equation $\tan^2 x = 1 - \sec x$, giving all solutions in the interval $-180° \leqslant x < 180°$.

15 Solve the equation $\cot^2 x + \operatorname{cosec} x = 11$, giving all solutions in the interval $0° \leqslant x < 360°$.

16 Solve the equation $4\tan^2 x + 12\sec x + 1 = 0$, giving all solutions in the interval $0° \leqslant x < 360°$.

17 Solve the equation $\cot^2 x + \operatorname{cosec}^2 x = 7$, giving all solutions in the interval $0 \leqslant x < 2\pi$.

18 Solve the equation $\tan^2 x + 3 = 3\sec x$, giving all solutions in the interval $0 \leqslant x < 2\pi$.

19 Solve the equation $\cot^2 x + 5\operatorname{cosec} x = 3$, giving all solutions in the interval $0 \leqslant x < 2\pi$.

20 Solve the equation $2\operatorname{cosec}^2 2x + \cot 2x = 3$, giving all solutions in the interval $-\pi \leqslant x < \pi$.

21 Given that $x = \sec A + \tan A$, show that $x + \dfrac{1}{x} = 2\sec A$.

22 Given that $x = \operatorname{cosec}\theta + \cot\theta$, show that $x + \dfrac{1}{x} = 2\operatorname{cosec}\theta$.

Key point summary

1 $\sin^{-1} x$ has domain $-1 \leqslant x \leqslant 1$ and range $-\dfrac{\pi}{2} \leqslant \sin^{-1} x \leqslant \dfrac{\pi}{2}$.

p43

The graph of $y = \sin^{-1} x$, $-1 \leqslant x \leqslant 1$ is obtained by reflecting the graph of $y = \sin x$, $-\dfrac{\pi}{2} \leqslant x \leqslant \dfrac{\pi}{2}$, in the line $y = x$.

Warning: $\sin^{-1} x \neq \dfrac{1}{\sin x}$.

2 $\sin^{-1}(\sin x) = x$ for $-\dfrac{\pi}{2} \leqslant x \leqslant \dfrac{\pi}{2}$,

p43

$\sin(\sin^{-1} x) = x$ for $-1 \leqslant x \leqslant 1$.

3 $\cos^{-1} x$ has domain $-1 \leqslant x \leqslant 1$ and range $0 \leqslant \cos^{-1} x \leqslant \pi$.

p44

The graph of $y = \cos^{-1} x$, $-1 \leqslant x \leqslant 1$ is obtained by reflecting the graph of $y = \cos x$, $0 \leqslant x \leqslant \pi$, in the line $y = x$.

Warning: $\cos^{-1} x \neq \dfrac{1}{\cos x}$.

4 $\cos^{-1}(\cos x) = x$ for $0 \leqslant x \leqslant \pi$,

p44

$\cos(\cos^{-1} x) = x$ for $-1 \leqslant x \leqslant 1$.

5 $\tan^{-1} x$ has domain all real numbers and range $-\dfrac{\pi}{2} < \tan^{-1} x < \dfrac{\pi}{2}$.

p45

The graph of $y = \tan^{-1} x$, is obtained by reflecting the graph of $y = \tan x$, $-\dfrac{\pi}{2} < x < \dfrac{\pi}{2}$, in the line $y = x$.

Warning: $\tan^{-1} x \neq \dfrac{1}{\tan x}$.

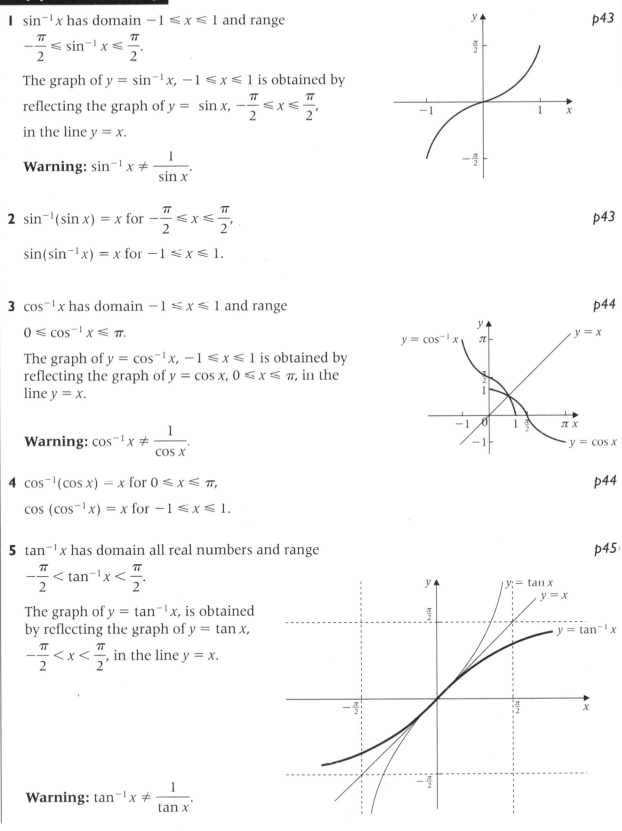

6 $\sec x = \dfrac{1}{\cos x}$

$\mathrm{cosec}\, x = \dfrac{1}{\sin x}$

$\cot x = \dfrac{1}{\tan x} = \dfrac{\cos x}{\sin x}$

p46

7

p50

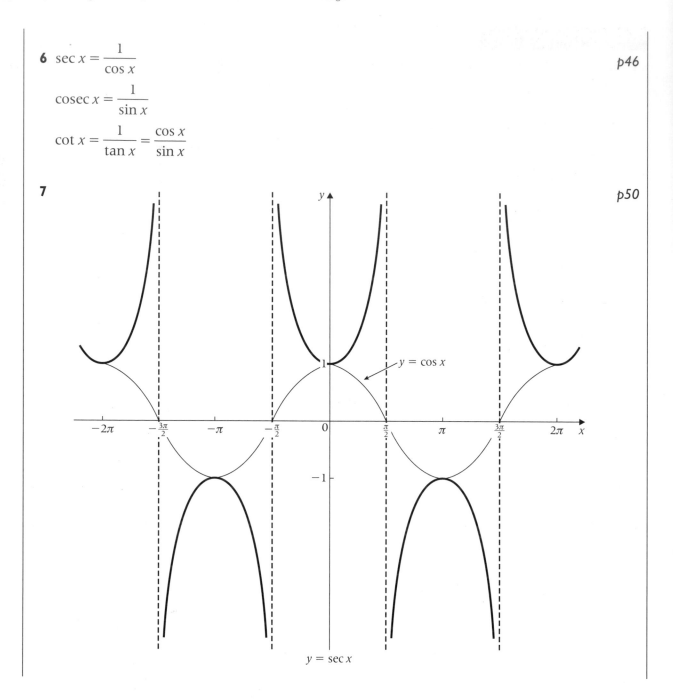

$y = \cos x$

$y = \sec x$

8

p51

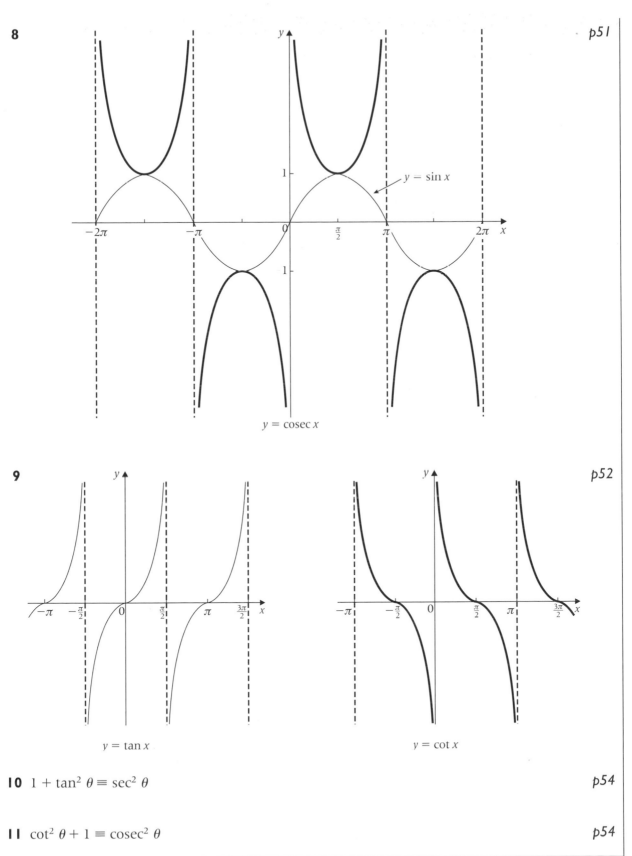

$y = \sin x$

$y = \operatorname{cosec} x$

9

p52

$y = \tan x$

$y = \cot x$

10 $1 + \tan^2 \theta \equiv \sec^2 \theta$

p54

11 $\cot^2 \theta + 1 \equiv \operatorname{cosec}^2 \theta$

p54

3

Test yourself	What to review
1 Show that the equation $\cos^{-1} x = \sin^{-1} x$ has only one real root and state its value to three significant figures.	*Section 3.1*
2 Solve the equation $\operatorname{cosec}(x + 30°) = 2.1$, for $0° \leqslant x < 360°$. Give your answers to one decimal place.	*Section 3.2*
3 By sketching the graphs of $y = \cot x$ and $y = \dfrac{x}{3}$ for $0 < x \leqslant \dfrac{\pi}{2}$, show that the equation $\cot x = \dfrac{x}{3}$ has one root and explain why this root must be in the interval $1 < x \leqslant \dfrac{\pi}{2}$.	*Section 3.3*
4 Solve the equation $3 \tan^2 x + 2 \sec x = 5$, giving all solutions in the interval $0 \leqslant x < 2\pi$.	*Section 3.4*

Test yourself **ANSWERS**

1 0.707.

2 121.6°, 358.4°.

4 $x = 0.723$, 2.09, 4.19, 5.56.

C3: The number e and calculus

Learning objectives

After studying this chapter, you should be able to:
- recognise the number e
- differentiate and integrate e^{kx}
- know what is meant by a natural logarithm
- realise that the inverse function of e^x is $\ln x$
- find the integral of $\frac{1}{x}$
- use exponential functions and logarithmic functions in calculus.

4.1 Gradients of exponential curves

Suppose you wish to find the gradient of the curve $y = 3^x$ at the point P where $x = 0$.

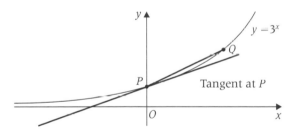

The point P has coordinates $(0, 1)$.

Take a point Q close to P, such as $(0.01, 3^{0.01})$.

The gradient of PQ is given by $\dfrac{3^{0.01} - 1}{0.01 - 0} \approx \dfrac{1.011\,047 - 1}{0.01} \approx 1.1047.$

By taking points even closer to P and using a suitable limiting argument, the gradient of the curve at P can be shown to be approximately $1.0986\ldots$.

For other exponential curves, the gradient at the point where $x = 0$ has a different value. The graph of $y = 2^x$, for example, has a gradient approximately equal to $0.693\,147\ldots$.

The ideal exponential curve would be of the form $y = a^x$, having gradient equal to 1 when $x = 0$.

Is it possible to find a value of the constant a that has this property?

4.2 A special limit

In order to find derivatives from first principles, you use the formula

$$\lim_{h \to 0} \left\{ \frac{f(x + h) - f(x)}{h} \right\}.$$

Consider the exponential function f where $f(x) = a^x$. The derived function is given by

$$\lim_{h \to 0} \left\{ \frac{a^{x+h} - a^x}{h} \right\}.$$

$$\lim_{h \to 0} \left\{ \frac{a^x(a^h - 1)}{h} \right\}.$$

But since the expression a^x is independent of h it can be brought outside the limit expression.

$$f'(x) = a^x \times \lim_{h \to 0} \left\{ \frac{(a^h - 1)}{h} \right\}$$

Hence the derivative of a^x is equal to $a^x \, L(a)$ where

$$L(a) = \lim_{h \to 0} \left\{ \frac{(a^h - 1)}{h} \right\}.$$

You can use your calculator to find the approximate value of this limit for different values of the constant a.

Some values of $L(a)$ are listed below with values given to three decimal places:

a	2	3	4	5	6
$L(a)$	0.693	1.099	1.386	1.609	1.792

This suggests there is a particular value of a, between 2 and 3 for which $L(a) = 1$.

Further calculations show that $L(2.718) = 0.999\,896\ldots$ and that $L(2.719) = 1.000\,264\ldots$.

Hence, there is a special number between 2.718 and 2.719 for which $L(a) = 1$.

This special number is denoted by the letter e. It is an irrational number and hence has a non-recurring decimal representation. The value of e is approximately $2.718\,281\,828\,458\,56 \ldots$

It is probably good to remember the first few decimal places of e rather like you have learned the first few decimal places of π.

> The number e is irrational and its value is approximately 2.718.

Throughout this chapter, a number of results will be proved using limits.

Your main concern should be how to use the proved results. The actual proofs will not be tested in the examination.

Since $a^{x+h} = a^x \times a^h$.

The gradient of the curve $y = a^x$ at the point where $x = 0$, mentioned in the first section, is in fact $L(a)$.

4.3 The derivative of e^x

The derivative of a^x was shown to be $a^x \, \mathrm{L}(a)$.
Substituting $a = e$, and since $\mathrm{L}(e) = 1$, it follows that

> The curve with equation $y = e^x$ has derivative $\dfrac{dy}{dx} = e^x$.

Worked example 4.1

Differentiate each of the following with respect to x.

(a) $3e^x + x^4$; **(b)** $2x - 7e^x + 3$

Solution

(a) $\dfrac{d}{dx}(3e^x + x^4) = 3e^x + 4x^3$; **(b)** $\dfrac{d}{dx}(2x - 7e^x + 3) = 2 - 7e^x$.

4.4 The derivative of e^{kx}, where k is a constant

You can use the formula

$$f'(x) = \lim_{h \to 0}\left\{\frac{f(x+h) - f(x)}{h}\right\} \text{ where } f(x) = e^{kx}$$

$$= \lim_{h \to 0}\left\{\frac{e^{k(x+h)} - e^{kx}}{h}\right\}$$

$$= \lim_{h \to 0}\left\{\frac{e^{kx}(e^{kh} - 1)}{h}\right\}$$

Since $e^{kx+kh} = e^{kx} \times e^{kh}$.

$$= e^{kx} \times \lim_{h \to 0}\left\{\frac{(e^{kh} - 1)}{h}\right\}$$

$$= e^{kx} \times \lim_{h \to 0}\left\{\frac{k(e^{kh} - 1)}{kh}\right\}$$

$$= k \, e^{kx} \times \lim_{h \to 0}\left\{\frac{(e^{kh} - 1)}{kh}\right\}$$

Let $s = kh$. Therefore as $h \to 0$, $s \to 0$.

$$f'(x) = k \, e^{kx} \times \lim_{s \to 0}\left\{\frac{(e^s - 1)}{s}\right\} = k \, e^{kx} \times 1$$

> $\dfrac{d}{dx}(e^{kx}) = k \, e^{kx}$

You need to remember this formula.

Worked example 4.2

Find $\dfrac{dy}{dx}$ for each of the following:

(a) $y = e^{-3x}$ (b) $y = x^5 - e^{4x}$

Solution

(a) $y = e^{-3x} \Rightarrow \dfrac{dy}{dx} = -3e^{-3x}$;

(b) $y = x^5 - e^{4x} \Rightarrow \dfrac{dy}{dx} = 5x^4 - 4e^{4x}$.

Worked example 4.3

A curve has equation $y = \dfrac{e^{5x} - 5}{e^{2x}}$. Find the gradient of the curve at the point where $x = 0$.

Solution

You need to rewrite the equation of the curve, making use of the fact that $\dfrac{1}{e^{2x}} = e^{-2x}$.

The curve has equation $y = \dfrac{e^{5x} - 5}{e^{2x}} = e^{-2x}(e^{5x} - 5) = e^{3x} - 5e^{-2x}$.

$$\dfrac{dy}{dx} = 3e^{3x} + 10e^{-2x}$$

When $x = 0$, $\dfrac{dy}{dx} = 3 + 10 = 13$

> Recall that $a^0 = 1$ whatever the value of a. Hence $e^0 = 1$.

EXERCISE 4A

1 Differentiate each of the following with respect to x:

     (a) e^{-x} (b) e^{3x} (c) e^{7x}

     (d) e^{-2x} (e) e^{4x} (f) e^{-6x}

2 Find the gradient of the following curves at the point where $x = 0$:

     (a) $y = x^5 - e^{2x}$ (b) $y = 5x^3 + 6x + e^{3x}$

     (c) $y = x\sqrt{x} - 2e^{-x}$

3 Calculate the gradient of the following curves at the point where $x = 1$:

     (a) $y = x^3 - 2e^{-3x}$ (b) $y = 4x^5 - 3e^{-x}$

     (c) $y = \dfrac{3}{x} + 5e^{2x}$

4 By writing e^{2x+3} as $e^{2x}.e^3$ show that $\dfrac{d}{dx}(e^{2x+3}) = 2e^{2x+3}$.

In a similar way, differentiate with respect to x:

(a) e^{x+4} (b) e^{2-x} (c) e^{3x+1}

(d) e^{7x-5} (e) e^{ax+b}

5 A curve has equation $y = \dfrac{e^{3x} - 4}{e^x}$. Find the gradient of the curve at the point where $x = 0$.

6 Find $\dfrac{dy}{dx}$ for each of the following:

(a) $y = (e^{2x} - 4)(e^{3x} + 2)$

(b) $y = (e^{3x} + 7)(e^{x} - 1)$

(c) $y = (e^{-x} - 2)^2$

7 A curve has equation $y = e^{2x} - 2x + 3$.

Find the value $\dfrac{dy}{dx}$ and $\dfrac{d^2y}{dx^2}$ at the point P where $x = 0$.

What can you deduce about P?

4.5 Integration of e^{kx}

Since the derivative of e^{kx} is equal to ke^{kx}, it follows that

$$\int e^{kx}dx = \frac{1}{k}e^{kx} + \text{constant}$$

You need to remember this formula.

Worked example 4.4

Find the area of the finite region bounded by the coordinate axes, the curve with equation $y = e^{3x}$ and the line $x = 4$.

Solution

The curve has been sketched and the required region shaded.

The region is bounded by the curve, the lines $x = 0$, $x = 4$ and the x-axis.

$$\text{Area} = \int_0^4 e^{3x}dx = \left[\frac{1}{3}e^{3x}\right]_0^4 = \frac{1}{3}e^{12} - \frac{1}{3}e^0 = \frac{(e^{12} - 1)}{3}.$$

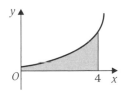

EXERCISE 4B

1 Integrate each of the following with respect to x:

(a) e^{-x} (b) e^{3x} (c) e^{7x}

(d) e^{-2x} (e) e^{4x} (f) e^{-6x}

2 Find:

(a) $\int e^{\frac{1}{2}x}\,dx$ **(b)** $\int x^2 - 3e^{2x}\,dx$ **(c)** $\int \sqrt{x} + 6e^{3x}\,dx$

3 Evaluate

(a) $\int_0^1 e^{-3x}\,dx$ **(b)** $\int_0^{\frac{1}{2}} 4e^{2x}\,dx$ **(c)** $\int_0^1 (4x - 5e^{-x})\,dx$

4 Calculate the area bounded by the curve with equation $y = 2x + e^{3x}$, the x-axis and the lines $x = 1$ and $x = 2$.

5 Calculate the area bounded by the curve with equation $y = 2 + x^2 + e^{\frac{x}{3}}$, the coordinate axes and the line $x = 3$.

6 Calculate the area bounded by the curve with equation $y = \dfrac{3 + e^{3x}}{e^x}$, the x-axis and the lines $x = 1$ and $x = 2$.

4.6 Natural logarithms

It was shown in the previous chapter that $N = a^x \Leftrightarrow x = \log_a N$. The special base being considered here is base e.

Therefore $N = e^x \Leftrightarrow x = \log_e N$. The logarithm to base e is referred to as a natural logarithm.

The French term is *logarithme naturel* which can be abbreviated to ln. That is why the button on your calculator to find natural logarithms is labelled $\boxed{\ln}$.

Hence $N = e^x \Leftrightarrow x = \ln N$.

It follows therefore that $\ln 1 = 0$ and that $\ln e = 1$.

Worked example 4.5

The function f is defined for all real values of x by

$$f(x) = e^x.$$

(a) Sketch the graph of f.   **(b)** State the range of f.

(c) Find $f^{-1}(x)$ and state the domain of f^{-1}.

Solution

(a) The graph of f is sketched on the right.

(b) Since e^x is always positive, the range is $f(x) > 0$.

(c) Letting $y = e^x$ and changing the subject gives $x = \ln y$.

Reflecting in the line $y = x$ is equivalent to interchanging x and y.
Hence $y = \ln x \Rightarrow f^{-1}(x) = \ln x$.

The domain of f^{-1} is equal to the range of f.
Hence the domain of f^{-1} is $x > 0$.

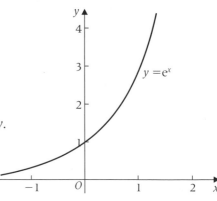

The natural logarithm $\ln x$ is defined for $x > 0$.

A useful relationship is

$$x = e^y \iff y = \ln x.$$

The graph of $y = \ln x$ is sketched below. You can see from its graph that $\ln x$ increases as x increases.

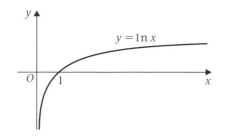

Note that $\ln 1 = 0$.

4

Since 'taking the natural logarithm' and 'finding e to the power of' are inverse operations, it follows that

$$e^{\ln b} = b \qquad \text{and} \qquad \ln(e^c) = c.$$

Also the rules of logarithms apply equally to natural logarithms so that

(1) $\ln a + \ln b = \ln ab$

(2) $\ln a - \ln b = \ln \dfrac{a}{b}$

(3) $n \ln a = \ln a^n$.

Worked example 4.6

Simplify the following:

(a) $e^{2\ln 3}$ **(b)** $\ln 12 - \ln 3 - 2 \ln 2$ **(c)** $\ln \sqrt{e}$

Solution

(a) Rewrite $2 \ln 3$ as $\ln 9$.
Hence $e^{2\ln 3} = e^{\ln 9} = 9$.

(b) Since $2 \ln 2 = \ln 4$, the expression can be written as

$$\ln 12 - \ln 3 - 2 \ln 2 = \ln \left(\frac{12}{3 \times 4} \right) = \ln 1 = 0.$$

(c) You need to realise that $\sqrt{e} = e^{\frac{1}{2}}$. Then

$$\ln \sqrt{e} = \ln e^{\frac{1}{2}} = \tfrac{1}{2}.$$

EXERCISE 4C

1 Simplify each of the following:

(a) $\ln e^3$ **(b)** $\ln e^{-2}$ **(c)** $\ln 2e$ **(d)** $e^{3\ln 2}$ **(e)** $e^{-\ln 5}$

2 Simplify:

(a) $\ln e^{\sin x}$ (b) $e^{2\ln(1+x)}$ (c) $e^{-\ln(5-x)}$

3 Solve the equations:

(a) $e^x = 5$ (b) $e^x = 7$ (c) $e^x = \frac{1}{2}$ (d) $e^{2x} = 4$

4 Solve the equations:

(a) $e^{3x} = 8$ (b) $e^{2x} = 5$ (c) $e^{-x} = \frac{1}{3}$ (d) $e^{-3x} = 27$

5 Solve the equations:

(a) $\ln x = 2$ (b) $\ln(2x) = 3$

(c) $\ln\left(\dfrac{x}{3}\right) = 4$ (d) $\ln\left(\dfrac{2}{x}\right) = 5$

6 Solve each of the following equations:

(a) $\ln x^2 = 8$ (b) $\ln x = -1$

(c) $\ln(x + 1) = 2\ln(x - 1)$ (d) $\ln(x + 3) + \ln(x - 1) = 0$

(e) $\ln(x + 2) - \ln(x + 1) = 1$

> You may need to reject some values of x for which the expressions are not defined.

7 State a geometrical transformation that maps the graph of $y = e^x$ onto the graph of the following:

(a) $y = 2e^x$ (b) $y = e^{-x}$ (c) $y = 2 + e^x$ (d) $y = e^{4-x}$

(e) $y = e^{\frac{x}{3}}$ (f) $y = e^{x-5}$ (g) $y = 3e^{\frac{x}{3}}$ (h) $y = \ln x$

8 Find a geometrical transformation that maps the graph of $y = \ln x$ onto the graph of the following:

(a) $y = 3\ln x$ (b) $y = 3 + \ln x$

(c) $y = 3 - \ln x$ (d) $y = 3\ln 3x$

9 Sketch the graphs of:

(a) $y = |\ln x|, x > 0$ (b) $y = \ln|x|, x \in \mathbb{R}, x \neq 0$

Worked example 4.7

A curve has equation $y = e^{4x} - 12x + 5$. Find the coordinates of its stationary point and determine its nature.

Solution

$$y = e^{4x} - 12x + 5 \Rightarrow \frac{dy}{dx} = 4e^{4x} - 12$$

Stationary points occur when $\dfrac{dy}{dx} = 0$.

Hence stationary points are given by

$$4e^{4x} - 12 = 0 \Rightarrow e^{4x} = 3 \Rightarrow 4x = \ln 3$$

$$\Rightarrow x = \frac{1}{4}\ln 3$$

Since $e^{4x} = 3$ the corresponding value of y is

$$3 - 12 \times \frac{1}{4} \ln 3 + 5 = 8 - 3 \ln 3.$$

$\dfrac{d^2y}{dx^2} = 16e^{4x}$. At the stationary point, since $e^{4x} = 3$,

$$\frac{d^2y}{dx^2} = 16 \times 3 = 48.$$

The second derivative is positive and so the point with coordinates $(\frac{1}{4} \ln 3, 8 - 3 \ln 3)$ is a minimum point.

Worked example 4.8

Find the area enclosed by the curve $y = e^{\frac{x}{2}}$, the x-axis and the lines $x = \ln 4$ and $x = \ln 9$.

Solution

$$\text{Area} = \int_{\ln 4}^{\ln 9} e^{\frac{x}{2}} dx = \left[2e^{\frac{x}{2}} \right]_{\ln 4}^{\ln 9}$$

$$= 2e^{\frac{\ln 9}{2}} - 2e^{\frac{\ln 4}{2}} = 2e^{\ln 3} - 2e^{\ln 2}$$

$$= 2 \times 3 - 2 \times 2 = 2$$

EXERCISE 4D

1 A curve has equation $y = e^x - 5x$. Find the coordinates of the stationary point and prove that it is a minimum point.

2 A curve has equation $y = e^{2x} - 6x + 3$.

   (a) Find $\dfrac{dy}{dx}$ and $\dfrac{d^2y}{dx^2}$.

   (b) Find the coordinates of the stationary point and determine its nature.

3 A curve has equation $y = 10x - e^{3x}$.

   (a) Find $\dfrac{dy}{dx}$ and $\dfrac{d^2y}{dx^2}$.

   (b) Find the coordinates of the stationary point and determine its nature.

4 A curve has equation $y = 5 + 4x + e^{-x}$.

   (a) Find $\dfrac{dy}{dx}$ and $\dfrac{d^2y}{dx^2}$.

   (b) Find the coordinates of the stationary point and determine its nature.

5 A curve has equation $y = e^{2x} - x^2 + x - 3$. Find the value of x for which $\dfrac{d^2y}{dx^2} = 0$.

6 Find the area of the region bounded by the curve with equation $y = e^{3x}$, the x-axis and the lines $x = \ln 2$ and $x = \ln 3$.

7 Find the area of the region enclosed by the curve with equation $y = e^{-x}$, the x-axis and the lines $x = \ln 2$ and $x = \ln 3$.

8 Find the area of the region bounded by the curve with equation $y = e^{2x}$, the x-axis and the lines $x = -\ln 3$ and $x = -\ln 2$.

9 Find the area of the region enclosed by the curve with equation $y = e^{\frac{x}{3}}$, the x-axis and the lines $x = -\ln 27$ and $x = \ln 8$.

4.7 The derivative of ln *x*

You can use the formula

$$f'(x) = \lim_{h \to 0} \left\{ \frac{f(x+h) - f(x)}{h} \right\} \text{ where } f(x) = \ln x.$$

This proof is included for completeness. It will not be tested in the C3 examination.

You only need to remember the key result below.

$$= \lim_{h \to 0} \left\{ \frac{\ln(x+h) - \ln x}{h} \right\}$$

$$= \lim_{h \to 0} \left\{ \frac{\ln\left(\dfrac{x+h}{x}\right)}{h} \right\}$$

$$= \lim_{h \to 0} \left\{ \frac{\ln\left(1 + \dfrac{h}{x}\right)}{h} \right\}$$

Making the substitution $s = \ln\left(1 + \dfrac{h}{x}\right)$ means

$e^s = 1 + \dfrac{h}{x} \Rightarrow e^s - 1 = \dfrac{h}{x}$ and so $h = x(e^s - 1)$.

When $h \to 0$ it means that $s \to 0$

Hence $f'(x) = \lim_{s \to 0} \left\{ \dfrac{s}{x(e^s - 1)} \right\}$

Since

$$\lim_{s \to 0} \left\{ \frac{(e^s - 1)}{s} \right\} = 1.$$

$$= \frac{1}{x} \times \lim_{s \to 0} \left\{ \frac{s}{(e^s - 1)} \right\}$$

$$= \frac{1}{x} \times 1 = \frac{1}{x}$$

The derivative of $\ln x$ is $\dfrac{1}{x}$.

You need to remember this formula.

4.8 The missing link

You have learned to integrate powers of x using

$$\int x^n dx = \frac{1}{n+1} x^{n+1} + \text{constant, for all values of } n \text{ except } n = -1.$$

You now have the missing link.
Using the result from the previous section,

since $\frac{d}{dx}(\ln x) = \frac{1}{x} = x^{-1}$, you can deduce that

$$\int \frac{1}{x} dx = \ln x + \text{constant}$$

You need to remember this formula.

4

Worked example 4.9

Find $\int_1^2 \frac{x^3 + 2}{x} dx.$

Solution

$$\int_1^2 \frac{x^3 + 2}{x} dx = \int_1^2 \left(x^2 + \frac{2}{x} \right) dx$$

$$= \left[\frac{x^3}{3} + 2 \ln x \right]_1^2 = \left(\frac{8}{3} + 2 \ln 2 \right) - \left(\frac{1}{3} + 2 \ln 1 \right)$$

$$= \frac{7}{3} + 2 \ln 2.$$

Worked example 4.10

A curve has equation $y = 5 + \ln 3x.$

(a) Find the gradient of the curve at the point where $x = 5$;

(b) Find the value of x at the point where the curve crosses the x-axis.

Solution

(a) $y = 5 + \ln 3x = 5 + \ln 3 + \ln x.$

Hence $\frac{dy}{dx} = 0 + 0 + \frac{1}{x} = \frac{1}{x}.$

When $x = 5$, gradient of curve is $\frac{1}{5}.$

(b) Solving $0 = 5 + \ln 3x.$

$$\Rightarrow \ln 3x = -5$$

$$\Rightarrow \quad 3x = e^{-5}$$

$$\Rightarrow \quad x = \frac{e^{-5}}{3}.$$

EXERCISE 4E

1 Find the derivative of each of the following:

(a) $\ln 2x$ (b) $x^2 - \ln 3x$ (c) $\ln x^4$ (d) $\ln \sqrt{x}$

2 Find: (a) $\int \dfrac{2}{x}\,dx$ (b) $\int_1^2 \dfrac{3}{x}\,dx$ (c) $\int_1^3 \dfrac{5 - x^2}{x}\,dx$

3 Find the area bounded by the curve with equation $y = \dfrac{6x^3 + 5}{x}$, the x-axis and the lines $x = 1$ and $x = 2$.

4 Find the value of the constant b such that $\int_e^b \dfrac{1}{x}\,dx = 2$.

5 The curve with equation $y = x^2 + 5 + \dfrac{2}{x}$ is defined for $x > 0$ and lies above the x-axis.

   (a) Find $\dfrac{dy}{dx}$ and hence find the coordinates of the stationary point M.

   (b) Calculate the value of $\dfrac{d^2y}{dx^2}$ at M.

   (c) Find the exact value of the area of the finite region bounded by the curve, the lines with equations $x = 1, x = 2$ and the x-axis.

6 The diagram shows a sketch of the curve $y = e^{2x} - 2$.

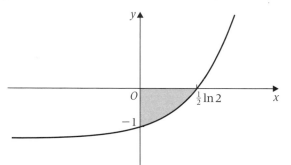

   (a) On separate diagrams, sketch the graphs of the following curves showing the coordinates of the points where the graph intersects the coordinate axes.
     (i) $y = |e^{2x} - 2|$,
     (ii) $y = e^{2x} - 5$.

   (b) (i) Find $\int (e^{2x} - 2)\,dx$.
     (ii) Hence show that the area of the shaded region bounded by the curve $y = e^{2x} - 2$ and the coordinate axes is $\ln 2 - \frac{1}{2}$.   [A]

7 A curve has equation $y = \dfrac{1}{4}x^3 - 6\ln x + 1$, $(x > 0)$

(a) (i) Find $\dfrac{dy}{dx}$.

(ii) Hence show that the gradient of the curve at the point where $x = \frac{2}{3}$ is $-8\frac{2}{3}$.

(b) (i) Given that the curve has just one stationary point, find the x-coordinate of this stationary point.

(ii) Find $\dfrac{d^2y}{dx^2}$.

(iii) Show that $\dfrac{d^2y}{dx^2} = 4.5$ at the stationary point and hence state the nature of the stationary point.

(c) The two points P and Q are on the curve $y = \dfrac{1}{4}x^3 - 6\ln x + 1$.

The x-coordinate of P is 4 and the x-coordinate of Q is 8. Find the gradient of the chord PQ in the form $a + b\ln 2$, where a and b are constants to be found. [A]

8 The function f is defined for all real **positive** values of x by

$$f(x) = e^{-2x} + \frac{3}{x} + 3.$$

(a) (i) Differentiate $f(x)$ with respect to x to find $f'(x)$.
(ii) Hence prove that f is a decreasing function.

(b) Find the range of f.

(c) Show that the area of the region bounded by the curve

$y = e^{-2x} + \dfrac{3}{x} + 3$, the x-axis and the lines $x = 1$ and $x = 2$ is

$$\frac{e^2 - 1}{2e^4} + 3(\ln 2 + 1).$$ [A]

9 A curve has equation $y = x^2 - 3x + \ln x + 2$, $(x > 0)$.

(a) (i) Find $\dfrac{dy}{dx}$.

(ii) Hence show that the gradient of the curve at the point where $x = 2$ is $1\frac{1}{2}$.

(b) (i) Show that the x-coordinates of the stationary points of the curve satisfy the equation $2x^2 - 3x + 1 = 0$.

(ii) Hence find the x-coordinates of each of the stationary points.

(iii) Find $\dfrac{d^2y}{dx^2}$.

(iv) Find the value of $\dfrac{d^2y}{dx^2}$ at each of the stationary points.

(v) Hence show that the y-coordinate of the maximum point is

$$\frac{3}{4} - \ln 2.$$ [A]

4.9 Polynomials involving exponentials and logarithms

Although an equation such as $e^{3y} - 2e^y + 1 = 0$ or $(\ln x)^2 - 3\ln x + 2 = 0$ might look rather frightening, it is only a disguised cubic or quadratic equation that you learned to solve earlier.

Worked example 4.11

Solve the equation $(\ln x)^2 - 3\ln x + 2 = 0$.

Solution

Firstly write $y = \ln x$. This gives a quadratic equation $y^2 - 3y + 2 = 0$ that can be factorised.

Hence $(y - 2)(y - 1) = 0$
$$\Rightarrow y = 2, y = 1$$

Therefore $\ln x = 2$ or $\ln x = 1$.
Hence $x = e^2$,   or $x = e^1 = e$.

Solutions are therefore $x = e, e^2$.

Worked example 4.12

Solve the equation $e^{3y} - 2e^{2y} - e^y + 2 = 0$.

Solution

You can write $e^y = x$. Hence $e^{3y} = (e^y)^3 = x^3$.

$$\Rightarrow e^{3y} - 2e^{2y} - e^y + 2 = x^3 - 2x^2 - x + 2 = 0.$$

Writing $p(x) = x^3 - 2x^2 - x + 2 = 0$, you can find a factor of $p(x)$ since $p(1) = 1 - 2 - 1 + 2 = 0$.

$$\Rightarrow (x - 1) \text{ is a factor.}$$

$p(x) = (x - 1)(ax^2 + bx + c)$
Comparing coefficients of x^3, $a = 1$.
Comparing constant terms: $-c = +2 \Rightarrow c = -2$
Therefore, $p(x) = (x - 1)(x^2 + bx - 2)$

Comparing coefficients of x: $-2 - b = -1 \Rightarrow b = -1$

The quadratic factor is $x^2 - x - 2 = (x - 2)(x + 1)$
$$\Rightarrow p(x) = (x - 1)(x + 1)(x - 2)$$
$$\Rightarrow x = 1, \quad x = -1, \quad x = 2$$

Hence $e^y = 1, \quad e^y = -1, \quad e^y = 2$

$$\Rightarrow y = 0, \quad y = \ln 2$$

> Alternatively, you could have used the factor theorem twice more and realised that $p(-1) = 0$ and $p(2) = 0$.

> Note that since $e^y > 0$, the equation $e^y = -1$ has no real solutions.

Worked examination question ─────────

The functions f and g are defined for all real values of x by
$$f(x) = 5x - 3$$
$$g(x) = e^{2x}.$$

(a) Write down the value of gf(0.6).

(b) State the range of g.

(c) Given that the domain of fg is the set of real numbers:
  (i) determine fg(x),
  (ii) state the range of fg.

(d) Show that g is an increasing function.

(e) The inverse of g is g^{-1}. Find $g^{-1}(x)$.   [A]

Solution

(a) $f(0.6) = 0 \Rightarrow gf(0.6) = g(0) = e^0 = 1$.

(b) Exponential functions are always positive so range of g is $g(x) > 0$.

(c) **(i)** $fg(x) = f(e^{2x}) = 5e^{2x} - 3$.
  (ii) Since $e^{2x} > 0$ it follows that $5e^{2x} - 3 > -3$.
    Hence the range of fg is $fg(x) > -3$.

(d) The derivative $g'(x) = 2\,e^{2x} > 0$ for all values of x.
Therefore g is an increasing function.

> This means that g is one-one and hence the inverse of g exists.

(e) If $y = g(x) = e^{2x}$ then for the inverse function
$x = g(y) = e^{2y}$.
Hence $2y = \ln x \Rightarrow y = \frac{1}{2} \ln x$.

Therefore $g^{-1}(x) = \frac{1}{2} \ln x$.

MIXED EXERCISE ─────────

1 Solve the following equations, where possible:
  (a) $e^{3x} = 2$ **(b)** $e^x = -3$ **(c)** $\ln 5x = 2$
  (d) $\ln 3x = -5$ **(e)** $\ln(5 - x) = e$
  (f) $\ln(2 - 3x) = 1$ **(g)** $\ln |3x| = 2$

2 Solve the following equations, where possible:
  (a) $e^{2x} - 9 = 0$ **(b)** $e^{2x} - 3e^x - 4 = 0$
  (c) $e^{2x} - 5e^x + 4 = 0$ **(d)** $e^{2x} + 5e^x + 6 = 0$
  (e) $e^{2x} - 4e^x + 2 = 0$

3 Solve the following equations:
  (a) $(\ln 2x)^2 = 16$ **(b)** $(\ln x)^2 - 5 \ln x + 6 = 0$
  (c) $(\ln x)^2 - \ln x - 20 = 0$ **(d)** $(\ln x)^2 - 4 \ln x - 1 = 0$

4 The function f is defined for $x > 0$ by $f(x) = \ln x$ and the function g is defined for all values of x by $g(x) = 2 - x$. The composite function fg has domain $x < 2$.

   (a) Find fg(1).

   (b) Sketch the graphs of $y = f(x)$ and $y = fg(x)$ on the same axes.

   (c) Describe a single geometric transformation which maps the graph of $y = f(x)$ onto the graph of $y = fg(x)$.   [A]

5 A curve has equation $y = 6e^{3x} + \sqrt{x}$, where $x \geqslant 0$. The region bounded by the curve, the coordinate axes and the line with equation $x = 1$ is R. Calculate the area of R.   [A]

6 The current, I amps, flowing through an electric circuit at time t seconds after closing a switch, is given by $I = 2e^{-\frac{t}{5}}$.

   (a) Write down the value of the current at the moment the switch is closed.

   (b) Determine the time in seconds for the current to decrease to 0.4 amps, giving your answer to three significant figures.

   (c) Calculate the rate of decrease of the current after exactly 3 seconds from the switch being closed, giving your answer to two significant figures. State clearly the units in your answer.   [A]

7 (a) Show that $(x + 3)$ is a factor of
$$p(x) = 2x^3 - 9x^2 - 38x + 21.$$

   (b) Express $p(x)$ as a product of three linear factors with integer coefficients.

   (c) Solve the equation $2e^{3y} - 9e^{2y} - 38e^y + 21 = 0$, leaving your answers in tcrms of natural logarithms.   [A]

8 A curve has equation $y = e^{2x} - 4x$.

   (a) Show that the x-coordinate of the stationary point on the curve is $\frac{1}{2} \ln 2$. Find the corresponding y-coordinate in the form $a + b \ln 2$, where a and b are integers to be determined.

   (b) Find an expression for $\dfrac{d^2y}{dx^2}$ and hence determine the nature of the stationary point.

   (c) Show that the area of the region enclosed by the curve, the x-axis and the lines $x = 0$ and $x = 1$ is $\frac{1}{2}(e^2 - 5)$.   [A]

9 The function f is defined for all positive values of x by

$$f(x) = 4 \ln x - \frac{1}{x}.$$

    (a) Find the derivative $f'(x)$.

    (b) Explain why f is an increasing function.

    (c) State, giving a reason, whether f^{-1} exists.   [A]

10 A curve has equation $y = e^{3x} - 24x$.

    (a) Determine $\dfrac{dy}{dx}$ and $\dfrac{d^2y}{dx^2}$ as functions of x.

    (b) The curve has a single stationary point, P. Find the x-coordinate of P in an exact form and show that its y-coordinate is $8(1 - 3 \ln 2)$.

        Find the value of $\dfrac{d^2y}{dx^2}$ at P and hence deduce the nature of the stationary point.

    (c) Find, in terms of e, the area of the region enclosed by the curve, the x-axis and the lines $x = 2$ and $x = 3$.   [A]

4

Key point summary

I The number e is irrational and its value is approximately 2.718.   *p62*

2 The curve with equation $y = e^x$ has derivative $\dfrac{dy}{dx} = e^x$.   *p63*

3 $\dfrac{d}{dx}(e^{kx}) = k\, e^{kx}$.   *p63*

4 $\displaystyle\int e^{kx} dx = \frac{1}{k}e^{kx} + \text{constant}.$ *p65*

5 The natural logarithm $\ln x$ is defined for $x > 0$. A useful equivalent statement is

    $x = e^y \Leftrightarrow y = \ln x.$ *p67*

6 The derivative of $\ln x$ is $\dfrac{1}{x}$.   *p70*

7 $\displaystyle\int \frac{1}{x} dx = \ln x + \text{constant}.$ *p71*

Test yourself	What to review
1 Find the gradient of the curve with equation $y = 5e^{3x} + 2e^{-x}$ at the point where $x = 0$.	*Section 4.4*
2 Calculate the area bounded by the curve with equation $y = 6x + 12e^{3x}$, the *x*-axis and the lines $x = 1$ and $x = 2$.	*Section 4.5*
3 Find the coordinates of the stationary point of the curve $y = 5 + 2x + e^{-2x}$ and determine its nature.	*Section 4.6*
4 Determine the exact value of $\int_{1}^{3} \dfrac{8x^4 - 5}{x}\, dx$.	*Section 4.8*
5 Solve the equation $(\ln x)^2 - 7 \ln x + 12 = 0$.	*Section 4.9*

Test yourself ANSWERS

1 13.

2 $9 + 4e^6 - 4e^3$.

3 $(0, 6)$, minimum.

4 $160 - 5 \ln 3$.

5 e^3, e^4.

CHAPTER 5

C3: Further differentiation and the chain rule

Learning objectives

After studying this chapter, you should be able to:
- find the derivatives of $\sin x$ and $\cos x$
- differentiate composite functions using the chain rule
- use the relationship $\dfrac{dy}{dx} = \dfrac{1}{\dfrac{dx}{dy}}$.

5

5.1 Review of applications of differentiation

The previous chapter introduced the functions $\ln x$ and e^x. You discovered how to differentiate and integrate these functions. In the C1 module you learnt how to find equations of tangents and normals. You also determined when a function was increasing or decreasing and when stationary points occurred on a curve. The following worked examples tie these ideas together.

Worked example 5.1

A curve is given by the equation $y = e^{3x} + x + 5$.

(a) Find the equation of the tangent to the curve at the point A where $x = 0$.

(b) The tangent at A crosses the x-axis at the point B. Find the coordinates of the point B.

(c) Find also the equation of the normal to the curve at A.

Solution

(a) A sketch of the curve and the tangent at A is shown below.

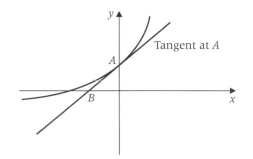

Firstly, $\dfrac{dy}{dx} = 3e^{3x} + 1$

$\Rightarrow$ the gradient of the tangent at $A = 3e^0 + 1 = 3 + 1 = 4$.
Also when $x = 0$, $y = e^0 + 0 + 5 = 6$.

So the tangent at A has gradient 4 and passes through the point $(0, 6)$.

$\Rightarrow$ the equation of the tangent at A is

$$y - 6 = 4(x - 0)$$

$\Rightarrow y = 4x + 6$

This is the equation of the tangent to the curve at $A(0, 6)$.

(b) To find where the tangent crosses the x-axis you put $y = 0$ in the equation of the tangent.

$$0 = 4x + 6$$

$\Rightarrow x = -1.5$

$\Rightarrow B$ is the point $(-1.5, 0)$

(c) For perpendicular lines, the product of the gradients $= -1$. The tangent at A has gradient 4.

Hence the normal has gradient $-\dfrac{1}{4}$.

It passes through the point $A(0, 6)$ and therefore the normal has equation $y = -\dfrac{1}{4}x + 6$.

> $\dfrac{d(e^{kx})}{dx} = ke^{kx}$
>
> $e^0 = 1$

Worked example 5.2

The function f is defined for all real values of x by
$f(x) = 2 - e^{3x} - x^3$.
Prove that f is a decreasing function.

Solution

Differentiating with respect to x:

$$f'(x) = 0 - 3e^{3x} - 3x^2 = -3(e^{3x} + x^2)$$

But e^{3x} is always positive. Also $x^2 \geqslant 0$ for all values of x. Hence $e^{3x} + x^2 > 0$ for all values of x.

Therefore $f'(x) < 0$ for all values of x.

So f is a decreasing function.

Worked example 5.3

The curve $y = 24 \ln x - x^3 + 8$ is defined for $x > 0$.

(a) Find the coordinates of the stationary point of the curve.

(b) Find the value of $\dfrac{d^2y}{dx^2}$ at the stationary point and hence determine whether it is a maximum or minimum point.

Solution

(a) $\dfrac{dy}{dx} = \dfrac{24}{x} - 3x^2$ and stationary points occur when $\dfrac{dy}{dx} = 0$.

$$\dfrac{24}{x} - 3x^2 = 0 \quad \Rightarrow \quad x^3 = 8$$

Hence $x = 2$ and therefore $y = 24\ln 2 - 8 + 8 = 24\ln 2$. So the coordinates of the stationary point are $(2, 24\ln 2)$.

(b) $\dfrac{d^2y}{dx^2} = -\dfrac{24}{x^2} - 6x$.

when $x = 2$, $\dfrac{d^2y}{dx^2} = -\dfrac{24}{4} - 12 = -18$.

Hence the stationary point is a maximum.

EXERCISE 5A

1 Find the equations of the tangent and the normal to the curve $y = 2e^{3x} + 4$ at the point where $x = 0$.

2 Show that the equation of the tangent to the curve $y = 7x^2 - 2e^x$ at the point where $x = 2$ is $y = (28 - 2e^2)x + 2e^2 - 28$.

3 A curve is defined for $x > 0$ by $y = 2\ln x + 5x$.

(a) Prove that the curve has no real stationary points.

(b) Find the equation of the tangent to this curve at the points:

    (i) P where $x = 1$,

    (ii) Q where $x = \frac{1}{2}$.

       (Leave your answer in terms of $\ln 2$.)

4 The function f is defined for $x > 0$ by $f(x) = 3\ln x - \dfrac{2}{x}$. Prove that f is an increasing function.

5 The function f is defined for all real **positive** values of x by

$$f(x) = e^{-2x} + \dfrac{3}{x} + 3.$$

(a) **(i)** Differentiate $f(x)$ with respect to x to find $f'(x)$.

    (ii) Hence prove that f is a decreasing function.

(b) Find the range of f. [A]

6 The diagram shows a sketch of the curve with equation $y = e^{2x} - 3$ which crosses the y-axis at the point A and the x-axis at the point B.

(a) Find the coordinates of A.

(b) Find the exact value of the x-coordinate of B.

(c) Show that the gradient of the curve at B is three times its gradient at A. [A]

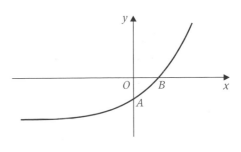

7 A curve has equation $y = \dfrac{6}{x} + 4 + 2\ln x$, $(x > 0)$.

(a) Find the stationary point, M, of the curve.

(b) Determine whether M is a maximum or minimum point.

(c) Find an equation of the normal to the curve at the point where $x = 1$.

5.2 Derivatives of sin x and cos x

The graph of $y = \sin x$, where x is measured in radians, is sketched below for $0 \leqslant x \leqslant 2\pi$.

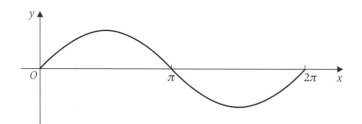

The gradient of the curve is equal to 1 at the origin and then decreases so that the gradient becomes zero when $x = \dfrac{\pi}{2}$ and then -1 when $x = \pi$.

Since $\cos^2\theta + \sin^2\theta = 1$, the point with coordinates $(\cos\theta, \sin\theta)$ always lies on the unit circle, i.e. the circle with its centre at the origin and with radius 1.

> No doubt you can think of a function g which is such that
> $g(0) = 1$, $g\!\left(\dfrac{\pi}{2}\right) = 0$ and
> $g(\pi) = -1$. If so you may be able to guess the derivative of the function $\sin x$.

The function f such that $f(x) = \sin x$ can therefore be thought of as the distance above the horizontal line through the centre of the circle, O, shown.

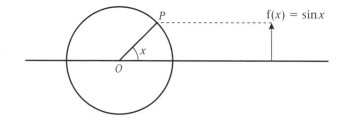

Suppose you choose another point Q on the unit circle close to P, where the angle the radius now makes with the horizontal is $x + h$, where h is small.

Since the radius of the circle is 1, the length of the arc PQ (using the formula that arc length $= r\theta$) is equal to $1 \times h = h$.

If the arc PQ is very small then the chord PQ is approximately of length h, and angle OPQ is approximately a right angle.

The point Q is at a distance $\sin(x + h)$ from the horizontal line and so this distance can be written as $\text{f}(x + h)$.

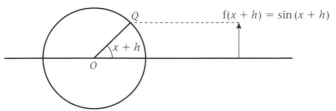

Drawing the vertical line QN and the horizontal line PN, the right-angled triangle PNQ is produced. Angle NPO must equal x, therefore angle NQP is equal to x.

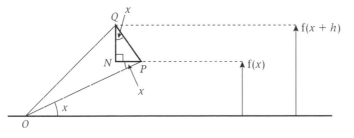

The difference in heights is QN and this is the same as $\text{f}(x + h) - \text{f}(x)$, where $\text{f}(x) = \sin x$.

Since the length of PQ is approximately equal to h, when h is small, the length of $QN \approx PQ \cos x \approx h \cos x$.

The derivative of $\text{f}(x) = \sin x$ is given by

$$\lim_{h \to 0}\left\{\frac{\text{f}(x + h) - \text{f}(x)}{h}\right\} = \lim_{h \to 0}\left\{\frac{h \cos x}{h}\right\} = \cos x.$$

Therefore the derivative of $\sin x$ can be seen to equal $\cos x$, where x is in radians.

$$\frac{\text{d}(\sin x)}{\text{d}x} = \cos x, \text{ where } x \text{ is in radians.}$$

This result must be learnt.

A similar argument can be made using $\text{g}(x) = \cos x$, to represent the horizontal distance of P from a vertical line through the centre of the circle.

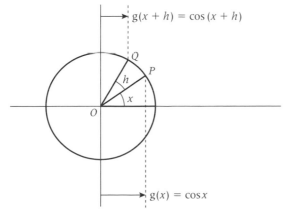

5

Using the triangle *NPQ* in the previous diagram:

The horizontal distance $NP = \cos x - \cos(x + h)$

But $NP \approx h \sin x$.

Hence $g(x + h) - g(x) = \cos(x + h) - \cos x \approx -h \sin x$.

The derivative of $g(x) = \cos x$ is given by

$$\lim_{h \to 0} \left\{ \frac{g(x + h) - g(x)}{h} \right\} \approx \lim_{h \to 0} \left\{ \frac{-h \sin x}{h} \right\} = -\sin x.$$

Hence the derivative of $\cos x$ is $-\sin x$.

> $$\frac{d(\cos x)}{dx} = -\sin x, \text{ where } x \text{ is in radians.}$$

Note the reason for the minus sign.

This result must be learnt. Remember the minus sign when you differentiate cos *x*.

Worked example 5.4

The curve $y = 3 \cos x + x + 2$ is defined for $0 \leqslant x \leqslant \dfrac{\pi}{2}$.

(a) Find the *x*-coordinate of the stationary point, *M*, of the curve.

    (b) (i) Find $\dfrac{d^2y}{dx^2}$.

        (ii) Hence determine whether *M* is a maximum or minimum point.

Solution

(a) $\dfrac{dy}{dx} = -3 \sin x + 1.$

The stationary point occurs when

$\sin x = \dfrac{1}{3} \quad \Rightarrow \quad x = \sin^{-1}\left(\dfrac{1}{3}\right).$

The *x*-coordinate of *M* is approximately 0.3398.

(b) (i) $\dfrac{d^2y}{dx^2} = -3 \cos x.$

When $x = \sin^{-1}\left(\dfrac{1}{3}\right)$, $\dfrac{d^2y}{dx^2} < 0.$

Hence *M* is a maximum point.

Worked example 5.5

The function f is defined for all real values of *x* by

$f(x) = 5 \sin x - 6x + 7.$

(a) Prove that f is a decreasing function.

(b) Find an equation for the tangent to the curve with equation $y = f(x)$ at the point where $x = 0$.

Solution

(a) Differentiating with respect to x:

$$f'(x) = 5\cos x - 6$$

But any value of $\cos x$ lies between -1 and $+1$.

Hence $-11 \leqslant f'(x) \leqslant -1$, so the derivative is always negative.

$$f'(x) < 0 \quad \Rightarrow \quad f \text{ is a decreasing function.}$$

(b) When $x = 0$, $f'(x) = 5\cos x - 6 = 5 - 6 = -1$, so the gradient of the tangent is -1.

Also when $x = 0$, $f(x) = 5\sin x - 6x + 7 = 0 - 0 + 7 = 7$, so the tangent passes through the point $(0, 7)$.

Hence the tangent has equation $y = -x + 7$ or $x + y = 7$.

EXERCISE 5B

1 Find $\dfrac{dy}{dx}$ for each of the following:

    (a) $y = 5\sin x + x^3$ **(b)** $y = 8\sin x + 2e^{3x}$

    (c) $y = 2\cos x + 5x - 4$ **(d)** $y = \ln x - 7\cos x$

    (e) $y = 4\sin x + 5\cos x$ **(f)** $y = 6\sin x - 8\cos x$

2 The curve $y = 4\sin x - 3x + 1$ is defined for $0 \leqslant x \leqslant \dfrac{\pi}{2}$.

    (a) Find the x-coordinate of the stationary point, M, of the curve.

    (b) (i) Find $\dfrac{d^2y}{dx^2}$.

          (ii) Hence determine whether M is a maximum or minimum point.

3 The curve $y = 5\cos x + 2x + 4$ is defined for $-\dfrac{\pi}{2} \leqslant x \leqslant \dfrac{\pi}{2}$.

    (a) Find the x-coordinate of the stationary point, P, of the curve.

    (b) (i) Find $\dfrac{d^2y}{dx^2}$.

          (ii) Hence determine whether P is a maximum or minimum point.

    (c) Find an equation for the tangent to the curve at the point where $x = 0$.

4 The curve $y = 5 - 6\sin x - 3x$ is defined for $0 \leqslant x \leqslant \pi$.

    (a) Find the x-coordinate of the stationary point, Q, of the curve.

    (b) (i) $\dfrac{d^2y}{dx^2}$.

          (ii) Hence determine whether Q is a maximum or minimum point.

    (c) Find an equation for the normal to the curve at the point where $x = \dfrac{\pi}{2}$.

5 Show that the curve with equation $y = 2\cos x + 8x - 3$ has no stationary points.

6 The function f is defined for all real values of x by

$$f(x) = 4\sin x + 5x - 7.$$

(a) Prove that f is an increasing function.

(b) Find an equation for the normal to the curve with equation $y = f(x)$ at the point where $x = 0$.

5.3 Chain rule for composite functions

Consider the expression $y = (2x - 3)^5$.
In order to differentiate with respect to x you will first have to use the binomial expansion and then differentiate term by term.

This is very time consuming and so fortunately there is a much quicker and easier method that relies upon noticing that the function is built up as a composite of two simpler functions.

By substituting $u = 2x - 3$, you can then write y in terms of u as

$$y = u^5.$$

Therefore,

$$\frac{dy}{du} = 5u^4 \quad \text{and} \quad \frac{du}{dx} = 2.$$

But how does this help to find $\dfrac{dy}{dx}$?

Consider the general composite function $y = fg(x)$.

Using the same notation as above,

$$u = g(x) \quad \text{and} \quad y = f(u).$$

Now making a small increase of δx in x will create a small increase of δu in u since u is a function of x. This will, in turn, create a small increase of δy in y since y is a function of u.

Clearly $\dfrac{\delta y}{\delta x} = \dfrac{\delta y}{\delta u} \times \dfrac{\delta u}{\delta x}$

> Since the δu's cancel each other out on the right-hand side.

You have seen previously that

as $\delta x \to 0$, $\dfrac{\delta y}{\delta x} \to \dfrac{dy}{dx}$ and similarly $\dfrac{\delta u}{\delta x} \to \dfrac{du}{dx}$.

> Since the small increase in u was created by the small increase in x.

It should also be clear that as $\delta x \to 0$, $\delta u \to 0$.

So as $\delta x \to 0$, $\dfrac{\delta y}{\delta u} \to \dfrac{dy}{du}$ giving the result

$$\frac{dy}{dx} = \frac{dy}{du} \times \frac{du}{dx}$$

or in function notation

$$\frac{dy}{dx} = f'(u) \times g'(x)$$

where $y = fg(x)$ and $u = g(x)$. This result is known as the **chain rule**.

The chain rule is an important and very useful technique.

Worked example 5.6

Differentiate $y = (2x - 3)^5$.

Solution

You have already seen above that you can write $u = 2x - 3$ and $y = u^5$ giving

$$\frac{du}{dx} = 2 \quad \text{and} \quad \frac{dy}{du} = 5u^4.$$

Using the chain rule gives

$$\frac{dy}{dx} = \frac{dy}{du} \times \frac{du}{dx} = 5u^4 \times 2 = 10u^4$$

but since the original function was given in terms of x and not u you should rewrite u in terms of x, so that

$$\frac{dy}{dx} = 10(2x - 3)^4$$

Worked example 5.7

Use the chain rule to differentiate the following composite functions with respect to x.

(a) $y = (3x + 5)^2$

(b) $y = e^{5x}$

Solution

(a) Here you can write

$$u = 3x + 5 \text{ and } y = u^2$$

$$\Rightarrow \frac{du}{dx} = 3 \quad \text{and} \quad \frac{dy}{du} = 2u$$

Using the chain rule gives,

$$\frac{dy}{dx} = \frac{dy}{du} \times \frac{du}{dx} = 2u \times 3 = 6u$$

$$\Rightarrow \frac{dy}{dx} = 6(3x + 5)$$

This result can be verified using the 'old' method of expanding brackets:
$$y = (3x + 5)(3x + 5)$$
$$= 9x^2 + 30x + 25$$
$$\Rightarrow \frac{dy}{dx} = 18x + 30 = 6(3x + 5)$$

(b) Here you can write

$$u = 5x \text{ and } y = e^u$$

$$\Rightarrow \quad \frac{du}{dx} = 5 \text{ and } \frac{dy}{du} = e^u$$

Using the chain rule gives,

$$\frac{dy}{dx} = \frac{dy}{du} \times \frac{du}{dx} = e^u \times 5 = 5e^u$$

$$\Rightarrow \quad \frac{dy}{dx} = 5e^{5x}$$

> Notice that the chain rule confirms the result you met in Chapter 4:
> $$\frac{d}{dx}(e^{kx}) = ke^{kx}$$

Worked example 5.8

Differentiate the following composite functions with respect to x:

(a) $y = \sin(4x - 7)$ **(b)** $y = 5\cos(2 - 3x)$

(c) $y = 3\ln(x^2 + 4x - 1)$

Solution

(a) Writing

$$u = 4x - 7 \text{ and } y = \sin u$$

$$\Rightarrow \quad \frac{du}{dx} = 4 \quad \text{and} \quad \frac{dy}{du} = \cos u$$

The chain rule gives,

$$\frac{dy}{dx} = \frac{dy}{du} \times \frac{du}{dx} = \cos u \times 4 = 4\cos u$$

$$\Rightarrow \quad \frac{dy}{dx} = 4\cos(4x - 7)$$

(b) You can write

$$u = 2 - 3x \quad \text{and} \quad y = 5\cos u$$

$$\Rightarrow \quad \frac{du}{dx} = -3 \quad \text{and} \quad \frac{dy}{du} = -5\sin u$$

The chain rule gives,

$$\frac{dy}{dx} = \frac{dy}{du} \times \frac{du}{dx} = -5\sin u \times -3 = 15\sin u$$

$$\Rightarrow \quad \frac{dy}{dx} = 15\sin(2 - 3x)$$

(c) Making the substitution

$$u = x^2 + 4x - 1 \quad \text{and} \quad y = 3\ln u$$

$$\Rightarrow \quad \frac{du}{dx} = 2x + 4 \quad \text{and} \quad \frac{dy}{du} = \frac{3}{u}$$

The chain rule gives,

$$\frac{dy}{dx} = \frac{dy}{du} \times \frac{du}{dx} = \frac{3}{u} \times (2x + 4) = \frac{3(2x + 4)}{u}$$

$$\Rightarrow \quad \frac{dy}{dx} = \frac{3(2x + 4)}{x^2 + 4x - 1}$$

With enough practice, it soon becomes possible to dispense with the formal approach of defining u in terms of x and then defining y in terms of u. You can simply visualise differentiating each 'layer' of the composite function and multiplying the derivatives together.

Worked example 5.9

Find $\dfrac{dy}{dx}$ for the following functions:

(a) $y = 5(2 - 6x)^3$ **(b)** $y = 2\sin(4x - 5)$

Solution

(a) $y = 5(2 - 6x)^3$

$\Rightarrow \dfrac{dy}{dx} = 15(2 - 6x)^2 \times -6 = -90(2 - 6x)^2$

(b) $y = 2\sin(4x - 5)$

$\Rightarrow \dfrac{dy}{dx} = 2\cos(4x - 5) \times 4 = 8\cos(4x - 5)$

Differentiating $5(\ )^3$ gives $15(\ )^2$ and differentiating $2 - 6x$ gives -6, then multiply the two together.

Differentiating $2\sin(\ )$ gives $2\cos(\ )$ and differentiating $4x - 5$ gives 4, then multiply the two together.

EXERCISE 5C

1 Use the chain rule to differentiate the following composite functions with respect to x:

(a) $y = (8x + 3)^4$ (use $u = 8x + 3$ and $y = u^4$)
(b) $y = (3 - 4x)^6$ (use $u = 3 - 4x$ and $y = u^6$)
(c) $y = 5(3x - 2)^{10}$ (use $u = 3x - 2$ and $y = 5u^{10}$)
(d) $y = 7(7 - x)^8$ (use $u = 7 - x$ and $y = 7u^8$)

In the first few questions you have been given u and $y = g(u)$, but as you gain confidence working through the exercise, you may wish to practise finding the derivative by simply visualising these.

2 Differentiate the following with respect to x:

(a) $y = e^{5x - 2}$ (use $u = 5x - 2$ and $y = e^u$)
(b) $y = 3\sin(6x + 2)$ (use $u = 6x + 2$ and $y = 3\sin u$)
(c) $y = 2\cos 8x$ (use $u = 8x$ and $y = 2\cos u$)
(d) $y = 5\ln(9x + 1)$ (use $u = 9x + 1$ and $y = 5\ln u$)
(e) $y = \dfrac{1}{2x + 7}$ (use $u = 2x + 7$ and $y = u^{-1}$)

3 Find $\dfrac{dy}{dx}$ for each of these:

(a) $y = 3\ln(3x^2 + 3x - 1)$ ($u = 3x^2 + 3x - 1$ and $y = 3\ln u$)
(b) $y = \dfrac{4}{3 - 2x}$ ($u = 3 - 2x$ and $y = 4u^{-1}$)
(c) $y = \dfrac{2}{(6x + 1)^2}$ ($u = 6x + 1$ and $y = 2u^{-2}$)
(d) $y = \dfrac{1}{e^x + 4}$ ($u = e^x + 4$ and $y = u^{-1}$)
(e) $y = e^{x^2}$ ($u = x^2$ and $y = e^u$).

4 Differentiate each of the following with respect to x:

(a) $y = 2(4x - 1)^6$

(b) $y = \dfrac{(8 - 3x)^5}{15}$

(c) $y = \dfrac{\sin(6x + 1)}{12}$

(d) $y = 4\ln(3 - x)$

(e) $y = 3e^{9 - 2x}$

(f) $y = -\cos(5 - 3x)$

(g) $y = e^{3x} + 6x^2$

(h) $y = 6x^3 - 2\ln 3x$

(i) $y = \dfrac{4\sin(6x - 5)}{3}$

(j) $y = e^{4 - 3x} + \sin 5x$

5 Find the gradient of the following curves at the specified point:

(a) $y = 3(2x - 1)^4$, where $x = 2$,

(b) $y = 4\cos 2\theta$, where $\theta = \dfrac{\pi}{4}$,

(c) $y = 4e^{3 - x}$, where $x = 2$,

(d) $y = 3\ln(x^2 + 2)$, where $x = -1$,

(e) $y = 5x^2 - (2x + 1)^3$, where $x = -2$.

6 Find the equation of the tangent to the curve given by $y = (2x - 3)^4$ at the point $P(1, 1)$.

7 A curve has equation $y = 4 - \dfrac{2}{(2x - 1)^2} - x$.

(a) Find $\dfrac{dy}{dx}$ and hence find the coordinates of the stationary point of the curve.

(b) Find $\dfrac{d^2y}{dx^2}$ and hence determine the nature of the stationary point.

8 A curve has equation $y = (3x - 2)^3 - 5x^2$.

(a) Find $\dfrac{dy}{dx}$ and hence find the coordinates of the two stationary points of the curve.

(b) Find $\dfrac{d^2y}{dx^2}$ and hence determine the nature of each of these stationary points.

5.4 Extending the chain rule

You saw in section 5.3 how the chain rule can be used to differentiate composite functions.

You may have noticed that all of the composite functions considered so far have been built up of **two** simpler functions. The basic principle, however, can be extended to differentiate composite functions built up of three or more simpler functions. For example, $\sin^2(2x + 1)$, $\ln(\sin 2x)$, etc.

The chain rule can be extended for three functions to

$$\frac{dy}{dx} = \frac{dy}{dv} \times \frac{dv}{du} \times \frac{du}{dx}, \text{ if } y = \text{fgh}(x)$$

> You can imagine the dvs and dus cancelling on the right-hand side.

where $u = \text{h}(x)$, $v = \text{g}(u)$ and $y = \text{f}(v)$.

Worked example 5.10

Differentiate the following functions with respect to x:

(a) $y = \sin^3(4x + 1)$ **(b)** $y = \ln(\cos 3x)$

Solution

(a) Here you can write

$$u = 4x + 1, \quad v = \sin u \quad \text{and} \quad y = v^3$$

$$\Rightarrow \frac{du}{dx} = 4, \frac{dv}{du} = \cos u \quad \text{and} \quad \frac{dy}{dv} = 3v^2$$

$$\Rightarrow \frac{dy}{dx} = \frac{dy}{dv} \times \frac{dv}{du} \times \frac{du}{dx} = 3v^2 \times \cos u \times 4 = 12v^2 \cos u$$

But you need to rewrite the expression in terms of x

$$\frac{dy}{dx} = 12 \sin^2(4x + 1) \cos(4x + 1)$$

(b) Here you can write

$$u = 3x; \quad v = \cos u; \quad y = \ln(v)$$

$$\frac{du}{dx} = 3; \quad \frac{dv}{du} = -\sin u; \quad \frac{dy}{dv} = \frac{1}{v}.$$

$$\frac{dy}{dx} = \frac{dy}{dv} \times \frac{dv}{du} \times \frac{du}{dx} = \frac{1}{\cos 3x} \times -\sin 3x \times 3$$

$$= -\frac{3 \sin 3x}{\cos 3x} = -3 \tan 3x$$

> Again, with practice, you can visualise differentiating one 'layer' at a time.

EXERCISE 5D

Differentiate each of the following functions using the chain rule:

1 $y = \sin^2 5x$ **2** $y = \cos^4 9x$

3 $y = \ln(\sin 4x)$ **4** $y = 2 \cos^5 4x$

5 $y = (e^{2x} - 5)^3$ **6** $y = \sin \sqrt{3x + 1}$

7 $y = (\ln \sqrt{x})^2$ **8** $y = e^{\sqrt{6x}}$

9 $y = \cos(\ln 8x)$ **10** $y = \cos^5 \sqrt{x}$

5.5 The reciprocal of the first derivative

You now have quite a few techniques for finding the derivative $\frac{dy}{dx}$ of a function. You may also occasionally find it necessary to find the derivative $\frac{dx}{dy}$.

Consider a small increase in x, δx, which produces a small increase in y, δy.

Now clearly $\dfrac{\delta y}{\delta x} = \dfrac{1}{\dfrac{\delta x}{\delta y}}$

and taking the limit as $\delta x \to 0$ gives

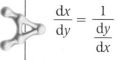

$$\frac{dy}{dx} = \frac{1}{\dfrac{dx}{dy}}$$

> While this result may appear obvious it cannot be taken for granted that similar results are also true. For example, the corresponding result for second (or any higher order) derivatives is **not** true.

Since by the chain rule $\dfrac{dy}{dx} \times \dfrac{dx}{dy} = \dfrac{dy}{dy} = 1$, it also follows immediately that $\dfrac{dy}{dx} = \dfrac{1}{\dfrac{dx}{dy}}$.

The key point above can also be written in the form:

$$\frac{dx}{dy} = \frac{1}{\dfrac{dy}{dx}}$$

Worked example 5.11

Verify the result above in the cases where:

(a) $y = \sqrt{x}$ (b) $y = \ln x$

Solution

(a) Firstly, $y = x^{\frac{1}{2}} \Rightarrow \dfrac{dy}{dx} = \dfrac{1}{2\sqrt{x}}$

Now $y = \sqrt{x} \Rightarrow x = y^2$

which gives $\dfrac{dx}{dy} = 2y = 2\sqrt{x}$.

Using the result gives

$\dfrac{dy}{dx} = \dfrac{1}{\left(\dfrac{dx}{dy}\right)} = \dfrac{1}{2\sqrt{x}}$ as expected.

(b) Firstly, $y = \ln x \Rightarrow \dfrac{dy}{dx} = \dfrac{1}{x}$.

Now $y = \ln x \Rightarrow x = e^y$,

which gives $\dfrac{dx}{dy} = e^y = e^{\ln x} = x$.

Using the result gives

$\dfrac{dy}{dx} = \dfrac{1}{\left(\dfrac{dx}{dy}\right)} = \dfrac{1}{x}$ as expected.

Worked example 5.12

Find $\dfrac{dy}{dx}$ for each of the following functions and hence write down an expression for $\dfrac{dx}{dy}$.

(a) $y = 5 \ln 3x$ **(b)** $y = (4x - 3)^3$ **(c)** $y = \dfrac{3}{5x - 1}$

Solution

(a) $y = 5 \ln 3x \Rightarrow \dfrac{dy}{dx} = \dfrac{5}{3x} \times 3 = \dfrac{5}{x}$

Using the result $\dfrac{dx}{dy} = \dfrac{1}{\dfrac{dy}{dx}}$ gives $\dfrac{dx}{dy} = \dfrac{x}{5}$.

> Notice the use of the chain rule to differentiate the composite function.

Instead of using the chain rule, you could have obtained this result using the laws of logarithms:

$$y = 5 \ln 3x = 5(\ln 3 + \ln x) = 5 \ln 3 + 5 \ln x$$

But $5 \ln 3$ is a constant and vanishes when you differentiate. Hence $\dfrac{dy}{dx} = \dfrac{5}{x}$ and therefore, as before,

$\dfrac{dx}{dy} = \dfrac{x}{5}$.

(b) $y = (4x - 3)^3 \Rightarrow \dfrac{dy}{dx} = 3(4x - 3)^2 \times 4 = 12(4x - 3)^2$

Using the reciprocal derivative rule gives

$$\dfrac{dx}{dy} = \dfrac{1}{12(4x - 3)^2}.$$

(c) $y = \dfrac{3}{5x - 1} = 3(5x - 1)^{-1}$

$$\Rightarrow \dfrac{dx}{dy} = -3(5x - 1)^{-2} \times 5 = -\dfrac{15}{(5x - 1)^2}$$

Using the result $\dfrac{dx}{dy} = \dfrac{1}{\dfrac{dy}{dx}}$ gives

$$\dfrac{dx}{dy} = -\dfrac{(5x - 1)^2}{15}.$$

Notice that you could have rearranged the original equation to confirm this result.

$$y = \frac{3}{5x - 1} \quad \Rightarrow \quad (5x - 1)y = 3 \quad \Rightarrow \quad 5xy - y = 3$$

Hence $5xy = 3 + y \quad \Rightarrow \quad x = \frac{3}{5}y^{-1} + \frac{1}{5}$

$$\frac{dx}{dy} = -\frac{3}{5}y^{-2}, \text{ but } y = \frac{3}{5x - 1}, \text{ so } y^{-2} = \frac{(5x - 1)^2}{9}.$$

Hence $\dfrac{dx}{dy} = -\dfrac{3}{5} \times \dfrac{(5x - 1)^2}{9} = -\dfrac{(5x - 1)^2}{15}.$

MIXED EXERCISE

1 Find $\dfrac{dy}{dx}$ for each of the following functions and hence write down an expression for $\dfrac{dx}{dy}$:

   (a) $y = 3\ln 4x$

   (b) $y = (2x + 4)^{-5}$

   (c) $y = \dfrac{4}{2x - 3}$

   (d) $y = \cos 2x$

   (e) $y = 3 + 4\sin 5x$

   (f) $y = e^{2x} + 5\cos 3x$

2 Given that $y = 3e^{\sin 2x}$, find $\dfrac{dy}{dx}$. Hence show that

$$\frac{dx}{dy} = \frac{\sec 2x}{2y}.$$

3 A curve has equation $y = 3\sin 2x + \cos 2x$.

   (a) (i) Find $\dfrac{dy}{dx}$.

      (ii) Hence show that any stationary points occur when $\tan 2x = 3$.

   (b) Find an equation of the tangent to the curve at the point where $x = \dfrac{\pi}{4}$.

4 A curve has equation $y = 4\sin 3x + 2\cos 3x + 7$.

   (a) (i) Find $\dfrac{dy}{dx}$.

      (ii) Hence show that any stationary points occur when $\tan 3x = 2$.

      (iii) Find the x-coordinates of all the stationary points for which $0 < x < \pi$.

   (b) Find an equation of the normal to the curve at the point where $x = \dfrac{\pi}{3}$.

5 Given that $x = 3 \sin y$, find $\dfrac{dx}{dy}$ in terms of y.

(a) Hence find $\dfrac{dy}{dx}$ in terms of y.

(b) Hence show that $\dfrac{dy}{dx} = \dfrac{1}{\sqrt{9 - x^2}}$.

> **Hint:** You may find the formula $\sin^2 y + \cos^2 y = 1$ useful.

(c) Use the result from **(b)** to write down a function of x that has derivative

$$\frac{1}{\sqrt{9 - x^2}}.$$

6 Given that $x = \ln(5 + y)$, find $\dfrac{dx}{dy}$ in terms of y.

(a) Hence find $\dfrac{dy}{dx}$ in terms of y.

(b) Show that $y = e^x - 5$ and find $\dfrac{dy}{dx}$ in terms of x. Show that this confirms your answer to **(a)**.

7 (a) Find $\dfrac{dy}{dx}$ when $y = \sin(x^2 + 3)$.

> This idea will be explored further in chapter 8.

(b) Hence find $\displaystyle\int x \cos(x^2 + 3)\, dx$.

8 The depth of water at the end of a pier is H metres at time t hours after high tide, where

$$H = 20 + 5 \cos(kt)$$

and k is a positive constant measured in radians per hour.

(a) Write down the greatest and least depths of water at the end of the pier.

(b) Sketch the graph of H against t from $t = 0$ to the time of the next high tide.

(c) The time interval between successive high tides is 11 hours and 15 minutes. Calculate the value of k, giving your answer to three significant figures.

(d) Calculate the rate of change of the depth of water exactly two hours after high tide, giving your answer in metres per hour. Interpret the sign of your answer.   [A]

Key point summary

1 $\dfrac{d(\sin x)}{dx} = \cos x$, where x is in radians. *p83*

2 $\dfrac{d(\cos x)}{dx} = -\sin x$, where x is in radians. *p84*

3 The chain rule can be used to differentiate composite functions. The chain rule is

p86 and p87

$$\frac{dy}{dx} = \frac{dy}{du} \times \frac{du}{dx}$$ or in function notation

$$\frac{dy}{dx} = f'(u) \times g'(x) \text{ where } y = fg(x) \text{ and } u = g(x).$$

4 The chain rule can be extended to

p91

$$\frac{dy}{dx} = \frac{dy}{dv} \times \frac{dv}{du} \times \frac{du}{dx}.$$

5 $$\frac{dy}{dx} = \frac{1}{\frac{dx}{dy}} \quad \text{or} \quad \frac{dx}{dy} = \frac{1}{\frac{dy}{dx}}$$

p92

Test yourself

What to review

1 Find an equation of the tangent to the curve $y = 3x + \ln x - 5$ at the point $P(1, -2)$.

Section 5.1

2 The function f is defined for all real values of x by

$$f(x) = 5 \cos x + 7x - 2.$$

Prove that f is an increasing function.

Section 5.2

3 Differentiate the following functions with respect to x:

(a) $y = 5(3x - 2)^4$ **(b)** $y = 4 \sin 7x$ **(c)** $y = 4e^{3x-2} + 3x^3$

Section 5.3

4 Find $\dfrac{dy}{dx}$ when:

(a) $y = \sin^4 2x$ **(b)** $y = e^{(2x+1)^3}$ **(c)** $y = \ln(\cos 4x)$.

Section 5.3

5 Find $\dfrac{dy}{dx}$ for each of the following and hence write down an expression for $\dfrac{dx}{dy}$:

(a) $y = 3(5 - 2x)^5$ **(b)** $y = \dfrac{1}{3x - 5}$

Section 5.4

Test yourself   ANSWERS

5 (a) $\dfrac{dy}{dx} = -\dfrac{1}{30(5 - 2x)^4}$;   **(b)** $\dfrac{dy}{dx} = -\dfrac{3}{(3x - 5)^2}$.

4 (a) $8 \sin^3 2x \cos 2x$;   **(b)** $6(2x + 1)^2 e^{(2x+1)^3}$;   **(c)** $-\dfrac{4 \sin 4x}{\cos 4x}$ or $-4 \tan 4x$.

3 (a) $60(3x - 2)^3$;   **(b)** $28 \cos 7x$;   **(c)** $12e^{3x-2} + 9x^2$.

2 $f'(x) = -5 \sin x + 7$. Since $\sin x \geqslant -1$ for all values of x, $f'(x) > 0$ and hence f is increasing.

1 $y = 4x - 6$.

CHAPTER 6

C3: Differentiation using the product rule and the quotient rule

Learning objectives

After studying this chapter, you should be able to:
■ differentiate using the product rule
■ differentiate using the quotient rule
■ differentiate $\tan x$, $\cot x$, $\sec x$ and $\csc x$.

6.1 Introduction

In this chapter you will be introduced to methods for differentiating expressions such as

$y = x^3 \sin x$ which is the *product* of x^3 and $\sin x$.

You will also learn how to differentiate *quotients* such as

$$y = \frac{x^3}{\sin x}.$$

6.2 Differentiation using the product rule

Consider the product

$$y = u \times v,$$

where u and v are each functions of x.

Making a small change δx in x produces a small change in each of the expressions u, v and y.

Let the small changes in u, v and y be δu, δv and δy, respectively.

Since

$$y = u \times v$$

making the small change in x gives

$$\begin{aligned}
y + \delta y &= (u + \delta u)(v + \delta v) \\
\Rightarrow y + \delta y &= uv + u\delta v + v\delta u + \delta u \delta v \\
\Rightarrow \delta y &= uv + u\delta v + v\delta u + \delta u \delta v - y \\
\Rightarrow \delta y &= u\delta v + v\delta u + \delta u \delta v
\end{aligned}$$

> **Common error:**
> You may think that if
> $y = x^3 \sin x$ then $\dfrac{dy}{dx} = 3x^2 \cos x$
> and if $y = \dfrac{x^3}{\sin x}$ then
> $\dfrac{dy}{dx} = \dfrac{3x^2}{\cos x}$.
>
> In other words, just differentiate each part of the product or quotient separately.
> Be warned! This approach **does not lead to the correct answer.**

> This is because u, v and y are all functions of x and therefore, their value depends on x.

> By multiplying out the brackets.

> $uv - y = 0$ since $y = uv$.

Dividing through by δx gives

$$\frac{\delta y}{\delta x} = u\frac{\delta v}{\delta x} + v\frac{\delta u}{\delta x} + \delta u\frac{\delta v}{\delta x}.$$

Now as $\delta x \to 0$ you have

$$\frac{\delta y}{\delta x} \to \frac{dy}{dx}, \frac{\delta u}{\delta x} \to \frac{du}{dx}, \frac{\delta v}{\delta x} \to \frac{dv}{dx} \text{ and } \delta u \to 0$$

giving

$$\frac{dy}{dx} = u\frac{dv}{dx} + v\frac{du}{dx} + 0 \times \frac{dv}{dx} = u\frac{dv}{dx} + v\frac{du}{dx}.$$

Thus,

> To differentiate the product $y = uv$ you can use the result:
>
> $$\frac{dy}{dx} = u\frac{dv}{dx} + v\frac{du}{dx}$$

This result is known as the **product rule**.

> Using function notation, the derivative of the product $f(x)\,g(x)$ is
>
> $$f(x)\,g'(x) + f'(x)\,g(x).$$

> The proof of the product rule is given here for the sake of completeness. It is very unlikely that you will be expected to prove it in an examination. It is extremely important, however, that you can use and apply the product rule with confidence.

> This rule is sometimes written using the 'dash' notation for derivatives, namely
> $$y' = uv' + u'v.$$

Worked example 6.1

Differentiate $y = x^3 \sin x$ with respect to x.

Solution

You can write $u = x^3 \Rightarrow \dfrac{du}{dx} = 3x^2$

and $v = \sin x \Rightarrow \dfrac{dv}{dx} = \cos x$.

The product rule gives

$$\frac{dy}{dx} = u\frac{dv}{dx} + v\frac{du}{dx}$$

$$= x^3 \cos x + \sin x(3x^2)$$

Leave the first alone.	Differentiate the second.	Leave the second alone.	Differentiate the first.

$$= x^3 \cos x + 3x^2 \sin x$$

$$\Rightarrow \frac{dy}{dx} = x^2(x \cos x + 3 \sin x)$$

Worked example 6.2

Find the derivative of $y = x^4 \cos x$.

Solution

Writing $f(x) = x^4 \Rightarrow f'(x) = 4x^3$

and $g(x) = \cos x \Rightarrow g'(x) = -\sin x$

the product rule for the derivative of $f(x) g(x)$ gives

$$f(x) g'(x) + f'(x) g(x)$$

$$= x^4(-\sin x) + \cos x(4x^3) = -x^4 \sin x + 4x^3 \cos x$$

$$\Rightarrow \frac{dy}{dx} = x^3(4 \cos x - x \sin x).$$

Worked example 6.3

Find the gradient of the curve $y = \sqrt{x} \ln x$ at the point $P(1, 0)$.

6

Solution

$$u = x^{\frac{1}{2}} \Rightarrow \frac{du}{dx} = \frac{1}{2\sqrt{x}}$$

and $v = \ln x \Rightarrow \dfrac{dv}{dx} = \dfrac{1}{x}$

The product rule gives

$$\frac{dy}{dx} = u\frac{dv}{dx} + v\frac{du}{dx}$$

$$= x^{\frac{1}{2}}\left(\frac{1}{x}\right) + \ln x \left(\frac{1}{2\sqrt{x}}\right) = \frac{1}{\sqrt{x}} + \frac{1}{2\sqrt{x}} \ln x$$

$$\Rightarrow \frac{dy}{dx} = \frac{1}{\sqrt{x}}(1 + \tfrac{1}{2} \ln x).$$

At P, $\dfrac{dy}{dx} = \dfrac{1}{\sqrt{1}}(1 + \tfrac{1}{2} \ln 1) = 1(1 + 0) = 1$,

so the gradient at P is equal to 1.

Worked example 6.4

Find the equation of the tangent to the curve $y = x^2 e^{3x}$ at the point $N(2, 4e^6)$.

Solution

$u = x^2 \Rightarrow \dfrac{du}{dx} = 2x$

and $v = e^{3x} \Rightarrow \dfrac{dv}{dx} = 3e^{3x}$

Notice that the chain rule has been used to differentiate e^{3x}.

The product rule gives

$$\frac{dy}{dx} = u\frac{dv}{dx} + v\frac{du}{dx}$$

$$= x^2(3e^{3x}) + e^{3x}(2x) = 3x^2e^{3x} + 2xe^{3x}$$

$$\Rightarrow \frac{dy}{dx} = xe^{3x}(3x + 2).$$

At N, $\dfrac{dy}{dx} = 2e^6(6 + 2) = 16e^6$:

The tangent at N has equation

$$y - 4e^6 = 16e^6(x - 2)$$
$$\Rightarrow y - 4e^6 = 16e^6 x - 32e^6$$
$$\Rightarrow y + 28e^6 = 16e^6 x$$

Worked example 6.5

A curve has equation $y = xe^{2x}$.

(a) Find the coordinates of the stationary point of the curve.

(b) Find $\dfrac{d^2y}{dx^2}$ and hence determine the nature of the stationary point.

Solution

(a) $u = x \Rightarrow \dfrac{du}{dx} = 1$

and $v = e^{2x} \Rightarrow \dfrac{dv}{dx} = 2e^{2x}$

The product rules gives

$$\frac{dy}{dx} = u\frac{dv}{dx} + v\frac{du}{dx}$$

$$= x(2e^{2x}) + e^{2x}(1) = 2xe^{2x} + e^{2x}$$

$$\Rightarrow \frac{dy}{dx} = e^{2x}(2x + 1).$$

At stationary points the gradient is zero.

$$e^{2x}(2x + 1) = 0$$
$$\Rightarrow e^{2x} = 0 \text{ or } 2x + 1 = 0$$

But $e^{2x} = 0$ has no solutions, so $x = -\dfrac{1}{2}$.

$x = -\dfrac{1}{2} \Rightarrow y = \left(-\dfrac{1}{2}\right)e^{-1} = -\dfrac{1}{2e}$

Hence, the curve has a single stationary point at $\left(-\dfrac{1}{2}, -\dfrac{1}{2e}\right)$.

(b) You already found that $\dfrac{dy}{dx} = 2xe^{2x} + e^{2x}$.

Notice that, in order to differentiate the first term, you have to use the product rule again.

$$\dfrac{d^2y}{dx^2} = [2x(2e^{2x}) + e^{2x}(2)] + 2e^{2x}$$

Using the product rule.

$$= 4xe^{2x} + 2e^{2x} + 2e^{2x} = 4xe^{2x} + 4e^{2x}$$

$$= 4e^{2x}(x + 1)$$

At the point $\left(-\dfrac{1}{2}, -\dfrac{1}{2e}\right)$, $\dfrac{d^2y}{dx^2} = 4e^{-1}\left(-\dfrac{1}{2} + 1\right) = \dfrac{2}{e} > 0$.

Since $\dfrac{d^2y}{dx^2}$ is positive, $\left(-\dfrac{1}{2}, -\dfrac{1}{2e}\right)$, is a minimum point.

EXERCISE 6A

1 Use the product rule to differentiate the following with respect to x:

(a) $y = x \sin x$ **(b)** $y = x^3 \cos x$

(c) $y = 4x \ln x$ **(d)** $y = \sqrt{x} \sin x$

(e) $y = 5x^3e^x$ **(f)** $y = \sin x \cos x$

(g) $y = e^x \ln x$ **(h)** $y = e^x \sin x$

(i) $y = 2\sqrt{x}(\ln x)$ **(j)** $y = 3x^5e^x$

2 Differentiate the following expressions:

(a) $y = x^3e^{2x}$ **(b)** $y = (6 - 3x)(2x - 1)^5$

(c) $y = e^x \sin 4x$ **(d)** $y = \sin 3x \cos 2x$

(e) $y = \sqrt{x + 1} \ln x$ **(f)** $y = 6x^2 \ln 5x$

(g) $y = e^{3x} \cos 2x$ **(h)** $y = 4x^3 \sqrt{2x - 1}$

(i) $y = 4x \ln(3x + 2)$ **(j)** $y = e^{5x} \sin x + 6x^2$

> Look out for the use of the chain rule also.

3 Find the gradient of the following curves at the specified point:

(a) $y = 3x^2e^x$ at the point $(1, 3e)$,

(b) $y = x \sin x$ at the point where $x = \dfrac{\pi}{6}$,

(c) $y = \sqrt{x} \ln x$ at the point where $x = 1$.

4 Find the equation of the tangent to the curve $y = xe^{3x}$ at the point $P(3, 3e^9)$.

5 A curve has equation $y = x \ln x$.

   (a) Find the coordinates of the stationary point of the curve, leaving your answer in terms of e.

   (b) Find $\dfrac{d^2y}{dx^2}$ and hence determine whether the stationary point is a maximum or a minimum.

6 A curve is defined for $x \geqslant -2$ by the equation
$$y = (x^2 + 3)\sqrt{(x + 2)}.$$

   (a) Show that $\dfrac{dy}{dx} = 0$ when $x = -1$ and find the x-coordinate of the other stationary point.

   (b) Find the value of $\dfrac{d^2y}{dx^2}$ when $x = -1$. Hence determine whether the stationary point when $x = -1$ is a maximum or minimum point.   [A]

7 The diagram below shows a sketch of the curve defined for $x > 0$ by the equation $y = x^2 \ln x$.

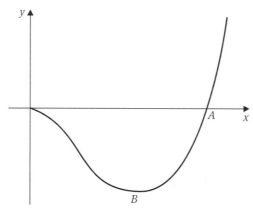

The curve crosses the x-axis at A and has a local minimum at B.

   (a) State the coordinates of A and calculate the gradient of the curve at A.

   (b) Determine the coordinates of B and determine the value of $\dfrac{d^2y}{dx^2}$ at B.   [A]

6.3 Differentiation using the quotient rule

Suppose you wish to differentiate a quotient of the form $\dfrac{f(x)}{g(x)}$.

You can write $y = \dfrac{u}{v}$, where u and v are functions of x.

A small change, δx, in x will produce small changes in the values of u, v and y.

Let these small changes in u, v and y be δu, δv and δy, respectively.

Now $y + \delta y = \dfrac{u + \delta u}{v + \delta v}$

$$\Rightarrow \delta y = \frac{u + \delta u}{v + \delta v} - y = \frac{u + \delta u}{v + \delta v} - \frac{u}{v}$$

$$\Rightarrow \delta y = \frac{v(u + \delta u) - u(v + \delta v)}{v(v + \delta v)}$$

By putting the fractions over a common denominator.

$$\Rightarrow \delta y = \frac{uv + v\delta u - uv - u\delta v}{v(v + \delta v)}$$

$$\Rightarrow \delta y = \frac{v\delta u - u\delta v}{v(v + \delta v)}$$

Now dividing both sides by δx gives

$$\frac{\delta y}{\delta x} = \frac{v\dfrac{\delta u}{\delta x} - u\dfrac{\delta v}{\delta x}}{v(v + \delta v)},$$

As with the product rule, the proof is included here for completeness.
You do not need to remember the proof but **you must learn this result** and feel confident when applying it.

and taking the limit as $\delta x \to 0$ gives

$$\frac{dy}{dx} = \frac{v\dfrac{du}{dx} - u\dfrac{dv}{dx}}{v^2}$$

Thus,

To differentiate a quotient $y = \dfrac{u}{v}$ you can use the result:

$$\frac{dy}{dx} = \frac{v\dfrac{du}{dx} - u\dfrac{dv}{dx}}{v^2}.$$

Note: Unlike the product rule the order in which the terms appear in the numerator is vital.

This result is known as the **quotient rule**.

This may also be written as
$$y' = \frac{vu' - uv'}{v^2}.$$

Using function notation, the derivative of the quotient $\dfrac{f(x)}{g(x)}$ is

$$\frac{f'(x)\,g(x) - f(x)\,g'(x)}{[g(x)]^2}.$$

Worked example 6.6

Differentiate $y = \dfrac{\sin x}{x^2}$ with respect to x.

Solution

You can write $f(x) = \sin x \Rightarrow f'(x) = \cos x$

and $g(x) = x^2 \Rightarrow g'(x) = 2x$.

The quotient rule for the derivative of $\dfrac{f(x)}{g(x)}$ is

$$\frac{f'(x)\,g(x) - f(x)\,g'(x)}{[g(x)]^2}$$

> This formula will be given in the formulae booklet for use in the examination.

$$= \frac{x^2(\cos x) - \sin x(2x)}{(x^2)^2} = \frac{x^2 \cos x - 2x \sin x}{x^4}$$

$$\Rightarrow \frac{dy}{dx} = \frac{x(x \cos x - 2 \sin x)}{x^4} = \frac{x \cos x - 2 \sin x}{x^3}$$

Worked example 6.7

Find the gradient of the curve with equation $y = \dfrac{\ln 2x}{x^3}$ at the point P where $x = 5$.

Solution

$u = \ln 2x \Rightarrow \dfrac{du}{dx} = \dfrac{2}{2x} = \dfrac{1}{x}$

> Notice the use of the chain rule. Or write
> $\ln 2x = \ln 2 + \ln x$.

and $v = x^3 \Rightarrow \dfrac{dv}{dx} = 3x^2$

The quotient rule gives

$$\frac{dy}{dx} = \frac{v\dfrac{du}{dx} - u\dfrac{dv}{dx}}{v^2}$$

> You may prefer to use this version of the quotient rule but you will have to learn it.

$$= \frac{x^3\left(\dfrac{1}{x}\right) - \ln 2x(3x^2)}{(x^3)^2} = \frac{x^2 - 3x^2 \ln 2x}{x^6}$$

$$\Rightarrow \frac{dy}{dx} = \frac{x^2(1 - 3 \ln 2x)}{x^6} = \frac{1 - 3 \ln 2x}{x^4}$$

at P, $\dfrac{dy}{dx} = \dfrac{1 - 3 \ln 10}{625}$.

Worked example 6.8

A curve has equation $y = \dfrac{(2-3x)}{(2x-1)^2}$, $\left(x \neq \dfrac{1}{2}\right)$.

Find the coordinates of the stationary point on the curve.

Solution

$f(x) = 2 - 3x \Rightarrow f'(x) = -3$

and $g(x) = (2x - 1)^2 \Rightarrow g'(x) = 4(2x - 1)$

The quotient rule gives

$$\frac{dy}{dx} = \frac{(2x-1)^2(-3) - 4(2-3x)(2x-1)}{(2x-1)^4}$$

$$= \frac{-3(2x-1) - 4(2-3x)}{(2x-1)^3}$$

$$= \frac{-6x + 3 - 8 + 12x}{(2x-1)^3}$$

$$\Rightarrow \frac{dy}{dx} = \frac{6x - 5}{(2x-1)^3}$$

It is important to realise that there is a common factor of $(2x - 1)$ in the numerator, that cancels one of the $(2x - 1)$ terms in the denominator.

If you had not cancelled the factor $(2x - 1)$, you would have obtained a quadratic in the numerator and been led to the conclusion $(2x - 1)(6x - 5) = 0$, suggesting incorrectly that there was also a turning point when $x = \frac{1}{2}$. This is a very common error that some students make when using the quotient rule.

At stationary points the gradient is zero

$$\frac{6x - 5}{(2x-1)^3} = 0$$

$$\Rightarrow 6x - 5 = 0 \Rightarrow x = \frac{5}{6}$$

When $x = \frac{5}{6}$, $y = \frac{(2 - \frac{15}{6})}{(\frac{10}{6} - 1)^2} = \frac{-\frac{1}{2}}{\frac{4}{9}} = -\frac{9}{8}$

$\Rightarrow$ the curve has one stationary point at $\left(\frac{5}{6}, -\frac{9}{8}\right)$.

Worked examination question

The diagram shows a sketch of the curve with equation

$$y = \frac{\ln x}{x} \qquad (x > 0).$$

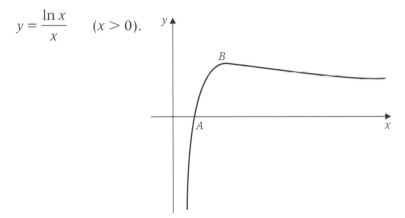

(a) State the coordinates of A, where the curve crosses the x-axis.

(b) Calculate, in terms of e, the coordinates of the maximum point B.

(c) Calculate, in terms of e, the value of $\frac{d^2y}{dx^2}$ at B. [A]

Solution

(a) The curve crosses the x-axis where $y = 0$,

$$\Rightarrow \frac{\ln x}{x} = 0 \Rightarrow \ln x = 0$$

$$\Rightarrow x = 1$$

So A has coordinates $(1, 0)$.

(b) $\dfrac{dy}{dx} = \dfrac{x\left(\dfrac{1}{x}\right) - \ln x}{x^2} = \dfrac{1 - \ln x}{x^2}$

At a stationary point the gradient is zero

$$\frac{1 - \ln x}{x^2} = 0 \Rightarrow 1 - \ln x = 0$$

$$\Rightarrow \ln x = 1$$

$$\Rightarrow x = e$$

When $x = e$, $y = \dfrac{1}{e}$

So B has coordinates $\left(e, \dfrac{1}{e}\right)$.

(c) Using the quotient rule again gives

$$\frac{d^2y}{dx^2} = \frac{x^2\left(-\dfrac{1}{x}\right) - 2x(1 - \ln x)}{x^4} = \frac{2 \ln x - 3}{x^3}$$

Cancelling the factor x in numerator and denominator.

At B, $\dfrac{d^2y}{dx^2} = \dfrac{2 \ln e - 3}{e^3} = -\dfrac{1}{e^3}$.

EXERCISE 6B

1 Differentiate the following with respect to x:

(a) $y = \dfrac{\cos x}{x}$

(b) $y = \dfrac{e^x}{x^2}$

(c) $y = \dfrac{\sin x}{\sqrt{x}}$

(d) $y = \dfrac{\ln x}{x}$

(e) $y = \dfrac{4x^2}{e^x}$

(f) $y = \dfrac{5 \ln x}{x^3}$

(g) $y = \dfrac{e^x}{3x^2}$

(h) $y = \dfrac{\sin x}{e^x}$

(i) $y = \dfrac{\sin x}{\cos x}$

(j) $y = \dfrac{\cos x}{\sin x}$

2 Find $\dfrac{dy}{dx}$ for each of the following expressions:

(a) $y = \dfrac{\sin 2x}{x^3}$

(b) $y = \dfrac{e^{-x}}{\sin x}$

(c) $y = \dfrac{4x^5}{e^{3x}}$

(d) $y = \dfrac{(1-3x)}{(2x+3)^2}$

(e) $y = \dfrac{\ln 5x}{x}$

(f) $y = \dfrac{x^3}{\cos 3x}$

(g) $y = \dfrac{4 \sin 2x}{2x^3}$

(h) $y = \dfrac{2e^{3x}}{x}$

(i) $y = \dfrac{7 \ln 3x}{x^2}$

(j) $y = \dfrac{e^{2-x}}{\sin 3x}$

3 By writing $y = \dfrac{u}{v}$ in the form $y = uv^{-1}$, use the product rule to establish the quotient rule.

4 A curve is defined by the equation

$$y = \frac{x+2}{(x-1)^2}, \; x \neq 1.$$

(a) Find the equation of the tangent to the curve at the point where $x = 0$.

(b) Find the coordinates of the point where this tangent crosses the curve again. [A]

5 The volume, $V \, \text{cm}^3$, in a container when the depth is x cm is given by

$$V = \frac{x^{\frac{1}{4}}}{(x+2)^{\frac{1}{2}}}, \; x > 0.$$

(a) Find $\dfrac{dV}{dx}$ and determine the value of x for which $\dfrac{dV}{dx} = 0$.

(b) Calculate the rate of change of volume when the depth is 1 cm and increasing at a rate of $0.01 \, \text{cm s}^{-1}$, giving your answer in $\text{cm}^3 \, \text{s}^{-1}$ to three significant figures. [A]

6 The diagram shows the curve with equation

$$y = \frac{\ln x}{x^2} \; \text{for} \; x > 0.$$

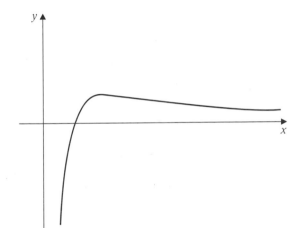

(a) State the x-coordinate of the point where the curve crosses the x-axis.

(b) Show that $\dfrac{dy}{dx} = \dfrac{1-2 \ln x}{x^3}$.

(c) Find the coordinates of the maximum point of the curve and calculate the value of $\dfrac{d^2y}{dx^2}$ there. [A]

6.4 Derivatives of other trigonometric functions

The derivatives of other trigonometric functions such as $\tan x$ and $\sec x$ can be found by using the derivatives of $\sin x$ and $\cos x$, together with the quotient rule.

Derivative of tan x

Suppose that you want to differentiate $y = \tan x$.
This can be rewritten as a quotient,

$$y = \frac{\sin x}{\cos x}.$$

Using the quotient rule, gives

$$\frac{dy}{dx} = \frac{\cos x(\cos x) - \sin x\,(-\sin x)}{\cos^2 x} = \frac{\cos^2 x + \sin^2 x}{\cos^2 x}$$

but $\cos^2 x + \sin^2 x \equiv 1$

$$\Rightarrow \frac{dy}{dx} = \frac{1}{\cos^2 x} = \sec^2 x$$

$\Rightarrow$

$$\frac{d}{dx}(\tan x) = \sec^2 x$$

Derivative of cot x

Suppose that you want to differentiate $y = \cot x$.
This can be rewritten as a quotient,

$$y = \frac{\cos x}{\sin x}.$$

The quotient rule gives

$$\frac{dy}{dx} = \frac{\sin x(-\sin x) - \cos x(\cos x)}{\sin^2 x} = \frac{-(\sin^2 x + \cos^2 x)}{\sin^2 x}$$

but $\sin^2 x + \cos^2 x \equiv 1$

$$\Rightarrow \frac{dy}{dx} = -\frac{1}{\sin^2 x} = -\operatorname{cosec}^2 x$$

$\Rightarrow$

$$\frac{d}{dx}(\cot x) = -\operatorname{cosec}^2 x$$

Derivative of sec x

In order to differentiate $y = \sec x$, you can rewrite it in the form

$$y = \frac{1}{\cos x}.$$

The quotient rule gives

$$\frac{dy}{dx} = \frac{\cos x(0) - 1(-\sin x)}{\cos^2 x} = \frac{\sin x}{\cos^2 x}$$

$$= \frac{1}{\cos x} \times \frac{\sin x}{\cos x} = \sec x \tan x$$

> You can also prove this result by writing sec x in the form $(\cos x)^{-1}$ and using the chain rule with $u = \cos x$ and $y = u^{-1}$.

$\Rightarrow$

$$\frac{d}{dx}(\sec x) = \sec x \tan x$$

Derivative of cosec x

To differentiate $y = \operatorname{cosec} x$, you can rewrite it in the form

$$y = \frac{1}{\sin x}.$$

The quotient rule then gives

$$\frac{dy}{dx} = \frac{\sin x(0) - 1(\cos x)}{\sin^2 x} = \frac{-\cos x}{\sin^2 x}$$

$$= -\frac{1}{\sin x} \times \frac{\cos x}{\sin x} = -\operatorname{cosec} x \cot x$$

> These standard results derived in this section will be given in the formulae booklet for use in the examination.

$\Rightarrow$

$$\frac{d}{dx}(\operatorname{cosec} x) = -\operatorname{cosec} x \cot x$$

Worked example 6.9

Differentiate the following expressions with respect to x.

(a) $y = \tan(3x - 4)$ **(b)** $y = \sin x \sec 2x$

Solution

(a) The chain rule gives

$$\frac{dy}{dx} = \sec^2(3x - 4) \times 3 = 3 \sec^2(3x - 4)$$

> The derivative of tan u is sec² u.

(b) The product rule gives

$$\frac{dy}{dx} = \sin x(2 \sec 2x \tan 2x) + \sec 2x(\cos x)$$

$$= \sec 2x(2 \sin x \tan 2x + \cos x)$$

> The derivative of sec u is sec u tan u.

Worked example 6.10

Differentiate the following expressions with respect to x:

(a) $y = \ln(\cot x)$

(b) $y = \dfrac{\operatorname{cosec} 3x}{\sin x}$

Notice the use of the chain rule in differentiating cosec $3x$.

Solution

(a) The chain rule gives

$$\frac{dy}{dx} = \frac{1}{\cot x} \times (-\operatorname{cosec}^2 x)$$

$$= -\frac{\sin x}{\cos x} \times \frac{1}{\sin^2 x} = -\frac{1}{\sin x \cos x} = -\operatorname{cosec} x \sec x$$

(b) The quotient rule gives

$$\frac{dy}{dx} = \frac{\sin x(-3\operatorname{cosec} 3x \cot 3x) - \operatorname{cosec} 3x(\cos x)}{\sin^2 x}$$

Notice the use of the chain rule in differentiating cosec $3x$.

$$= \frac{-\operatorname{cosec} 3x (3\sin x \cot 3x + \cos x)}{\sin^2 x}$$

EXERCISE 6C

1 Differentiate the following with respect to x:

(a) $y = 5x^3 + \tan 4x$

(b) $y = \sec(6x - 3)$

(c) $y = \cot(1 - x)$

(d) $y = 5\operatorname{cosec} 3x$

2 Differentiate the following expressions with respect to the appropriate variable:

(a) $y = 3t^4 \sec t$

(b) $y = \dfrac{\cot x}{x^3}$

(c) $y = \ln(\sec t)$

(d) $y = \sin x + x^2 \tan x$

3 A curve has equation $y = \tan 3x$. Find the equation of the tangent to the curve at the point P on the curve where $x = \dfrac{\pi}{4}$.

4 Show that the gradient of the curve with equation $y = \sec x \tan x$ can be written in the form

$$\sec x(2\sec^2 x - 1).$$

5 A curve is given by the equation $y = e^{-x} \sec 2x$. Find an expression for the gradient of the curve.

6 By writing $y = \sec x$ as $y = (\cos x)^{-1}$, use the chain rule to establish the result $\dfrac{d}{dx}(\sec x) = \sec x \tan x$.

7 By writing $y = \operatorname{cosec} x$ as $y = (\sin x)^{-1}$, use the chain rule to establish the result $\dfrac{d}{dx}(\operatorname{cosec} x) = -\operatorname{cosec} x \cot x$.

8 Find an equation for the tangent to the curve $y = \sec^2 3x$ at the point where $x = \dfrac{\pi}{12}$.

9 A curve has equation $y = \dfrac{\tan 3x + \cos 2x}{\tan x - \sec 3x}$. Find the gradient of the curve at the point where $x = 0$.

10 The function f is defined for $-\dfrac{\pi}{4} < x < \dfrac{\pi}{4}$ by

$f(x) = \sec 2x + \tan 2x$.

   (a) Find $f'(x)$ and hence show that f is an increasing function.

   (b) Find the equation of the normal to the curve $y = f(x)$ at the point where $x = 0$.

The key to success in differentiation is to appreciate which of the various techniques need to be used.

The following exercise contains questions on basic differentiation, the chain rule, the product rule and the quotient rule. It is left to you to decide which technique, or combination of techniques, is most appropriate.

6

MIXED EXERCISE

1 Differentiate the following expressions with respect to x:

   (a) $y = (x - 3)(x + 2)$ **(b)** $y = 5x^3 - 2x + e^{2x}$

   (c) $y = \dfrac{4x^3 - 2}{x}$ **(d)** $y = x(x - 2)(x + 5)$

   (e) $y = e^{-x} \sin x$ **(f)** $y = 3 \tan 5x$

   (g) $y = \dfrac{2 \ln x}{x^3}$ **(h)** $y = 4 \cot(2 - 3x)$

2 Find the gradients of the following curves at the specified points:

   (a) $y = 5 \sin 2\theta$ at the point where $\theta = \dfrac{\pi}{6}$,

   (b) $y = e^{2x + 7}$ at the point where $x = -3$,

   (c) $y = \sqrt{x}(x + 1)$ at the point where $x = 9$.

3 Find the equation of the tangent to the curve with equation $y = \sqrt{x}e^x$ at the point where $x = 4$.

4 Find the equation of the tangent to the curve $y = 2 \tan 2x$ at the point where $x = \dfrac{\pi}{8}$.

5 A curve has equation $y = x^2 e^{-x}$.

(a) Find the coordinates of its stationary points.

(b) Find the value of $\dfrac{d^2y}{dx^2}$ at each of the stationary points and hence determine their nature.

6 A curve is defined for $x > 0$ by the equation

$$y = (10 - x)\ln 2x.$$

Find the gradient of the curve when $x = 1$, leaving your answer in terms of $\ln 2$.

7 The radius, r cm, of a circular ink spot, t seconds after it first appears, is given by

$$r = \frac{1 + 4t}{2 + t}.$$

Calculate:

(a) the time taken for the radius to double its initial size,

(b) the rate of increase of the radius in cm s^{-1} when $t = 3$,

(c) the value to which r tends as t tends to infinity.　　[A]

Hint: The initial size is found by letting $t = 0$.

8 The curve with equation $y = (x + 2)e^{-x}$ is sketched below.

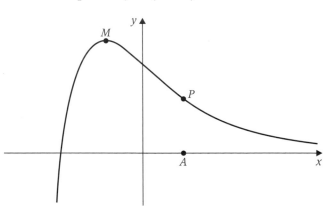

The maximum point M and the point P lie on the curve and A lies on the x-axis. The points A and P each have x-coordinate equal to 3.

(a) Find the value of $\dfrac{dy}{dx}$ at the point P, leaving your answer in terms of e.

(b) Determine the equation of the tangent to the curve at P and find the x-coordinate of the point B where this tangent crosses the x-axis.

(c) Calculate the coordinates of the maximum point M. [A]

9 The curve with equation $y = (x^2 - 4x + 1)e^{-x}$ is sketched below.

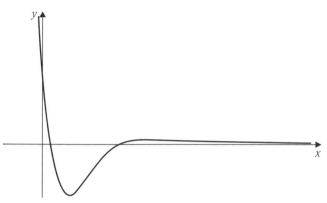

(a) Determine the exact surd values of x where the curve crosses the x-axis.

(b) Calculate the exact values of the coordinates of the stationary points of the curve. [A]

10 A model plane moves so that its height y metres above horizontal ground is given by

$$y = \frac{8x}{x^3 + 1}, \quad x \geq 0,$$

when its horizontal distance is x metres from the take-off point on the ground.

(a) Find the value of $\dfrac{\mathrm{d}y}{\mathrm{d}x}$ when $x = 1$.

(b) (i) Find the rate of change of y in m s^{-1} when $x = 1$ and x is increasing at a rate of 0.8 m s^{-1}.

(ii) Interpret the sign in your answer to **(i)**. [A]

6

Key point summary

1 To differentiate a product $y = u \times v$, you can use the **product rule**:   *p98*

$$\frac{\mathrm{d}y}{\mathrm{d}x} = u\frac{\mathrm{d}v}{\mathrm{d}x} + v\frac{\mathrm{d}u}{\mathrm{d}x}$$

2 Using function notation, the derivative of the product   *p98* $f(x)g(x)$ is

$$f(x)g'(x) + f'(x)g(x).$$

3 To differentiate a quotient $y = \dfrac{u}{v}$, you can use the **quotient rule**: *p103*

$$\frac{dy}{dx} = \frac{v\dfrac{du}{dx} - u\dfrac{dv}{dx}}{v^2}$$

4 Using function notation, the derivative of the *p103*

quotient $\dfrac{f(x)}{g(x)}$ is $\dfrac{f'(x)\,g(x) - f(x)\,g'(x)}{[g(x)]^2}$.

This formula is given in the formulae booklet for use in examinations.

5 $\dfrac{d}{dx}(\tan x) = \sec^2 x$ *p108*

$\dfrac{d}{dx}(\cot x) = -\text{cosec}^2\, x$ *p108*

$\dfrac{d}{dx}(\sec x) = \sec x \tan x$ *p109*

$\dfrac{d}{dx}(\text{cosec}\, x) = -\text{cosec}\, x \cot x$ *p109*

These formulae are given in the formulae booklet for use in the examination.

Test yourself	**What to review**

1 Differentiate the following with respect to x: *Section 6.2*

   (a) $y = x^2(x + 1)^5$ **(b)** $y = x \sin x$

   (c) $y = e^{-2x} \cos x$ **(d)** $y = x^2 \ln(3x - 1)$

---

2 Find $\dfrac{dy}{dx}$ for each of the following expressions: *Section 6.3*

   (a) $y = \dfrac{x^2}{x + 1}$ **(b)** $y = \dfrac{x}{\cos x}$

   (c) $y = \dfrac{\sin 3x}{x}$ **(d)** $y = \dfrac{\cos(1 - 2x)}{e^x}$

---

3 Differentiate each of the following with respect to x: *Section 6.4*

   (a) $y = 5 \tan 2x$ **(b)** $y = 2 \sec(1 - 4x)$

   (c) $y = 7 \cot 5x$ **(d)** $y = \text{cosec}(x^2 + 3)$.

1 **(a)** $5x^2 (x + 1)^4 + 2x(x + 1)^5$;

(b) $x \cos x + \sin x$;

(c) $-e^{-2x} (\sin x + 2 \cos x)$;

(d) $\dfrac{3x^2}{3x - 1} + 2x \ln (3x - 1)$.

2 **(a)** $\dfrac{2x (x + 1) - x^2}{(x + 1)^2} = \dfrac{x(2 + x)}{(x + 1)^2}$;

(b) $\dfrac{\cos x + x \sin x}{\cos^2 x}$

(c) $\dfrac{3x \cos 3x - \sin 3x}{x^2}$;

(d) $\dfrac{2 \sin (1 - 2x) - \cos (1 - 2x)}{e^x}$

3 **(a)** $10 \sec^2 2x$;

(b) $-8 \sec (1 - 4x) \tan (1 - 4x)$;

(c) $-35 \operatorname{cosec}^2 5x$;

(d) $-2x \operatorname{cosec} (x^2 + 3) \cot (x^2 + 3)$.

Test yourself ANSWERS

CHAPTER 7

C3: Numerical solution of equations and iterative methods

Learning objectives

After studying this chapter, you should be able to:
- locate roots of equations by change of sign
- use numerical methods to find solutions of equations
- understand the principle of iteration
- appreciate the need for convergence in iterations
- understand how cobweb and staircase diagrams demonstrate convergence or divergence for equations of the form $x = g(x)$.

7.1 Exact solutions to equations

You have learned how to solve linear equations of the form $ax + b = 0$, and quadratic equations of the form $ax^2 + bx + c = 0$, where a, b and c are constants.

In the sixteenth century in Italy various formulae were developed to solve cubic equations.
Cardano derived a formula to solve equations of the form

$$x^3 + cx = d.$$

His solution was

$$x = \sqrt[3]{u} - \sqrt[3]{v}$$

where $u = \sqrt{\left(\frac{d}{2}\right)^2 + \left(\frac{c}{3}\right)^3} + \frac{d}{2}$ and $v = \sqrt{\left(\frac{d}{2}\right)^2 + \left(\frac{c}{3}\right)^3} - \frac{d}{2}.$

> Look at a History of Maths website and consider the interesting contest between Fiore and Tartaglia.

> This section is included for interest but will not be tested in the C3 examination.

Worked example 7.1

Use Cardano's formula to find the exact root of $x^3 - 6x = 6$.

Solution

$c = -6$, $d = 6$ and so $\left(\frac{d}{2}\right)^2 + \left(\frac{c}{3}\right)^3 = (3)^2 + (-2)^3 = 9 - 8 = 1$

$x = \sqrt[3]{1 + 3} - \sqrt[3]{1 - 3} = \sqrt[3]{4} - \sqrt[3]{(-2)}$

$\quad = \sqrt[3]{4} + \sqrt[3]{2}$

> You can check this answer because $\sqrt[3]{4} + \sqrt[3]{2} \approx 2.847$ and $2.847^3 - 6 \times 2.847 \approx 6$.

Sometimes, the answer produced by this method is rather complicated. For example, using Cardano's method, the solution to $x^3 + 6x = 20$ is $\sqrt[3]{\sqrt{108} + 10} - \sqrt[3]{\sqrt{108} - 10}$.

> Evaluate this surd expression on your calculator and you should get a whole number. Check that this integer satisfies the original equation.

Cardano's method and other similar formulae do not always provide all the solutions to a cubic equation and as equations become as complicated as $x^5 - 3x^2 + 2x - 3 = 0$, there is no known analytical solution and so numerical methods have to be used.

7.2 Numerical methods

Suppose the graph of $y = f(x)$ is continuous and crosses the x-axis between $x = a$ and $x = b$.

The situation may be as in diagram (1) or as in diagram (2).

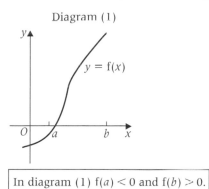
Diagram (1)

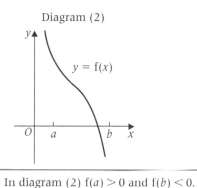
Diagram (2)

In diagram (1) $f(a) < 0$ and $f(b) > 0$. | In diagram (2) $f(a) > 0$ and $f(b) < 0$.

In each case there is a change in sign, either going from negative to positive or from positive to negative.

> If the graph of $y = f(x)$ is continuous over the interval $a \leqslant x \leqslant b$, and $f(a)$ and $f(b)$ have different signs, then at least one root of the equation $f(x) = 0$ must lie in the interval $a < x < b$.

Worked example 7.2

Verify that a root of the equation $x^5 - 3x^2 + 2x - 3 = 0$ lies between 1.44 and 1.45.

Solution

Let $f(x) = x^5 - 3x^2 + 2x - 3$.

$f(1.44) = 1.44^5 - 3(1.44)^2 + 2 \times 1.44 - 3 = -0.149...$

$f(1.45) = 1.45^5 - 3(1.45)^2 + 2 \times 1.45 - 3 = 0.002\,234...$

> You could check this on your graphics calculator by drawing the graph and using the trace facility to see where the curve crosses the x-axis.

7

Since $f(1.44) < 0$ and $f(1.45) > 0$, the change in sign indicates that at least one root lies between 1.44 and 1.45.

A sketch of the graph of $y = x^5 - 3x^2 + 2x - 3$ reveals that it crosses the x-axis only once and so you have actually located the approximate value of the only real root of the equation $x^5 - 3x^2 + 2x - 3 = 0$.

Worked example 7.3

Find two consecutive negative integers between which the root of the equation $x^3 - 17x + 32 = 0$ lies.

Solution

Let $f(x) = x^3 - 17x + 32$.

The graph of $y = f(x)$ is continuous for all values of x.

> Always set up a function first so the equation is in the form $f(x) = 0$.

$f(-1) = -1 + 17 + 32 = 48$
$f(-5) = -125 + 85 + 32 = -8$

Therefore the root must lie between -1 and -5 and is likely to be nearer -5.

$f(-4) = -64 + 68 + 32 = 36$

Since $f(-5) < 0$ and $f(-4) > 0$ there is a change of sign to indicate that a root lies between -5 and -4. The question indicates that the equation has a single real root.

Therefore two consecutive negative integers between which the root lies are -5 and -4.

In each of the examples above, the functions considered were polynomials and the graphs of all polynomials are continuous. You cannot apply the change of sign test when the function you are considering is not continuous over the interval concerned.

Worked example 7.4

Given that $f(x) = \dfrac{1}{x}$. Find $f(-1)$ and $f(0.5)$.

Can you deduce that a root of $\dfrac{1}{x} = 0$ lies between -1 and 0.5?

Solution

$f(-1) = -1$; $f(0.5) = 2$.

Although there is a change in sign, you need to consider the graph of $y = f(x)$.

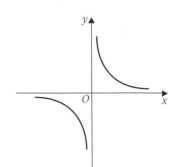

The graph is **not** continuous between -1 and 0.5. It has a discontinuity when $x = 0$. You can **not** deduce that the equation has a root between -1 and 0.5.

Worked example 7.5

Prove that the real root of the equation $x^3 - 3x + 4 = 0$ is -2.196 correct to three decimal places.

Solution

Your graphics calculator will show that the graph of $y = x^3 - 3x + 4$ is continuous and that it crosses the x-axis once only.

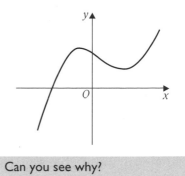

Let $f(x) = x^3 - 3x + 4$.

In order to prove that the root is -2.196 correct to three decimal places, it is necessary to show that the root lies between -2.1955 and -2.1965.

Can you see why?

$f(-2.1955) = 0.003\,706\,4\ldots$ and $f(-2.1965) = -0.007\,760\,8\ldots$

The change of sign shows that the root lies between -2.1955 and -2.1965. Hence the root is equal to -2.196 correct to three decimal places.

Worked example 7.6

Verify that the equation $e^{-x} = x$ has a root between 0.56 and 0.57 and find this root correct to two decimal places.

Solution

Let $f(x) = e^{-x} - x$.
Since $f(0.56) = 0.0112\ldots$ and $f(0.57) = -0.004\,47\ldots$ and the graph of $y = f(x)$ is continuous, then the equation $f(x) = 0$ has a root between 0.56 and 0.57.

Because the value of $f(0.57)$ is nearer to zero than $f(0.56)$, it suggests that the root is nearer to 0.57 than 0.56.

A sensible next guess might be 0.567 or 0.568.
$f(0.567) = 0.000\,245\,6\ldots$ whereas $f(0.568) = -0.001\,34\ldots$.
Hence, the root is likely to be 0.567 to three decimal places.

In order to verify this, you need to calculate $f(0.5675)$ which equals $-0.000\,558\,9\ldots$ and because $f(0.567)$ is positive, the root must lie between 0.567 and 0.5675.

Therefore, the root is 0.567 correct to three decimal places.

EXERCISE 7A

1 Verify that the equation $x^5 + 7x - 10 = 0$ has a root between 1.1 and 1.2.

2 Verify that the equation $x^3 - 3x + 5 = 0$ has a root between -2.3 and -2.1.

3 Verify that the equation $x^5 + 2x - 7 = 0$ has a root between 1.3 and 1.4.

4 Verify that the equation $x^3 + 12x - 39 = 0$ has a root between 2.2 and 2.3.

5 Find two consecutive integers between which the root of the equation $x^7 + 2x^2 + 3 = 0$ lies.

6 Prove that one root of the equation $x^4 - 23x + 19 = 0$ lies between 0 and 1 and find two consecutive integers between which the other root lies.

7 The equation $x^3 + 7x - 19 = 0$ has a root close to 1.8. Determine whether the root lies between 1.75 and 1.80 or between 1.80 and 1.85.

8 Consider the equation $x^5 - 9x - 7 = 0$.
   (a) Show that the equation has a root between -1.5 and -1.4.
   (b) Given that $g(x) = x^5 - 9x - 7$ show that $g(-1)$ and $g(2)$ have the same sign.
      (i) Does that mean there are no roots of the equation between -1 and 2?
      (ii) Find $g(-\frac{1}{2})$ and $g(1\frac{1}{2})$. What can you deduce about roots of the equation?

> You may wish to plot $y = g(x)$ on your graphics calculator to help you answer **(b)**.

9 Use Cardano's formula given in Section 7.1 to find the exact solutions to the equations:
   (a) $x^3 + 12x = 12$
   (b) $x^3 + 6x = 2$
   (c) $x^3 - 6x = 40$

> This question is included for interest only. Nothing similar will be tested in the C3 examination.

10 Given that $f(x) = \dfrac{1}{2x - 1}$, find $f(0)$ and $f(1)$.
   A student claims that the change in sign proves that the equation $\dfrac{1}{2x - 1} = 0$ has a root between 0 and 1. Is she correct?

7.3 Approximate values for points of intersection

In earlier units, you learnt that the points of intersection of two curves can often be found by solving simultaneous equations. After eliminating one of the variables the resulting equation may not always have exact solutions, and so you may have to use the numerical methods introduced in the previous section.

Worked example 7.7

The x-coordinate of the point of intersection of the line $y = 3x - 5$ and the curve $y = x^3 - 5x + 7$ is α. Show that $-3.4 < \alpha < -3.3$.

Solution

The point of intersection is given by equating the two expressions for y.

$\Rightarrow 3x - 5 = x^3 - 5x + 7$

$\Rightarrow x^3 - 8x + 12 = 0$ where $x = \alpha$ is the root of the equation.

Let $f(x) = x^3 - 8x + 12$

$\quad\quad \Rightarrow f(-3.4) = (-3.4)^3 - 8 \times (-3.4) + 12 = -0.104$

$\quad\quad \Rightarrow f(-3.3) = (-3.3)^3 - 8 \times (-3.3) + 12 = 2.463$

Hence $f(-3.4) < 0$ and $f(-3.3) > 0$.

The graph of $y = x^3 - 8x + 12$ is a polynomial graph and so it is continuous for all values of x, so the change in sign means that $-3.4 < \alpha < -3.3$.

Worked example 7.8

The two curves with equations $y = 2^x$ and $y = x^3 - 5$ intersect at the point (α, β). Determine two consecutive integers between which α lies.

Solution

You could plot the graphs on your graphics calculator and verify that the curves have a single point of intersection where the x-coordinate is less than 5.

It is important to follow this procedure, though.

For the point of intersection $2^x = x^3 - 5$

$$\Rightarrow 2^x - x^3 + 5 = 0$$

Create a new function, f, such that $f(x) = 2^x - x^3 + 5$

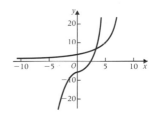

Rearrange the equation so it is of the form $f(x) = 0$.

Then $f(1) = 2^1 - 1^3 + 5 = 6$
$\quad\quad\quad f(2) = 2^2 - 2^3 + 5 = 1$
and $f(3) = 2^3 - 3^3 + 5 = -14$.

Since $y = f(x)$ is continuous for all values of x and there is a change of sign between $x = 2$ and $x = 3$, α must lie between the consecutive integers 2 and 3.

EXERCISE 7B

1 Show that the curve with equation $y = x^3$ and the curve with equation $y = x^2 + 1$ intersect at a point whose x-coordinate lies between 1.4 and 1.5.

2 Show that the line with equation $y = 2x - 7$ and the curve with equation $y = x^5 + 3$ intersect at a point whose x-coordinate lies between -1.8 and -1.6.

3 Verify that the line with equation $x + 3y = 1$ and the curve with equation $y = x^3 - x + 2$ intersect at a point whose x-coordinate lies between -1.4 and -1.3.

4 Prove that the curve with equation $y = x^5$ and the curve with equation $y = x^2 - 5$ intersect at a point whose x-coordinate lies between -1.3 and -1.2.

5 The two curves with equations $y = x^5 - x^3 + 1$ and $y = x^2 + 3$ intersect at the point (p, q). Show that $1.50 < p < 1.51$.

6 The two curves with equations $y = 3^x$ and $y = x^7$ intersect at the point (r, s). Prove that $1.20 < r < 1.21$.

7 The two curves with equations $y = 19 - 2^x$ and $y = 5^x$ intersect at the point (t, u). Show that $1.71 < t < 1.72$.

8 The curves $y = x^4 - 3$ and $y = 5x - x^6$ intersect at the points (a, b) and (c, d), where $a \approx -0.57$ and $c \approx 1.36$.

   (a) Determine whether $-0.58 < a < -0.57$ or $-0.57 < a < -0.56$.

   (b) Determine whether $1.35 < c < 1.36$ or $1.36 < c < 1.37$.

7.4 Iterative methods

Suppose you wish to find the square root of 17 but do not have a square root function on your calculator.

This problem is essentially the same as saying, what would be the length of the side of a square with area 17 square units?

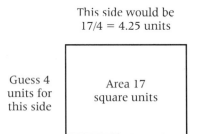

This side would be
17/4 = 4.25 units

Guess 4 units for this side

Area 17 square units

Thinking geometrically, if you guess that one side of a rectangle has length 4 units and the area is known to be 17 square units, then the other side must have length 17/4 = 4.25 units.

Since you are hoping to produce a square, a better guess for the side of the square is the average of these two lengths, namely

$$\frac{1}{2}(4 + 4.25) = 4.125 \text{ units.}$$

Repeating the process, if one side has length 4.125 units then the other side is $17/4.125 = 4.1212\ldots$ units. Therefore a better guess for the side of the square is

$$\frac{1}{2}(4.125 + 4.1212\ldots) = 4.123\ 106\ 06\ldots \text{ units.}$$

This is an example of an **iterative method** where the previous answer is fed into a formula to produce the next result.

If you have a graphics calculator with an ANS key, this procedure can be speeded up considerably.

Try typing in 4 followed by $\boxed{\text{EXE}}$ (on a Casio calculator) or $\boxed{\text{ENTER}}$ (on a Texas or Sharp calculator).

Then type in 0.5 (ANS + 17 ÷ ANS) followed by $\boxed{\text{EXE}}$ or $\boxed{\text{ENTER}}$ and you will see the value 4.125.

Now keep pressing the $\boxed{\text{EXE}}$ or $\boxed{\text{ENTER}}$ key and you will get the sequence

> 4.123 106 061...
> 4.123 105 626...
> 4.123 105 626...

We say that the sequence **converges** to the value 4.1231, to four decimal places.

In general the square root of any positive number N can be found by this method using 0.5(ANS + N ÷ ANS) after inputting your first guess.

> You can check this is correct by pressing $\sqrt{17}$ on your calculator.

> Try it with a few values and you will see that your first guess does not even have to be close to the value of the square root of N. You should see that after a few key presses the values converge to $\sqrt{N}$.

7

7.5 Recurrence relations

A sequence can be generated by a simple recurrence relation of the form $x_{n+1} = f(x_n)$.

For example, when $x_1 = 2$ and $x_{n+1} = x_n^2 - 1$, you can find

$$x_2 = 2^2 - 1 = 3$$

$x_3 = 8$, $x_4 = 63$, $x_5 = 3968$, etc.

A recurrence relation, or iterative formula, to find the square root of N (discussed in the previous section) would be

$$x_{n+1} = \frac{1}{2}\left(x_n + \frac{N}{x_n}\right)$$ with a suitable starting value x_1.

Worked example 7.9

A sequence is defined by $x_{n+1} = \dfrac{x_n - 3}{1 + 2x_n}$, $x_1 = 2$.

Find the values of x_2, x_3 and x_4.

Solution

You can find x_2 by using $x_2 = \dfrac{x_1 - 3}{1 + 2x_1} = -\dfrac{1}{5}$.

Then $x_3 = \dfrac{x_2 - 3}{1 + 2x_2} = \dfrac{\dfrac{-1}{5} - 3}{1 + \dfrac{-2}{5}} = \dfrac{\dfrac{-16}{5}}{\dfrac{3}{5}} = \dfrac{-16}{3} = -5\dfrac{1}{3}$.

Finally, $x_4 = \dfrac{x_3 - 3}{1 + 2x_3} = \dfrac{\dfrac{-16}{3} - 3}{1 + \dfrac{-32}{3}} = \dfrac{\dfrac{-25}{3}}{\dfrac{-29}{3}} = \dfrac{25}{29}$.

> You could have obtained decimal equivalents from your calculator using 4 ENTER
>
> (ANS − 3) ÷ (1 + 2ANS)
>
> ENTER

7.6 Convergent and divergent sequences

A sequence may be defined inductively by a recurrence relation such as in the previous worked example, or by a defining formula such as $u_n = \dfrac{1 + 3n}{n + 2}$.

In this case, you need to substitute $n = 1, 2, \dots$ to obtain

$$u_1 = \frac{1 + (3 \times 1)}{1 + 2} = \frac{4}{3}, \; u_2 = \frac{1 + (3 \times 2)}{2 + 2} = \frac{7}{4}, \text{ etc.}$$

As n gets larger, $u_{100} = \dfrac{1 + (3 \times 100)}{100 + 2} = \dfrac{301}{102} \approx 2.951$

and $u_{1000} = \dfrac{1 + (3 \times 1000)}{1000 + 2} = \dfrac{3001}{1002} \approx 2.995$

> An intuitive idea of whether a sequence converges or diverges is very useful before you try to find its limit.

so that the values are getting closer and closer to 3.
This sequence is said to **converge** to the value 3.

Worked example 7.10

For each of the following sequences, find x_2, x_3 and x_4, and state whether the sequence is convergent or divergent.

(a) $x_{n+1} = 1 + 2x_n, \quad x_1 = 5$,

(b) $x_{n+1} = \dfrac{5x_n + 1}{1 + 2x_n}, \quad x_1 = 2$.

Solution

(a) $x_{n+1} = 1 + 2x_n$ and $x_1 = 5$, so $x_2 = 1 + 2x_1 = 1 + 10 = 11$.

Similarly, $x_3 = 1 + 2x_2 = 1 + 22 = 23$ and
$x_4 = 1 + 2x_3 = 1 + 46 = 47$.

The values will continue to increase and so the sequence diverges to ∞.

(b) Since $x_{n+1} = \dfrac{5x_n + 1}{1 + 2x_n}$ and $x_1 = 2$, you can find

$$x_2 = \frac{5x_1 + 1}{1 + 2x_1} = \frac{10 + 1}{1 + 4} = \frac{11}{5} = 2.2.$$

Similarly, $x_3 = \dfrac{5x_2 + 1}{1 + 2x_2} = \dfrac{11 + 1}{1 + 4.4} = \dfrac{12}{5.4} = 2.2222\ldots$ and

also $x_4 = \dfrac{5x_3 + 1}{1 + 2x_3} = \dfrac{12.111\ldots}{5.444\ldots} = 2.224\ldots\,.$

It is difficult to see from just a few values, but this sequence is converging to a limit.

If you use the ANS key on a graphics calculator with

2 [EXE]

$(5\,\text{ANS} + 1) \div (1 + 2\,\text{ANS})$

[EXE]

[EXE] , etc., you find it quickly settles down to the value

2.224 744 871... .

> You may have to press [ENTER] rather than [EXE] on your calculator.

Alternatively, if you know the sequence tends to a limit L, you know that $x_n \to L$ and $x_{n+1} \to L$.

Since $x_{n+1} = \dfrac{5x_n + 1}{1 + 2x_n}$ you can write down an equation involving L.

$$L = \frac{5L + 1}{1 + 2L} \Rightarrow L + 2L^2 = 5L + 1$$

or $\quad 2L^2 - 4L - 1 = 0.$

Hence $L = \dfrac{4 \pm \sqrt{24}}{4} \Rightarrow L = 1 + \dfrac{\sqrt{6}}{2}$ or $L - 1 - \dfrac{\sqrt{6}}{2}.$

Since $x_1 = 2$, all the subsequent terms must be positive.

Hence the exact value of the limit is $1 + \dfrac{\sqrt{6}}{2}.$

EXERCISE 7C

1 A sequence is defined by $x_{n+1} = \dfrac{x_n + 5}{1 - 3x_n}$, $x_1 = 0$. Find the values of x_2, x_3 and x_4.

2 A sequence is defined by $u_{n+1} = \dfrac{u_n^2 + 1}{3 + u_n}$, $u_1 = 2$. Find the values of u_2, u_3 and u_4.

3 Use the iterative formula $x_{n+1} = \dfrac{1}{2}\left(x_n + \dfrac{N}{x_n}\right)$ with $x_1 = 3$ to find x_2, x_3 and x_4, giving your answers to six significant figures, in the cases where:

(a) $N = 10$, **(b)** $N = 13$.

Verify that these successive values are getting closer to:

(a) $\sqrt{10}$, **(b)** $\sqrt{13}$.

4 An iterative formula to find the cube root of the number V is given by $x_{n+1} = \frac{1}{3}\left(2x_n + \frac{V}{x_n^2}\right)$.

(a) With $V = 30$ and $x_1 = 3$, find the values of x_2, x_3 and x_4, giving your answers to six significant figures.

(b) With $V = 100$ and $x_1 = 5$, find the values of x_2, x_3 and x_4, giving your answers to six significant figures.

> Can you justify this formula by thinking of a cube, making a guess at the length of two equal sides and a suitable averaging technique to obtain the next approximation?

5 The iteration $x_{n+1} = x_n(2 - Nx_n)$ can be used to find the reciprocal of N without needing to use division.

(a) Use $N = 7$ and $x_1 = 0.1$ to find three iterations and verify that the values are approaching $1/7$.

(b) Use $N = 53$ and $x_1 = 0.02$ to find three iterations and verify that the values are approaching $1/53$.

6 For each of the following sequences, find the values of x_2, x_3 and x_4, and ascertain whether the sequence converges or diverges.

(a) $x_{n+1} = 4 + 5x_n$, $x_1 = -1$.

(b) $x_{n+1} = 4 + 5x_n$, $x_1 = -2$.

(c) $x_{n+1} = 3 - \frac{1}{2}x_n$, $x_1 = -2$.

(d) $x_{n+1} = \frac{7x_n - 1}{3 + 2x_n}$, $x_1 = 1.7$.

7.7 Cobweb and staircase diagrams

It is sometimes possible to rearrange an equation into the form $x = g(x)$ and then an iterative formula of the form $x_{n+1} = g(x_n)$ can be used to try to find a solution to the equation. This all depends on whether the iteration converges and you can draw diagrams to ascertain whether this is the case.

Consider the iterative formula $x_{n+1} = \frac{1}{2}x_n + 1$.

You can draw the graphs of $y = x$ and $y = \frac{1}{2}x + 1$ on the same axes.

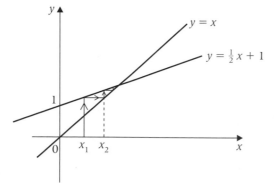

> On your calculator, try 0.5
> $\boxed{\text{EXE}}$ followed by 0.5 ANS + 1
> $\boxed{\text{EXE}}$ and you will see convergence to the value 2.

Suppose $x_1 = \frac{1}{2}$. The next value is $x_2 = \frac{1}{2}x_1 + 1 = \frac{5}{4}$. A vertical line can be drawn to $y = \frac{1}{2}x + 1$ to obtain the y-value equivalent to x_2. In order to get the next x-value as x_2 you draw a horizontal line to meet $y = x$ and then a vertical line to $y = \frac{1}{2}x + 1$ again and the process is repeated.

A diagram of the form shown is produced which converges to the point of intersection and this is known as **staircase convergence**.

Because it looks like a staircase.

Suppose the limit exists and let it be L. Then as $n \to \infty$, $x_n \to L$ and $x_{n+1} \to L$.

So $L = \frac{1}{2}L + 1 \Rightarrow \frac{1}{2}L = 1 \Rightarrow L = 2$.

The iteration converges to the value 2, which is the x-coordinate of the point of intersection of the two graphs.

Try using the diagram with $x_1 = 3$ and you should see that convergence takes place with a staircase coming downwards to the limit this time. Remember to draw the vertical line to meet $y = \frac{1}{2}x + 1$ **first** then the horizontal line to $y = x$, etc.

Another situation occurs with the following recurrence relation $x_{n+1} = 2 - \frac{1}{3}x_n$ with $x_1 = \frac{1}{2}$.

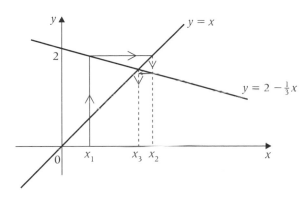

This diagram is known as a **cobweb diagram** and for this iteration there is convergence. Again, try different values of x_1 and you will see it still converges. This will not always be the case, however.

The limit L can be found as in the previous example.
$x_n \to L$ and $x_{n+1} \to L$.
So $L = 2 - \frac{1}{3}L \Rightarrow \frac{4}{3}L = 2 \Rightarrow L = 1.5$.
The limit of the sequence is 1.5.

Look carefully at a spider's web and you will see a clear spiral.

Convergence does not always take place as you can see below.

Suppose $x_{n+1} = 5 - 3x_n$ with $x_1 = 1$.

The terms of the sequence are $1, 2, -1, 8, -19, \ldots$ and the cobweb diagram below shows the spiralling outwards with clearly no convergence, whatever the starting value of x_1.

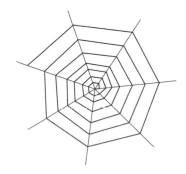

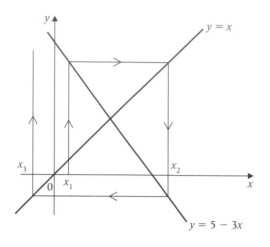

Worked example 7.11

Show that the equation $x^3 - 5x + 7 = 0$ can be rearranged to form the following iterative formulae:

(a) $x_{n+1} = \dfrac{1}{5}(x_n^3 + 7),$ **(b)** $x_{n+1} = \sqrt[3]{(5x_n - 7)}.$

Using $x_1 = -2$ in each case, find the values of x_2 and x_3. Show on separate diagrams whether or not convergence takes place.

Solution

(a) $x^3 - 5x + 7 = 0 \Rightarrow 5x = x^3 + 7 \Rightarrow x = \dfrac{1}{5}(x^3 + 7)$

Hence, an iterative formula is $x_{n+1} = \dfrac{1}{5}(x_n^3 + 7).$

$x_1 = -2 \Rightarrow x_2 = 0.2(-8 + 7) = -0.2$
$\qquad \Rightarrow x_3 = 0.2(-0.008 + 7) = 1.3984$

The graphs of $y = x$ and $y = \dfrac{1}{5}(x^3 + 7)$ are drawn and the

diagram below shows that the iteration is **not** convergent.

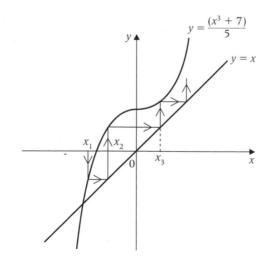

(b) $x^3 - 5x + 7 = 0 \Rightarrow x^3 = 5x - 7 \Rightarrow x = \sqrt[3]{(5x - 7)}$

Hence, an iterative formula is $x_{n+1} = \sqrt[3]{(5x_n - 7)}$.

$x_1 = -2 \Rightarrow x_2 = \sqrt[3]{-17} = -\sqrt[3]{17} = -2.571\,28\ldots$

$\Rightarrow x_3 = \sqrt[3]{-19.856\ldots} = -\sqrt[3]{19.856\ldots} = -2.7079\ldots$

The graphs of $y = x$ and $y = \sqrt[3]{(5x - 7)}$ are drawn and the diagram below shows that the iteration is convergent.

> **Note:** Many calculators cannot find cube roots of negative numbers and so you have to input a positive number and deal with the minus sign yourself.

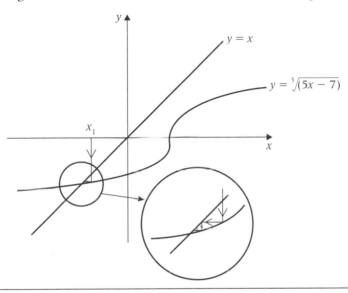

An iteration of the form $x_{n+1} = g(x_n)$ converges when the gradient of $y = g(x)$ at the point of intersection with the line $y = x$ satisfies the condition $|g'(x)| < 1$, provided a suitable value for x_1 is chosen.

Worked example 7.12

A sequence is defined by $x_{n+1} = \sqrt{(28 - 3x_n)}$, $x_1 = 3$.

(a) Find the values of x_2, x_3 and x_4.

(b) Explain why all the values of x_n are positive.

(c) Given that the sequence has limit L:

    (i) show that L must satisfy the equation $L^2 + 3L - 28 = 0$,

    (ii) find the limit of the sequence.

(d) Draw an appropriate diagram to show how convergence takes place.

Solution

(a) $x_2 = 4.358\,898\,9\ldots$, $x_3 = 3.863\,069\ldots$, $x_4 = 4.051\,02\ldots$

(b) Since x_1 is positive and x_2 is the positive square root of a positive number, x_2 is also positive. Similarly, if x_n is positive then x_{n+1} is positive. Therefore, all the values of x_n are positive.

(c) (i) Since the limit exists and it equals L. Then as $n \to \infty$,
$x_n \to L$ and $x_{n+1} \to L$.
So, $L = \sqrt{(28-3L)} \Rightarrow L^2 = 28 - 3L \Rightarrow L^2 + 3L - 28 = 0$.

(ii) $L^2 + 3L - 28 = 0 \Rightarrow (L + 7)(L - 4) = 0 \Rightarrow L = -7, L = 4$
But since all the terms of the sequence are positive, the value of L must be positive. You can reject the negative root.
Hence $L = 4$.
The limit of the sequence is 4.

(d) The diagram below shows how convergence takes place.

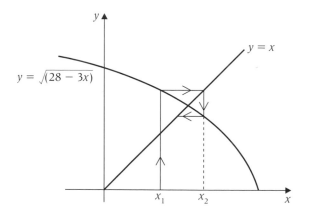

EXERCISE 7D

1 Draw diagrams to illustrate whether convergence takes place for each of the following recurrence relations with $x_1 = 0$.

(a) $x_{n+1} = \frac{1}{2}x_n - 2$

(b) $x_{n+1} = 2x_n + 1$

(c) $x_{n+1} = 3 - \frac{1}{4}x_n$,

(d) $x_{n+1} = 4 - 2x_n$

(e) $x_{n+1} = 5 + \frac{3}{4}x_n$

2 Using the results from question **1** and any further sketches you may need, determine the condition on the constant k for the iterative formula $x_{n+1} = kx_n + c$ to be convergent.

3 The equation $x = \sin 2x$, where x is in radians, has a single positive root β.

(a) Use the iterative formula $x_{n+1} = \sin 2x_n$ with $x_1 = 1$ to find x_2 and x_3, giving your answers to four decimal places.

(b) On a copy of the diagram below, indicate whether the sequence of iterations starting with $x_1 = 1$ (indicated by the mark on the x-axis) converges to β or diverges.

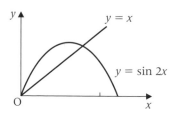

4 Show that the equation $\theta = 1 + \sin\theta$ has a root between 1.8 and 2.0.
Draw a diagram to show whether the iteration
$\theta_{n+1} = 1 + \sin\theta_n$ converges to this root with $\theta_1 = 1.9$ and find the values of θ_2, θ_3, θ_4 and θ_5, to four decimal places.

5 The equation $x^5 - 3x + 7 = 0$ has a single root α.

(a) Find two integers between which α lies.

(b) Rearrange the equation $x^5 - 3x + 7 = 0$ in different ways and form two iterative equations of the form
$x_{n+1} = g(x_n)$. By means of appropriate sketches of cobweb or staircase diagrams, determine whether the iteration converges to α in each case.

6 The graph of $y = x^3 - 4x^2 + x + 6$ intersects the line with equation $y = 10 + x$ at the point where $x = \alpha$. Show that α satisfies the equation $x = 4 + \dfrac{4}{x^2}$ and that α lies between 4 and 5.

Use the iterative formula $x_{n+1} = 4 + \dfrac{4}{x_n^2}$ with $x_1 = 4$ to find x_5,
giving your answer to four significant figures.
Draw a diagram to show how these iterations converge to α.

7 Sketch the curve with equation $y = e^x$ and on the same axes draw an appropriate straight line to show that the equation
$e^x + x - 3 = 0$ has exactly one root α.

(a) Prove that α lies between 0.7 and 0.8.

(b) Show that the equation $e^x + x - 3 = 0$ can be rearranged in the form $x = \ln[f(x)]$. Use an iteration of the form
$x_{n+1} = g(x_n)$ based on this rearrangement with $x_1 = 0.8$ to find the values of x_2, x_3 and x_4, giving your answers to three decimal places. Draw a diagram to show how convergence takes place.   [A]

8 (a) Sketch the graph of $y = 3^x - 5$ and an appropriate straight line to show that the equation $x = 3^x - 5$ has two real roots.

(b) Explain by means of an appropriate cobweb or staircase diagram how the iteration $x_{n+1} = 3^{x_n} - 5$ behaves for all possible values of x_1.

(c) By means of an appropriate value of x_1, determine one of the roots correct to three decimal places.

(d) Determine a possible rearrangement of the formula which would enable you to converge to the other root with an appropriate starting value.

7

9 Find a cubic equation whose single root can be found using

the iterative formula $x_{n+1} = 3 + \dfrac{2}{x_n^2}$.

Using any non-zero value for x_1, evaluate the iterations until they agree to five decimal places. Hence, find the value of the root to four decimal places.

Show on a diagram how convergence takes place.

10 A sequence is defined by $x_{n+1} = \sqrt{2x_n + 3}$, $x_1 = 2$.

(a) Find the values of x_2, x_3 and x_4.

(b) Explain why all the values of x_n are positive.

(c) Given that the limit of the sequence is L:

    (i) show that L must satisfy the equation
        $L^2 - 2L - 3 = 0$,

    (ii) find the limit of the sequence.

(d) Draw an appropriate diagram to show how convergence takes place.

11 A sequence is defined by $x_{n+1} = \dfrac{1}{5}(x_n^2 + 6)$, $x_1 = 2.5$.

(a) Find the values of x_2, x_3 and x_4.

(b) Show that if $x_n < 3$ then $x_{n+1} < 3$.

(c) Given that the limit of the sequence is L:

    (i) show that L must satisfy the equation
        $L^2 - 5L + 6 = 0$,

    (ii) find the limit of the sequence.

(d) Draw an appropriate diagram to show how convergence takes place.

12 A sequence is defined by $x_{n+1} = -\sqrt{3x_n + 10}$, $x_1 = -1$.

(a) Find the values of x_2, x_3 and x_4.

(b) Explain why all the values of x_n are negative.

(c) Given that the limit of the sequence is L:

    (i) show that L must satisfy the equation
        $L^2 - 3L - 10 = 0$,

    (ii) find the limit of the sequence.

(d) Draw an appropriate diagram to show how convergence takes place.

13 A sequence is defined by $x_{n+1} = \sqrt{x_n + 12}$, $x_1 = 2$.

(a) Find the values of x_2, x_3 and x_4, giving your answers to three decimal places.

(b) Given that the limit of the sequence is L:

    (i) show that L must satisfy the equation
        $L^2 - L - 12 = 0$,

    (ii) find the value of L.

(c) The graphs of $y = \sqrt{x + 12}$ and $y = x$ are sketched below.

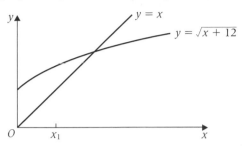

On a copy of the sketch, draw a cobweb or staircase diagram to show how convergence takes place.   [A]

Key point summary

1 If the graph of $y = f(x)$ is continuous over the interval $a \leqslant x \leqslant b$, and $f(a)$ and $f(b)$ have different signs, then at least one root of the equation $f(x) = 0$ must lie in the interval $a < x < b$.   *p117*

2 Cobweb and/or staircase diagrams can be drawn to illustrate whether convergence takes place or not for iterations of the form $x_{n+1} = g(x_n)$.   *p127*

3 An iteration of the form $x_{n+1} = g(x_n)$ converges when the gradient of $y = g(x)$ at the point of intersection with the line $y = x$ satisfies the condition $|g'(x)| < 1$, provided a suitable value for x_1 is chosen.   *p129*

7

Test yourself   What to review

1 Verify that the equation $x^3 - 7x^2 - 1 = 0$ has a root between 6.8 and 7.2.

Section 7.2

2 Prove that the x-coordinate of the point of intersection of $y = \sin x$ and $x + y = 2$ is 1.106, correct to three decimal places.

Section 7.3

3 A sequence is defined by $u_{n+1} = \dfrac{2 - u_n^2}{3 + u_n}$, $u_1 = 0$. Find the values of u_2 and u_3.

Section 7.5

4 For each of the following sequences, find the values of x_2, x_3 and x_4 and state whether the sequence is convergent or not:

(a) $x_{n+1} = 3 - 4x_n$, $x_1 = 2$

(b) $x_{n+1} = \dfrac{x_n - 3}{1 - 2x_n}$, $x_1 = -1$.

Section 7.6

Test yourself (continued)	What to review

5 (a) Sketch on the same axes the graphs of $y = \ln x$ and $y = 3 - x$.

(b) Use your graph from **(a)** to explain why the equation
$$\ln x - 3 + x = 0$$
has a single root, α.

(c) Prove that α lies between 2.20 and 2.21.

Section 7.3

6 The iterative formula $x_{n+1} = \frac{1}{4} \cos x_n$ is to be used with $x_1 = 0$. Find the values of x_2, x_3 and x_4 to four significant figures, and draw a diagram to show whether the sequence converges or diverges.

Section 7.7

Test yourself ANSWERS

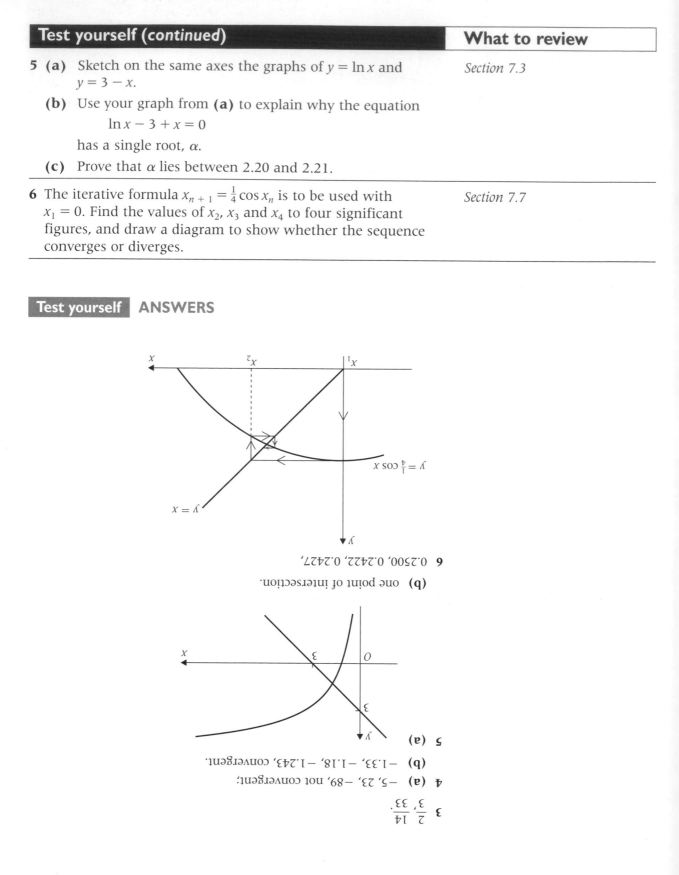

6 0.2500, 0.2422, 0.2427,

(b) one point of intersection.

5 (a)

4 (b) −1.33, −1.18, −1.243, convergent.

4 (a) −5, 23, −89, not convergent;

3 $\frac{2}{3}, \frac{14}{33}$.

CHAPTER 8

C3: Integration by inspection and substitution

Learning objectives

After studying this chapter, you should be able to:

■ integrate expressions using the reverse idea of the chain rule
■ integrate trigonometric expressions

■ integrate expressions of the form $\dfrac{f'(x)}{f(x)}$

■ integrate expressions using a suitable substitution.

8.1 Review

In this chapter you will learn how to integrate more functions to add to those you have already considered.

You will recall from C1 and C2, the basic result

$$\int x^n \, dx = \frac{x^{n+1}}{n+1} + c, \quad n \neq -1,$$

> c is the arbitrary constant.

and how, in Chapter 4, the missing link was established to give

$$\int \frac{1}{x} \, dx = \ln x + c.$$

The exponential function was also integrated in Chapter 4 to give

$$\int e^{kx} \, dx = \frac{1}{k} e^{kx} + c.$$

Worked example 8.1

Find $\displaystyle\int \left(10\sqrt{x^3} + \frac{1}{5x} + \frac{4}{e^{2x}} \right) dx.$

Solution

$$\int\left(10\sqrt{x^3} + \frac{1}{5x} + \frac{4}{e^{2x}}\right) dx = \int\left(10x^{\frac{3}{2}} + \frac{1}{5} \times \frac{1}{x} + 4e^{-2x}\right) dx$$

$$= 10\int x^{\frac{3}{2}} dx + \frac{1}{5}\int\frac{1}{x} dx + 4\int e^{-2x} dx$$

$$= \frac{10x^{\frac{5}{2}}}{\frac{5}{2}} + \frac{1}{5}\ln x + 4 \times \frac{e^{-2x}}{-2} + c$$

$$= 4\sqrt{x^5} + \frac{1}{5}\ln x - 2e^{-2x} + c$$

EXERCISE 8A

Find each of the following:

1 $\displaystyle\int\left(6x^2 + \frac{2}{x} + 7\right) dx$

2 $\displaystyle\int\left(\sqrt{x} + e^{2x}\right) dx$

3 $\displaystyle\int\left(\frac{1}{2x} - \frac{2}{e^{4x}}\right) dx$

4 $\displaystyle\int\left(\frac{x^2 e^{3x} + x + 3}{x^2}\right) dx$

5 $\displaystyle\int\left(e^{2x} - 1\right)^2 dx$

6 $\displaystyle\int_0^1\left(2x + 1\right)^4 dx$

> Hint: Multiply out the brackets.

8.2 Integrating functions of the form $(ax + b)^n$ and $e^{cx + d}$ with respect to x

In chapter 5 you used the chain rule to differentiate many functions. Because integration is the reverse process to differentiation you can use the results of section 5.3 to evaluate many more integrals.

Worked example 8.2

Find $\displaystyle\int(2x + 7)^6 dx$.

Solution

Consider $\dfrac{d}{dx}(2x + 7)^7$.

> Increase the power by 1.

$$\frac{d}{dx}(2x + 7)^7 = 7(2x + 7)^6 \times 2$$

> Use the chain rule.

$$\Rightarrow 14(2x + 7)^6 = \frac{d}{dx}(2x + 7)^7$$

Integrating both sides with respect to x gives

$$\int 14(2x + 7)^6 dx = \int\frac{d}{dx}(2x + 7)^7 dx$$

$$14\int(2x + 7)^6 dx = (2x + 7)^7 + k$$

> Since 14 is a constant.

$$\Rightarrow \int(2x + 7)^6 dx = \frac{1}{14}(2x + 7)^7 + c.$$

> Where the constant $c = \dfrac{k}{14}$.

In general,

> For constants a, b and n, where $n \neq -1$,
> $$\int (ax + b)^n \, dx = \frac{1}{a(n + 1)}(ax + b)^{n + 1} + c.$$

In words, to integrate $(ax + b)^n$ with respect to x,

- increase index by 1
- divide at front by new index
- divide at front by derivative of $(ax + b)$, namely a.

Worked example 8.3

Evaluate $\displaystyle\int_{-1}^{0} \frac{1}{(4 + 3x)^2} \, dx$.

Solution

$$\int_{-1}^{0} \frac{1}{(4 + 3x)^2} \, dx = \int_{-1}^{0} (3x + 4)^{-2} \, dx = \left[\frac{1}{(3)(-2 + 1)}(3x + 4)^{-2 + 1} \right]_{1}^{0}$$

new index $= -2 + 1$
divide by $(-2 + 1)$
divide by 3, the derivative of $(3x + 4)$.

$$= \left[-\frac{1}{3}(3x + 4)^{-1} \right]_{-1}^{0}$$

$$-\frac{1}{-12} - \frac{1}{-3} = \frac{1}{4}$$

Worked example 8.4

Find $\displaystyle\int \frac{2}{\sqrt{4x + 9}} \, dx$.

Solution

$$\int \frac{2}{\sqrt{4x + 9}} \, dx = 2\int (4x + 9)^{-\frac{1}{2}} \, dx$$

$$= 2\left[\frac{(4x + 9)^{\frac{1}{2}}}{(4)(\frac{1}{2})} \right] + c$$

$-\dfrac{1}{2} + 1 = \dfrac{1}{2}$

$$\Rightarrow \int \frac{2}{\sqrt{4x + 9}} \, dx = \sqrt{4x + 9} + c$$

Integrals of the type $\int e^{ax + b} \, dx$ can also be found by applying the reverse process to differentiation by the chain rule or by the alternative method illustrated by the next worked example.

8

Worked example 8.5

Find $\int e^{3x + 5}\,dx$.

Solution

$$\int e^{3x + 5}\,dx = \int e^5 e^{3x}\,dx$$

By the law of indices.

$$= e^5 \int e^{3x}\,dx = e^5\left[\frac{1}{3}e^{3x}\right] + c$$

Since e^5 is a constant and $\int e^{kx}\,dx = \frac{1}{k}e^{kx} + c.$

$$= \frac{1}{3}e^{3x + 5} + c$$

Try using the reverse chain rule approach to find the integral in the previous worked example.

Start by considering $\dfrac{d}{dx}\,(e^{3x + 5})$.

In general,

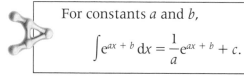

For constants a and b,

$$\int e^{ax + b}\,dx = \frac{1}{a}e^{ax + b} + c.$$

In words, to integrate the exponential function $e^{ax + b}$ with respect to x,

- write down the exponential function
- divide at the front by the derivative of $(ax + b)$, namely a.

EXERCISE 8B

1 Find:

(a) $\int (x + 2)^6\,dx$

(b) $\int (5x + 1)^8\,dx$

(c) $\int (3 - 2x)^5\,dx$

(d) $\int 4 - (1 + 2x)^6\,dx$

(e) $\int \sqrt{4x + 1}\,dx$

(f) $\int \sqrt[3]{8x - 27}\,dx$

(g) $\int \dfrac{1}{(x - 1)^2}\,dx$

(h) $\int \dfrac{1}{(2 - 5x)^3}\,dx$

(i) $\int \dfrac{1}{2(5 + 4x)^4}\,dx$

(j) $\int \sqrt{\dfrac{4}{2x - 1}}\,dx$

(k) $\int e^{2x}\,dx$

(l) $\int \dfrac{3}{e^{3x}}\,dx$

(m) $\int (e^{5x + 6} - 1)\,dx$

(n) $\int \dfrac{4}{e^{3 - 2x}}\,dx$

2 Evaluate:

(a) $\int_0^1 (2x + 1)^4\,dx$

(b) $\int_{-1}^2 (1 + 3x)^5\,dx$

(c) $\int_1^2 \dfrac{1}{(1 - 2x)^2}\,dx$

(d) $\int_0^1 \dfrac{1}{\sqrt[3]{7x + 1}}\,dx$

(e) $\int_{\frac{1}{2}}^2 e^{2x - 1}\,dx$

(f) $\int_0^{0.25} \dfrac{1}{2e^{4x - 1}}\,dx$

3 (a) Write $x^2 - 2x + 3$ in the form $(x - a)^2 + b$, where a and b are constants to be found.

(b) Hence, show that $\dfrac{x^2 - 2x + 3}{(x - 1)^2} \equiv 1 + \dfrac{2}{(x - 1)^2}$.

(c) Hence, evaluate $\displaystyle\int_2^3 \dfrac{x^2 - 2x + 3}{(x - 1)^2}\, dx$.

4 Evaluate $\displaystyle\int_0^1 \dfrac{x^2 + 2x}{(x + 1)^2}\, dx$.

8.3 Further integration of trigonometric functions

In sections 5.2 and 6.4 of this book you were shown the three results:

$\dfrac{d}{dx}(\sin x) = \cos x$,

$\dfrac{d}{dx}(\cos x) = -\sin x$, and

$\dfrac{d}{dx}(\tan x) = \sec^2 x$,

where x is in radians.

Integrating both sides of each of the three results leads to:

> For x in radians,
>
> $\displaystyle\int \cos x \, dx = \sin x + c$
>
> $\displaystyle\int \sin x \, dx = -\cos x + c$
>
> $\displaystyle\int \sec^2 x \, dx = \tan x + c$.

These three integrals can be generalised by applying the reverse of the chain rule.

Worked example 8.6

Find $\displaystyle\int \sin 4x \, dx$.

Solution

Consider $\dfrac{d}{dx}(\cos 4x) = -4 \sin 4x$.

Integrating both sides with respect to x leads to

$\displaystyle\int \dfrac{d}{dx}(\cos 4x)\, dx = \int -4 \sin 4x \, dx$

$\Rightarrow \quad \cos 4x + k = -4 \int \sin 4x \, dx$

$\Rightarrow \quad \displaystyle\int \sin 4x \, dx = -\dfrac{1}{4} \cos 4x + c$.

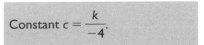

Constant $c = \dfrac{k}{-4}$.

Worked example 8.7

Find $\int \cos(2x + 3)\,dx$.

Solution

Consider $\dfrac{d}{dx}(\sin(2x + 3)) = 2\cos(2x + 3)$.

Integrating both sides with respect to x leads to

$$\int 2\cos(2x + 3)\,dx = \sin(2x + 3) + k$$

$$\Rightarrow \int \cos(2x + 3)\,dx = \frac{1}{2}\sin(2x + 3) + c.$$

In a similar way you can obtain the following general results:

For constants a and b and x in radians,

$$\int \cos(ax + b)\,dx = \frac{1}{a}\sin(ax + b) + c$$

$$\int \sin(ax + b)\,dx = -\frac{1}{a}\cos(ax + b) + c$$

$$\int \sec^2(ax + b)\,dx = \frac{1}{a}\tan(ax + b) + c.$$

These integrals should be memorised as they do not appear in the formulae booklet in this form.

Worked example 8.8

Show that $\displaystyle\int_0^{\frac{\pi}{6}} \sec^2\left(3x - \frac{\pi}{4}\right) dx = \frac{2}{3}$.

Solution

$$\int_0^{\frac{\pi}{6}} \sec^2\left(3x - \frac{\pi}{4}\right) dx = \left[\frac{1}{3}\tan\left(3x - \frac{\pi}{4}\right)\right]_0^{\frac{\pi}{6}}$$

$$= \frac{1}{3}\tan\left(\frac{\pi}{2} - \frac{\pi}{4}\right) - \frac{1}{3}\tan\left(0 - \frac{\pi}{4}\right)$$

$$= \frac{1}{3}\tan\left(\frac{\pi}{4}\right) - \frac{1}{3}\tan\left(-\frac{\pi}{4}\right)$$

$$= \frac{1}{3}(1) - \frac{1}{3}(-1)$$

$$= \frac{2}{3}$$

The following three results, which you were shown in section 6.4, are given in the examination formulae booklet:

$$\frac{d}{dx}(\operatorname{cosec} x) = -\operatorname{cosec} x \cot x,$$

$$\frac{d}{dx}(\sec x) = \sec x \tan x, \text{ and}$$

$$\frac{d}{dx}(\cot x) = -\operatorname{cosec}^2 x,$$

where x is in radians.

Applying the chain rule you can obtain three more general

results, for example $\dfrac{d}{dx}(\cosec ax) = -a\cosec ax \cot ax$.

Integrating both sides of each of the three results with respect to x leads to:

> For x in radians and a constant a,
>
> $$\int \cosec ax \cot ax \, dx = -\frac{1}{a}\cosec ax + c$$
>
> $$\int \sec ax \tan ax \, dx = \frac{1}{a}\sec ax + c$$
>
> $$\int \cosec^2 ax \, dx = -\frac{1}{a}\cot ax + c.$$

Worked example 8.9

Given that $\tan\dfrac{\pi}{6} = \dfrac{1}{\sqrt{3}}$, find the exact value of $\displaystyle\int_{\frac{\pi}{12}}^{\frac{\pi}{8}} \cosec^2 2x \, dx$.

Solution

$$\int_{\frac{\pi}{12}}^{\frac{\pi}{8}} \cosec^2 2x \, dx = \left[-\frac{1}{2}\cot 2x\right]_{\frac{\pi}{12}}^{\frac{\pi}{8}}$$

$$= -\frac{1}{2}\cot\frac{\pi}{4} - \left(-\frac{1}{2}\cot\frac{\pi}{6}\right) = -\frac{1}{2}(1) + \frac{1}{2}(\sqrt{3})$$

$$= \frac{\sqrt{3}-1}{2}$$

$$\cot\frac{\pi}{6} = \frac{1}{\tan\frac{\pi}{6}} = \frac{1}{\left(\frac{1}{\sqrt{3}}\right)} = \sqrt{3}$$

Sometimes you may have to rewrite the integrand in a more suitable form before integrating. The next example illustrates the point.

Worked example 8.10

Find $\displaystyle\int\left(\frac{2\sin x}{\cos^2 x}\right)dx$.

Solution

$$\int\left(\frac{2\sin x}{\cos^2 x}\right)dx = 2\int\frac{1}{\cos x}\left(\frac{\sin x}{\cos x}\right)dx = 2\int\sec x \tan x \, dx$$

$$\Rightarrow \int\frac{2\sin x}{\cos^2 x}\,dx = 2\sec x + c.$$

EXERCISE 8C

1 (a) Find $\int \sin x \, dx$.

(b) Hence show that $\int_0^{\frac{\pi}{3}} \sin x \, dx = 0.5$.

2 Evaluate:

(a) $\int_0^{\frac{\pi}{2}} \cos x \, dx$
    (b) $\int_0^{\frac{\pi}{4}} \sec^2 \theta \, d\theta$

3 Find:

(a) $\int \cos 2x \, dx$
    (b) $\int \sin 3x \, dx$

(c) $\int \sec^2 \frac{\theta}{2} \, d\theta$
    (d) $\int 1 - \cos(2x + 4) \, dx$

(e) $\int x - \sin(4x - 1) \, dx$
    (f) $\int \sec^2(1 - 3\theta) \, d\theta$

(g) $\int \csc^2 4x \, dx$
    (h) $\int \tan 3x \sec 3x \, dx$

(i) $\int \frac{\cos 2x}{\sin^2 2x} \, dx$

4 By using the identity $1 + \tan^2 A \equiv \sec^2 A$, find $\int \tan^2 x \, dx$.

5 (a) Find $\int \tan^2 2x \, dx$.

(b) Hence evaluate $\int_0^{\frac{\pi}{8}} \tan^2 2x \, dx$.

6 Evaluate:

(a) $\int_0^{\frac{\pi}{12}} \sec^2 3\theta \, d\theta$
    (b) $\int_0^{\frac{\pi}{8}} \cos 4x \, dx$

(c) $\int_{\frac{\pi}{6}}^{\frac{\pi}{2}} \sin 2x \, dx$
    (d) $\int_0^{\frac{\pi}{9}} \sin\left(6x + \frac{\pi}{3}\right) dx$

(e) $\int_0^{\frac{\pi}{6}} \cos\left(4x + \frac{\pi}{6}\right) dx$
    (f) $\int_0^{\frac{\pi}{4}} (1 + \cos 2\theta) \, d\theta$

(g) $\int_0^{\frac{\pi}{4}} (1 + \sin 2\theta) \, d\theta$
    (h) $\int_{\frac{\pi}{6}}^{\frac{\pi}{4}} 2 \csc^2 2x \, dx$

(i) $\int_0^{\frac{\pi}{9}} \frac{\sin 3x}{\cos^2 3x} \, dx$

7 By using the identity $1 + \cot^2 A = \csc^2 A$, find $\int \cot^2 x \, dx$.

8 Evaluate $\int_{\frac{\pi}{8}}^{\frac{\pi}{4}} \cot^2 2x \, dx$.

8.4 Integrals of the form $\int \dfrac{f'(x)}{f(x)}\, dx$

Here are four examples of integrals of the form $\int \dfrac{f'(x)}{f(x)}\, dx$:

$$\int \frac{5}{5x+7}\, dx, \quad \int \frac{2x}{x^2+4}\, dx, \quad \int \frac{3e^x+2}{3e^x+2x+1}\, dx, \quad \int \frac{2\cos x}{5+2\sin x}\, dx.$$

If you can find an expression for the general integral

$\int \dfrac{f'(x)}{f(x)}\, dx$, where $f(x)$ is any non-zero function of x, you will be

able to write down the answers for the four integrals given above.

Again you can make use of the chain rule for differentiation.

Consider $\dfrac{d}{dx}\,[\ln f(x)]$.

$$\frac{d}{dx}\,[\ln f(x)] = \frac{1}{f(x)} \times f'(x)$$

Integrating both sides with respect to x leads to

$$\int \frac{f'(x)}{f(x)}\, dx = \ln|f(x)| + c$$

The modulus sign stresses the fact that you cannot take the natural logarithm of a negative value.

In words, the integral of a rational function, in which the numerator is the derivative of the denominator, is the natural logarithm of the modulus of the denominator.

Returning to the list of the four integrals, you are now able to write down the following:

$$\int \frac{5}{5x+7}\, dx, = \ln|5x+7| + c,$$

$$\int \frac{2x}{x^2+4}\, dx = \ln|x^2+4| + c,$$

$$\int \frac{3e^x+2}{3e^x+2x+1}\, dx = \ln|3e^x+2x+1| + c,$$

$$\int \frac{2\cos x}{5+2\sin x}\, dx = \ln|5+2\sin x| + c.$$

As a first step, to integrate a rational function you check to see if the numerator is the same as the derivative of the denominator. If it is then the integral is just the natural logarithm of the modulus of the denominator.

Worked example 8.11

Find $\int \dfrac{x}{x^2+1}\, dx$.

8

Solution

Differentiating the denominator $x^2 + 1$ gives $2x$. Although this is not **exactly** the same as the numerator, x, since 2 is a constant you can write $\int \dfrac{x}{x^2 + 1}\, dx$ as $\dfrac{1}{2} \int \dfrac{2x}{x^2 + 1}\, dx$.

So $\int \dfrac{x}{x^2 + 1}\, dx = \dfrac{1}{2} \int \dfrac{2x}{x^2 + 1}\, dx = \dfrac{1}{2} \ln|x^2 + 1| + c$

$$= \dfrac{1}{2} \ln (x^2 + 1) + c$$

> **Warning:** You can only take constants outside of an integral sign. For example,
> $$\int \dfrac{1}{x^2 + 1}\, dx \neq \dfrac{1}{2x} \int \dfrac{2x}{x^2 + 1}\, dx.$$

> For real x, $x^2 + 1$ is always positive so you do not need the modulus sign.

Worked example 8.12

Show that the area bounded by the curve $y = \tan 2x$, the x-axis and the line $x = \dfrac{\pi}{6}$ is $\dfrac{1}{2} \ln 2$.

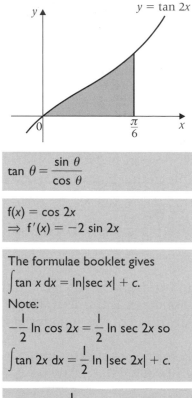

Solution

The required area is $\displaystyle\int_0^{\frac{\pi}{6}} \tan 2x\, dx$.

$$\int_0^{\frac{\pi}{6}} \tan 2x\, dx = \int_0^{\frac{\pi}{6}} \dfrac{\sin 2x}{\cos 2x}\, dx$$

$$= -\dfrac{1}{2} \int_0^{\frac{\pi}{6}} \dfrac{-2 \sin 2x}{\cos 2x}\, dx$$

$$= -\dfrac{1}{2} \Big[\ln|\cos 2x|\Big]_0^{\frac{\pi}{6}}$$

$$= -\dfrac{1}{2} \left(\ln \cos \dfrac{\pi}{3} - \ln \cos 0\right)$$

$$= -\dfrac{1}{2} \left(\ln \dfrac{1}{2} - \ln 1\right)$$

$$= \dfrac{1}{2} \left(-\ln \dfrac{1}{2}\right) = \dfrac{1}{2} \ln 2$$

> $\tan \theta = \dfrac{\sin \theta}{\cos \theta}$

> $f(x) = \cos 2x$
> $\Rightarrow f'(x) = -2 \sin 2x$

> The formulae booklet gives
> $\int \tan x\, dx = \ln|\sec x| + c.$
> Note:
> $-\dfrac{1}{2} \ln \cos 2x = \dfrac{1}{2} \ln \sec 2x$ so
> $\int \tan 2x\, dx = \dfrac{1}{2} \ln|\sec 2x| + c.$

> $-\ln k = \ln \dfrac{1}{k}$

Worked example 8.12 illustrates how we can generalise integrals given in the formulae booklet to obtain

For a constant a and x in radians,
$$\int \tan ax\, dx = \dfrac{1}{a} \ln|\sec ax| + c$$
$$\int \cot ax\, dx = \dfrac{1}{a} \ln|\sin ax| + c$$
$$\int \sec ax\, dx = \dfrac{1}{a} \ln|\sec ax + \tan ax| + c$$
$$\int \operatorname{cosec} ax\, dx = -\dfrac{1}{a} \ln|\operatorname{cosec} ax + \cot ax| + c$$

> In the next exercise you will be asked to prove some of these results for particular values of a.

Worked example 8.13

Find $\int (1 + \sec 3x)^2 \, dx$.

Solution

$$\int (1 + \sec 3x)^2 \, dx = \int 1 + 2 \sec 3x + \sec^2 3x \, dx$$

$$= \int 1 \, dx + 2 \int \sec 3x \, dx + \int \sec^2 3x \, dx$$

$$= x + \frac{2}{3} \ln |\sec 3x + \tan 3x| + \frac{1}{3} \tan 3x + c$$

Worked example 8.14

Find $\int \dfrac{x + e^{2x}}{x^2 + e^{2x}} \, dx$.

Solution

Firstly differentiate $(x^2 + e^{2x})$ with respect to x to get $2x + 2e^{2x}$ which can be factorised to $2(x + e^{2x})$.

So $\int \dfrac{x + e^{2x}}{x^2 + e^{2x}} \, dx = \dfrac{1}{2} \int \dfrac{2(x + e^{2x})}{x^2 + e^{2x}} \, dx$

$$= \frac{1}{2} \ln (x^2 + e^{2x}) + c.$$

> Since, for real x, $x^2 + e^{2x} > 0$, you may omit the modulus sign.

EXERCISE 8D

1 Find:

(a) $\displaystyle\int \frac{3}{3x + 5} \, dx$

(b) $\displaystyle\int \frac{6x}{3x^2 + 5} \, dx$

(c) $\displaystyle\int \frac{3x^2}{x^3 + 8} \, dx$

(d) $\displaystyle\int \frac{x}{2x^2 + 3} \, dx$

(e) $\displaystyle\int \frac{1}{x} - \frac{2}{x + 3} \, dx$

(f) $\displaystyle\int \frac{6x}{2x^2 + 5} \, dx$

(g) $\displaystyle\int \frac{x}{(2 - x)(2 + x)} \, dx$

(h) $\displaystyle\int \frac{e^{3x}}{4 - e^{3x}} \, dx$

(i) $\displaystyle\int \frac{\cos 2x}{1 + \sin 2x} \, dx$

(j) $\displaystyle\int (\tan x + 1)^2 \, dx$

(k) $\displaystyle\int \frac{3}{x \ln x} \, dx$ $\left[\textbf{Hint for (k):} \dfrac{1}{x \ln x} = \dfrac{\frac{1}{x}}{\ln x} \right]$

2 Show that $\int \cot x \, dx = \ln |\sin x| + c$.

3 By writing $\sec 4x$ as $\dfrac{\sec 4x (\sec 4x + \tan 4x)}{\sec 4x + \tan 4x}$, find $\int \sec 4x \, dx$.

4 By writing cosec $2x$ as $\dfrac{\text{cosec } 2x(\text{cosec } 2x + \cot 2x)}{\text{cosec } 2x + \cot 2x}$,

find $\displaystyle\int \text{cosec } 2x\, dx$.

5 Evaluate:

(a) $\displaystyle\int_0^2 \left(\frac{1}{x+3}\right) dx$

(b) $\displaystyle\int_1^2 \left(\frac{1}{x} + \frac{x}{x^2+1}\right) dx$

(c) $\displaystyle\int_0^{\frac{\pi}{3}} \tan x\, dx$

(d) $\displaystyle\int_1^2 \left(\frac{1}{4x} - \frac{x^2}{x^3+1}\right) dx$

(e) $\displaystyle\int_0^{\frac{\pi}{2}} \left(\frac{\cos x}{3 - \sin x}\right) dx$

(f) $\displaystyle\int_2^3 \frac{4x}{(x-1)(x+1)} dx$

(g) $\displaystyle\int_0^1 \left(\frac{x+1}{x^2+2x+3}\right) dx$

(h) $\displaystyle\int_0^1 \left(\frac{2e^x}{e^x + e^{-x}}\right) dx$

[**Hint for (h)**: Rewrite the integrand with denominator $e^{2x} + 1$.]

6 Show that the area bounded by the curve $y = \dfrac{2x(2x^2+1)}{2x^4 + 2x^2 + 5}$,

the x-axis and the line $x = 2$ is $\ln 3$.

So far in this chapter you have been expected to recognise some standard integrals. If you found this difficult, the work which you will now learn in Section 8.5 offers an alternative approach for finding some of the integrals discussed earlier.

8.5 Integration by substitution

The method of integration by substitution involves a change of variable chosen so as to simplify the integral to be found. Everything in the integral must be changed to the new variable, including the 'dx'.

Consider $y = \displaystyle\int f(x)\, dx \;\Rightarrow\; \dfrac{dy}{dx} = f(x)$.

Introducing the variable u, and integrating both sides of the

chain rule $\dfrac{dy}{du} = \dfrac{dy}{dx} \times \dfrac{dx}{du}$ with respect to u, gives

$y = \displaystyle\int \frac{dy}{dx} \times \frac{dx}{du}\, du \;\Rightarrow\; y = \int f(x) \times \frac{dx}{du}\, du.$

Comparing this to $\longrightarrow y = \displaystyle\int f(x)\, dx$, you can see that

$$\ldots dx = \ldots \frac{dx}{du}\, du. \text{ Similarly } \ldots du = \frac{du}{dx}\, dx.$$

These results are used in integration by substitution.

The first worked example in this section is the same as Worked example 8.2 but is used here to illustrate the method of integration by substitution.

Worked example 8.15

Find $\int (2x + 7)^6 \, dx$.

Solution

An obvious change of variable is to

let $u = 2x + 7 \implies \dfrac{du}{dx} = 2 \implies \dfrac{dx}{du} = \dfrac{1}{2}$.

$$\int (2x + 7)^6 \, dx = \int u^6 \left(\frac{1}{2} \, du \right) = \frac{u^7}{14} + c$$

$$\implies \int (2x + 7)^6 \, dx = \frac{(2x + 7)^7}{14} + c$$

> Used $dx = \dfrac{dx}{du} \, du$.

> The final answer must be in terms of the original variable, x.

In examination questions, the substitution to be used is sometimes given.

Worked example 8.16

Use the substitution $s = \sin x$ to find $\int \sin^4 x \cos^3 x \, dx$.

> To integrate odd powers of $\cos x$, the substitution is usually $s = \sin x$.

Solution

Let $s = \sin x \implies \dfrac{ds}{dx} = \cos x \implies ds = \cos x \, dx$.

> Always check the integrand before isolating dx.

$$\int \sin^4 x \cos^3 x \, dx = \int \sin^4 x \cos^2 x \, (\cos x \, dx)$$

$$= \int \sin^4 x \, (1 - \sin^2 x) \, (\cos x \, dx)$$

> Used $\cos^2 x = 1 - \sin^2 x$.

$$= \int s^4 (1 - s^2) \, ds$$

$$= \int s^4 - s^6 \, ds = \frac{s^5}{5} - \frac{s^7}{7} + c$$

$$\int \sin^4 x \cos^3 x \, dx - \frac{1}{5} \sin^5 x - \frac{1}{7} \sin^7 x + c$$

> The final answer must be in terms of the original variable.

The previous two worked examples have been for indefinite integrals. Although you could solve definite integrals in the same way, by reaching the final line before substituting the values for x, there is a slightly shorter approach as illustrated in the next worked examples.

Worked example 8.17

Find $\displaystyle\int_0^4 \frac{2x}{\sqrt{9 + x^2}} \, dx$.

> You will recall that the limits 4 and 0 are the values of x, since dx.

8

Solution

Noting that the derivative of $9 + x^2$ is the numerator, you use the substitution $u = 9 + x^2$.

Differentiating with respect to x gives $\dfrac{du}{dx} = 2x$

$$\Rightarrow du = 2x\,dx.$$

You now only need to change the limits from values of x to values of u.

Since $u = 9 + x^2$, when $x = 4$, $u = 9 + 4^2 = 25$ and when $x = 0$, $u = 9$.

You are now ready to change the integral from the variable x to the variable u.

$$\int_0^4 \frac{2x}{\sqrt{9+x^2}}\,dx = \int_0^4 \frac{1}{\sqrt{9+x^2}}\,(2x\,dx) = \int_9^{25} \frac{1}{\sqrt{u}}\,du = \int_9^{25} u^{-\frac{1}{2}}\,du$$

> The position of the limits for u must match the position of the corresponding limits for x.

$$= \left[2u^{\frac{1}{2}}\right]_9^{25}$$

$$= 10 - 6 = 4$$

$$\int_0^4 \frac{2x}{\sqrt{9+x^2}}\,dx = 4$$

Worked example 8.18

$O(0, 0)$ and $P(1, 2)$ are two points on the curve with equation $y = \dfrac{6x^5}{4 - x^3}$. By using the substitution $u = 4 - x^3$, or otherwise, find the area of the region bounded by the arc OP of the curve, the x-axis and the line $x = 1$.

Solution

The required area is given by $\displaystyle\int_0^1 \frac{6x^5}{4 - x^3}\,dx$.

Let $u = 4 - x^3 \Rightarrow \dfrac{du}{dx} = -3x^2 \Rightarrow du = -3x^2\,dx$.

When $x = 1$, $u = 3$ and when $x = 0$, $u = 4$.

$$\int_0^1 \frac{6x^5}{4-x^3}\,dx = \int_0^1 \frac{-2x^3(-3x^2\,dx)}{4-x^3} = \int_4^3 \frac{2(u-4)(du)}{u}$$

> The position of the limits for u must match the position of the corresponding limits for x.

$$= 2\int_4^3 \frac{u}{u} - \frac{4}{u}\,du = 2\int_4^3 1 - \frac{4}{u}\,du$$

$$= 2\left[u - 4\ln u\right]_4^3$$

$$= 2[(3 - 4\ln 3) - (4 - 4\ln 4)]$$

$$\int_0^1 \frac{6x^5}{4-x^3}\,dx = 2\left[4\ln\frac{4}{3} - 1\right]$$

EXERCISE 8E

1 Use the given substitution, or otherwise, to find these integrals:

(a) $\int (4x + 1)^8 \, dx, \quad u = 4x + 1$

(b) $\int \cos\left(5x + \dfrac{\pi}{3}\right) dx, \quad u = 5x + \dfrac{\pi}{3}$

(c) $\int 4x(2x + 1)^5 \, dx, \quad u = 2x + 1$

(d) $\int \dfrac{5x}{5x^2 + 7} \, dx, \quad u = 5x^2 + 7$

(e) $\int \dfrac{2x}{(4x + 5)^3} \, dx, \quad u = 4x + 5$

(f) $\int \dfrac{4x}{(x^2 + 3)^3} \, dx, \quad u = x^2 + 3$

(g) $\int \dfrac{x^2}{x + 3} \, dx, \quad u = x + 3$

(h) $\int \cos^3 x \, dx, \quad s = \sin x$

(i) $\int \cos^2 x \sin^3 x \, dx, \quad c = \cos x$

(j) $\int \sin^2 2x \tan 2x \, dx, \quad u = \cos 2x$

(k) $\int \dfrac{4x}{\sqrt{x^2 - 1}} \, dx, \quad u = x^2 - 1$

(l) $\int 4x \, e^{x^2} \, dx, \quad u = x^2$

(m) $\int \dfrac{2}{e^{2x} + 4} \, dx, \quad u = e^{2x} + 4$

(n) $\int \dfrac{1}{\sqrt{x}} \sin \sqrt{x} \, dx, \quad x = u^2$

2 Using the given substitution, or otherwise, evaluate these integrals:

(a) $\int_1^2 x(x - 1)^4 \, dx, \quad u = x - 1$

(b) $\int_0^1 (2x - 1)^6 \, dx, \quad u = 2x - 1$

(c) $\int_0^1 \dfrac{3x^2 + 2}{2x^3 + 4x + 3} \, dx, \quad u = 2x^3 + 4x + 3$

8

(d) $\displaystyle\int_0^3 \frac{2x}{(4x+1)^2}\, dx, \quad u = 4x + 1$

(e) $\displaystyle\int_{-1}^0 \frac{x}{(x^2+2)^2}\, dx, \quad u = x^2 + 2$

(f) $\displaystyle\int_4^5 \frac{x^2}{x-3}\, dx, \quad u = x - 3$

(g) $\displaystyle\int_0^{\frac{\pi}{2}} \cos^5 x\, dx, \quad s = \sin x$

(h) $\displaystyle\int_0^{\frac{\pi}{3}} \cos^5 x \tan x\, dx, \quad c = \cos x$

(i) $\displaystyle\int_0^{\frac{\pi}{4}} \sin^3 2x \cos^2 2x\, dx, \quad u = \cos 2x$

(j) $\displaystyle\int_0^2 3x\sqrt{4 - x^2}\, dx, \quad u = 4 - x^2$

(k) $\displaystyle\int_0^1 6x^2\, e^{x^3}\, dx, \quad u = x^3$

(l) $\displaystyle\int_{-\ln 2}^0 \frac{2}{e^x + 1}\, dx, \quad u = e^x + 1$

3 Find:

(a) $\displaystyle\int 2x(x^2+1)^8\, dx$ **(b)** $\displaystyle\int \frac{4x}{\sqrt{x^2+1}}\, dx$ **(c)** $\displaystyle\int \cos^3 2x\, dx$

4 Evaluate:

(a) $\displaystyle\int_0^1 \frac{1}{(1+x)^2}\, dx$ **(b)** $\displaystyle\int_0^{\frac{\pi}{3}} \frac{\sin x}{\cos^4 x}\, dx$ **(c)** $\displaystyle\int_1^4 \frac{1}{2\sqrt{x}(1+\sqrt{x})^2}\, dx$

5 Evaluate: $\displaystyle\int_1^5 x\sqrt{2x-1}\, dx$.

MIXED EXERCISE

1 Find:

(a) $\displaystyle\int \frac{3}{4x} - \frac{1}{e^{5x}}\, dx$ **(b)** $\displaystyle\int_1^2 \frac{1}{(1-2x)^3}\, dx$

(c) $\displaystyle\int_{\frac{1}{4}}^2 e^{4x-1}\, dx$ **(d)** $\displaystyle\int \cos 2x + \sin 4x\, dx$

(e) $\displaystyle\int \frac{4}{4x+5}\, dx$ **(f)** $\displaystyle\int \frac{e^{2x}}{6 - e^{2x}}\, dx$

(g) $\displaystyle\int_0^{\frac{\pi}{2}} \frac{\cos x}{2 - \sin x}\, dx$

2 (a) Show that $\displaystyle\int_0^6 \frac{1}{2+u}\,\mathrm{d}u = \ln 4$.

(b) Use the substitution $x = u^2$ to show that

$$\int_0^{36} \frac{1}{\sqrt{x}(2+\sqrt{x})}\,\mathrm{d}x = \ln 16. \qquad \text{[A]}$$

3 Use the substitution $u = \cos t$ to evaluate

$$\int_{\frac{\pi}{3}}^{\frac{\pi}{2}} \cos^2 t \sin t\,\mathrm{d}t.$$

4 The equation of a curve C is given by $y = \dfrac{\cos x}{2 - \sin x}$, $0 \leqslant x \leqslant \pi$.
Find the exact area of the region bounded by the curve C, the
x- and y-axes and the line $x = \dfrac{\pi}{6}$. Express your answer in its
simplest form. [A]

5 Use the substitution $u = x + 1$ to express the integral

$$I = \int_0^1 \frac{x^2}{(x+1)^2}\,\mathrm{d}x \text{ in the form } \int_a^b \left(1 - \frac{c}{u} + \frac{1}{u^2}\right)\mathrm{d}u, \text{ stating the}$$

value of each of the constants a, b and c.
Hence find the exact value of I. [A]

6 By means of the substitution $u = 3 + e^x$, or otherwise, evaluate

$$\int_0^{\ln 6} e^{2x}(3 + e^x)^{\frac{1}{2}}\,\mathrm{d}x. \qquad \text{[A]}$$

Key point summary

1 For constants a, b and n, where $n \neq -1$,   *p137*

$$\int (ax+b)^n\,\mathrm{d}x = \frac{1}{a(n+1)}\,(ax+b)^{n+1} + c.$$

2 For constants a and b,   *p138*

$$\int e^{ax+b}\,\mathrm{d}x = \frac{1}{a}\,e^{ax+b} + c.$$

3 For x in radians,   *p139*

$$\int \cos x\,\mathrm{d}x = \sin x + c$$

$$\int \sin x\,\mathrm{d}x = -\cos x + c$$

$$\int \sec^2 x\,\mathrm{d}x = \tan x + c.$$

4 For constants a and b and x in radians,   *p140*

$$\int \cos(ax+b)\,\mathrm{d}x = \frac{1}{a}\sin(ax+b) + c$$

$$\int \sin(ax+b)\,\mathrm{d}x = -\frac{1}{a}\cos(ax+b) + c$$

$$\int \sec^2(ax+b)\,\mathrm{d}x = \frac{1}{a}\tan(ax+b) + c.$$

8

5 For constants a and x in radians, $p141$

$$\int \operatorname{cosec} ax \cot ax \, dx = -\frac{1}{a} \operatorname{cosec} ax + c$$

$$\int \sec ax \tan ax \, dx = \frac{1}{a} \sec ax + c$$

$$\int \operatorname{cosec}^2 ax \, dx = -\frac{1}{a} \cot ax + c.$$

6 $\displaystyle\int \frac{f'(x)}{f(x)} \, dx = \ln|f(x)| + c$ $p143$

7 For a constant a, and x in radians, $p144$

$$\int \tan ax \, dx = \frac{1}{a} \ln|\sec ax| + c$$

$$\int \cot ax \, dx = \frac{1}{a} \ln|\sin ax| + c$$

$$\int \sec ax \, dx = \frac{1}{a} \ln|\sec ax + \tan ax| + c$$

$$\int \operatorname{cosec} ax \, dx = -\frac{1}{a} \ln|\operatorname{cosec} ax + \cot ax| + c.$$

8 These results are used in integration by substitution: $p146$

$$\dots dx = \dots \frac{dx}{du} \, du \quad \text{and} \quad \dots du = \frac{du}{dx} \, dx.$$

Test yourself   What to review

1 Find $\displaystyle\int_1^4 \frac{1 + \sqrt{x}}{x} \, dx$. *Section 8.1*

2 Find: *Section 8.2*

(a) $\displaystyle\int \left(\frac{x}{4} + 3\right)^7 dx$ (b) $\displaystyle\int e^{4x-3} \, dx$ (c) $\displaystyle\int \frac{6}{\sqrt{9x+1}} \, dx$

3 Find $\displaystyle\int (\sec 2x + \tan 2x)^2 \, dx$. *Section 8.3*

4 Find $\displaystyle\int \frac{2e^x}{e^x + 1} + \cot 5x + \frac{x(x-1)}{2x^3 - 3x^2 + 7} \, dx$. *Section 8.4*

5 Use the substitution $u = 4 + \sin 2x$ to find: *Section 8.5*

(a) $\displaystyle\int \frac{\cos 2x}{(4 + \sin 2x)} \, dx$

(b) $\displaystyle\int_0^{\frac{3\pi}{4}} \cos 2x \, (4 + \sin 2x)^4 \, dx$

(c) $\displaystyle\int 4 \cos 2x \, e^{4 + \sin 2x} \, dx$

Test yourself ANSWERS

1 $2 + \ln 4.$

2 **(a)** $\frac{1}{2}\left(\frac{x}{4} + 3\right)^8 + c;$ **(b)** $\frac{1}{4}e^{4x - 3} + c;$ **(c)** $\frac{4}{3}\sqrt{9x + 1} + c;$

3 $\tan 2x - x + \sec 2x + c.$

4 $2 \ln (e^x + 1) + \frac{1}{5} \ln |\sin 5x| + \frac{1}{6}\ln|2x^3 - 3x^2 + 7| + c.$

5 **(a)** $\frac{1}{2} \ln(4 + \sin 2x) + c;$

(b) $-78.1;$

(c) $2e^4 + \sin 2x + c.$

CHAPTER 9

C3: Integration by parts and standard integrals

Learning objectives

After studying this chapter, you should be able to:
- integrate expressions using integration by parts
- use relevant standard integrals quoted in the formulae booklet.

9.1 Introduction

In chapter 8 you learned how to use the method of 'integration by substitution'. In this chapter you will learn another method for integrating some functions; the method is known as 'integration by parts'. Later in the chapter you will use the standard integrals

$$\int \frac{1}{a^2 + x^2}\, dx \quad \text{and} \quad \int \frac{1}{\sqrt{a^2 - x^2}}\, dx,\ \text{where } a \text{ is a constant, and which}$$

are both given in the formulae booklet.

9.2 Integration by parts

In section 6.2 you were shown how to differentiate product of functions of x by using the product rule

$$\frac{d}{dx}(uv) = u\frac{dv}{dx} + v\frac{du}{dx}.$$

> For example,
> $$\frac{d}{dx}(x^2 \ln x) = x^2\left(\frac{1}{x}\right) + \ln x\,(2x).$$

In this section you will be shown how to adapt this result to enable you to solve some different integrals.
Rewrite the product rule in the form

$$u\frac{dv}{dx} = \frac{d}{dx}(uv) - v\frac{du}{dx}.$$

Integrating both sides with respect to x gives

$$\int u\frac{dv}{dx}\, dx = \int \frac{d}{dx}(uv)\, dx - \int v\frac{du}{dx}\, dx$$

$$\Rightarrow \quad \int u\frac{dv}{dx}\, dx = uv - \int v\frac{du}{dx}\, dx.$$

$$\int u \frac{dv}{dx} \, dx = uv - \int v \frac{du}{dx} \, dx$$

is known as the integration by parts formula and is given, in this form, in the examination formulae booklet.

The process of integration by parts is used to change a relatively complicated integral (normally involving the product of two functions of x) into a more manageable integral.

In the examination, the method of integration by parts is most commonly, although not exclusively, used to solve integrals of the form

$$\int x^n \sin mx \, dx$$

$$\int x^n \cos mx \, dx$$

$$\int x^n e^{mx} \, dx$$

$$\int x^n \ln mx \, dx.$$

Worked example 9.1

Find $\int x \sin 2x \, dx$.

Solution

To find
$$\int x \sin 2x \, dx$$

use integration by parts, $\int u \dfrac{dv}{dx} \, dx = uv - \int v \dfrac{du}{dx} \, dx$

> The aim is to simplify the integral to one in which the integrand involves one function rather than the product of two, so you choose $u = x$ in this example.

with $u = x$ and $\dfrac{dv}{dx} = \sin 2x$

$\Rightarrow \dfrac{du}{dx} = 1$ and $v = -\dfrac{1}{2} \cos 2x$.

$$\int x \sin 2x \, dx = \left[x \left(-\frac{1}{2} \cos 2x \right) \right] - \int \left(-\frac{1}{2} \cos 2x \right)(1) \, dx$$

> $\int \sin ax \, dx = -\dfrac{1}{a} \cos ax + c$

$$\int x \sin 2x \, dx = -\frac{1}{2} x \cos 2x + \frac{1}{2} \int (\cos 2x) \, dx$$

$$= -\frac{1}{2} x \cos 2x + \frac{1}{4} \sin 2x + c$$

> $\int \cos ax \, dx = \dfrac{1}{a} \sin ax + c$

9

Worked example 9.2

Find $\int_1^2 \ln x \, dx$.

Solution

To find $\int_1^2 \ln x \, dx$ rewrite it as $\int_1^2 (1) \ln x \, dx$.

Use integration by parts, $\quad \int u \dfrac{dv}{dx} \, dx = uv - \int v \dfrac{du}{dx} \, dx$

with $u = \ln x$ and $\dfrac{dv}{dx} = 1$

> Since you cannot integrate ln x directly, for integrals of the form $\int x^n \ln mx \, dx$ you use $u = \ln mx$ and $\dfrac{dv}{dx} = x^n$.

$\Rightarrow \dfrac{du}{dx} = \dfrac{1}{x}$ and $v = x$.

$\int_1^2 \ln x \, dx = \left[(\ln x)(x) \right]_1^2 - \int_1^2 (x)\left(\dfrac{1}{x}\right) dx$

$\qquad = \left[x(\ln x) \right]_1^2 - \int_1^2 1 \, dx$

$\qquad = \left[x(\ln x) - x \right]_1^2$

$\qquad = (2\ln 2 - 2) - (\ln 1 - 1)$

$\int_1^2 \ln x \, dx = 2\ln 2 - 1$

> Simplify $x\left(\dfrac{1}{x}\right)$ to 1 before integrating.

Integrals of the form $\int (ax + b)e^{kx} \, dx$, $\int (ax + b) \sin kx \, dx$ can be solved by multiplying out the brackets and dealing with each term separately or directly, as illustrated by the next worked example.

Worked example 9.3

Find $\int_0^{\frac{\pi}{5}} (5x + 3) \cos 5x \, dx$.

Solution

To find $\qquad \int_0^{\frac{\pi}{5}} (5x + 3) \cos 5x \, dx$

use integration by parts, $\int u \dfrac{dv}{dx} \, dx = uv - \int v \dfrac{du}{dx} \, dx$

with $u = (5x + 3)$ and $\dfrac{dv}{dx} = \cos 5x$

$\Rightarrow \dfrac{du}{dx} = 5$ and $v = \dfrac{1}{5} \sin 5x$.

$\displaystyle\int_0^{\frac{\pi}{5}} (5x + 3) \cos 5x \, dx = \left[(5x + 3)(\tfrac{1}{5} \sin 5x) \right]_0^{\frac{\pi}{5}} - \int_0^{\frac{\pi}{5}} (\tfrac{1}{5} \sin 5x)(5) \, dx$

$\qquad\qquad = \dfrac{\pi + 3}{5} \sin \pi - \dfrac{3}{5} \sin 0 - \left[-\dfrac{1}{5} \cos 5x \right]_0^{\frac{\pi}{5}}$

$\qquad\qquad = 0 - 0 + \dfrac{1}{5} (\cos \pi - \cos 0)$

$\qquad\qquad = -\dfrac{2}{5}$

Sometimes it is necessary to apply integration by parts more than once to solve an integral. The next worked example illustrates the method.

> In the exam, you will not need to use integration by parts more than twice in solving an integral.

Worked example 9.4

Find $\displaystyle\int x^2 \, e^x \, dx$.

Solution

To find $\displaystyle\int x^2 \, e^x \, dx$

use integration by parts, $\displaystyle\int u \dfrac{dv}{dx} \, dx = uv - \int v \dfrac{du}{dx} \, dx$

with $u = x^2$ and $\dfrac{dv}{dx} = e^x$

> As before, to reduce the power of x, use $u = x^2$ and $\dfrac{dv}{dx} = e^x$.

$\Rightarrow \dfrac{du}{dx} = 2x$ and $v = e^x$.

$\displaystyle\int x^2 \, e^x \, dx = [x^2 \, e^x] - \int e^x (2x) \, dx$

> $\displaystyle\int e^x \, dx = e^x + c$

$\displaystyle\int x^2 \, e^x \, dx = [x^2 \, e^x] - 2\int xe^x \, dx \qquad [*]$

Now $\displaystyle\int xe^x \, dx = xe^x - \int e^x \, dx$

$\displaystyle\int xe^x \, dx = xe^x - e^x + k$

> $\displaystyle\int xe^x \, dx$ cannot be integrated directly. You have to use integration by parts again with $u = x$ and $\dfrac{dv}{dx} = e^x$.

Substitute in [*] gives

$\displaystyle\int x^2 \, e^x \, dx = [x^2 \, e^x] - 2[xe^x - e^x + k]$.

$\displaystyle\int x^2 \, e^x \, dx = x^2 \, e^x - 2xe^x + 2e^x + c$

> You can check the answer by differentiating the right-hand side to give $x^2 e^x$. Remember to use the product rule when differentiating the first two terms.

EXERCISE 9A

1 Find these integrals:

(a) $\displaystyle\int x\, e^{2x}\, dx$ **(b)** $\displaystyle\int x\, e^{\frac{x}{2}}\, dx$ **(c)** $\displaystyle\int \frac{x}{e^x}\, dx$

(d) $\displaystyle\int \theta \sin \theta\, d\theta$ **(e)** $\displaystyle\int x \cos 4x\, dx$ **(f)** $\displaystyle\int x \cos 3x\, dx$

(g) $\displaystyle\int x \ln x\, dx$ **(h)** $\displaystyle\int x^6 \ln x\, dx$ **(i)** $\displaystyle\int \frac{\ln 4x}{x^3}\, dx$

2 Evaluate these integrals:

(a) $\displaystyle\int_0^2 x\, e^x\, dx$ **(b)** $\displaystyle\int_0^1 x\, e^{4x}\, dx$ **(c)** $\displaystyle\int_0^4 \frac{x}{e^{2x}}\, dx$

(d) $\displaystyle\int_0^\pi x \sin \frac{1}{2} x\, dx$ **(e)** $\displaystyle\int_0^{\frac{\pi}{2}} \theta \cos \theta\, d\theta$ **(f)** $\displaystyle\int_0^{\frac{\pi}{2}} \theta^2 \cos \theta\, d\theta$

(g) $\displaystyle\int_1^2 x^3 \ln x\, dx$ **(h)** $\displaystyle\int_1^e \ln x^2\, dx$ **(i)** $\displaystyle\int_1^{\frac{e}{2}} \frac{\ln 2x}{x^4}\, dx$

> Hint **(h)**: $\ln x^2 = 2 \ln x$.

3 Find these integrals:

(a) $\displaystyle\int (x + 3)e^x\, dx$ **(b)** $\displaystyle\int (3x - 2)e^{3x}\, dx$

(c) $\displaystyle\int x(1 + \cos 2x)\, dx$ **(d)** $\displaystyle\int (3 + 2x) \ln x\, dx$

4 Evaluate these integrals:

(a) $\displaystyle\int_0^1 (2x + 1)e^{2x}\, dx$ **(b)** $\displaystyle\int_0^{\frac{\pi}{2}} (4\theta - 3) \sin \theta\, d\theta$

(c) $\displaystyle\int_{\frac{1}{2}}^1 (2x + 1) \ln 2x\, dx$

5 Show that $\displaystyle\int_{-\frac{\pi}{2}}^{\frac{\pi}{2}} x \sin 2x\, dx = \frac{\pi}{2}.$

6 Show that, for $n \geqslant 0$, $\displaystyle\int_1^e x^n \ln x\, dx = \frac{n\, e^{n+1} + 1}{(n+1)^2}.$

7 Find $\displaystyle\int \theta \sec^2 \theta\, d\theta$

9.3 Standard integrals

The following two integrals are in the formulae booklet and may be quoted in the examination.

$$\int \frac{1}{a^2 + x^2}\, dx = \frac{1}{a} \tan^{-1}\left(\frac{x}{a}\right) + c$$

$$\int \frac{1}{\sqrt{a^2 - x^2}}\, dx = \sin^{-1}\left(\frac{x}{a}\right) + c$$

> The second result can be shown by using the substitution $x = a \sin \theta$.

The next worked example shows you how the first of these two results is derived.

Worked example 9.5

Use the substitution $x = a \tan \theta$ to show that

$$\int \frac{1}{a^2 + x^2} \, dx = \frac{1}{a} \tan^{-1} \left(\frac{x}{a} \right) + c.$$

Solution

$x = a \tan \theta \Rightarrow dx = a \sec^2 \theta \, d\theta$

$$\int \frac{1}{a^2 + x^2} \, dx = \int \frac{a \sec^2 \theta}{a^2 (1 + \tan^2 \theta)} \, d\theta = \frac{1}{a} \int 1 \, d\theta = \frac{1}{a} \theta + c$$

$$= \frac{1}{a} \tan^{-1} \left(\frac{x}{a} \right) + c$$

Worked example 9.6

Evaluate $\int_0^1 \frac{2x + 1}{9 + x^2} \, dx$, giving your answer to three significant figures.

Solution

$$\int_0^1 \frac{2x + 1}{9 + x^2} \, dx = \int_0^1 \frac{2x}{9 + x^2} + \frac{1}{9 + x^2} \, dx$$

> Differentiating the denominator does not give the **full** numerator so you have to split the fraction.

$$= \int_0^1 \frac{2x}{9 + x^2} \, dx + \int_0^1 \frac{1}{3^2 + x^2} \, dx$$

$$= \left[\ln (9 + x^2) \right]_0^1 + \left[\frac{1}{3} \tan^{-1} \left(\frac{x}{3} \right) \right]_0^1$$

$$= \ln 10 - \ln 9 + \frac{1}{3} \left(\tan^{-1} \frac{1}{3} - \tan^{-1} 0 \right)$$

$$= \ln \frac{10}{9} + \frac{1}{3} \tan^{-1} \left(\frac{1}{3} \right)$$

$$= 0.105 \, 36\ldots + 0.107 \, 25\ldots = 0.213 \text{ (to 3 s.f.)}$$

> Set your calculator to radian mode.

9

Worked example 9.7

Find $\int \frac{2x + 1}{\sqrt{4 - x^2}} \, dx$.

Solution

Splitting the fraction leads to

$$\int \frac{2x + 1}{\sqrt{4 - x^2}} \, dx = \int \frac{2x}{\sqrt{4 - x^2}} \, dx + \int \frac{1}{\sqrt{4 - x^2}} \, dx$$

> You may prefer to use the substitution $u = 4 - x^2$ to find the first integral.

$$= \int -(-2x)(4 - x^2)^{-\frac{1}{2}} \, dx + \int \frac{1}{\sqrt{2^2 - x^2}} \, dx$$

$$= -2(4 - x^2)^{\frac{1}{2}} + \sin^{-1} \left(\frac{x}{2} \right) + c$$

EXERCISE 9B

1 Find:

(a) $\displaystyle\int \frac{1}{\sqrt{25 - x^2}}\, dx$

(b) $\displaystyle\int \frac{1}{25 + x^2}\, dx$

(c) $\displaystyle\int \frac{1}{\sqrt{0.25 - x^2}}\, dx$

(d) $\displaystyle\int \frac{1}{0.01 + x^2}\, dx$

2 By writing $\dfrac{1}{1 + 4x^2}$ in the form $\dfrac{1}{4(\frac{1}{4} + x^2)}$, find $\displaystyle\int \frac{1}{1 + 4x^2}\, dx$.

3 By writing $\dfrac{1}{\sqrt{4 - 9x^2}}$ in the form $\dfrac{1}{3\left(\sqrt{\frac{4}{9} - x^2}\right)}$, find $\displaystyle\int \frac{1}{\sqrt{4 - 9x^2}}\, dx$.

4 Find:

(a) $\displaystyle\int \frac{1}{\sqrt{1 - 4x^2}}\, dx$

(b) $\displaystyle\int \frac{1}{9 + 4x^2}\, dx$

(c) $\displaystyle\int \frac{1}{\sqrt{0.25 - 16x^2}}\, dx$

(d) $\displaystyle\int \frac{1}{16 + 9x^2}\, dx$

5 Evaluate the following integrals, giving your answers to three significant figures:

(a) $\displaystyle\int_{0.2}^{0.5} \frac{1}{\sqrt{1 - x^2}}\, dx$

(b) $\displaystyle\int_{1}^{3} \frac{1}{1 + x^2}\, dx$

(c) $\displaystyle\int_{1}^{4} \frac{1}{\sqrt{25 - x^2}}\, dx$

(d) $\displaystyle\int_{1}^{2} \frac{1}{4 + x^2}\, dx$

(e) $\displaystyle\int_{1}^{3} \frac{1}{25 + 4x^2}\, dx$

(f) $\displaystyle\int_{0.2}^{0.4} \frac{1}{\sqrt{1 - 4x^2}}\, dx$

6 Find:

(a) $\displaystyle\int \frac{2x + 1}{4 + x^2}\, dx$

(b) $\displaystyle\int \frac{x + 2}{9 + x^2}\, dx$

(c) $\displaystyle\int \frac{1 - 2x}{\sqrt{25 - x^2}}\, dx$

(d) $\displaystyle\int \frac{3 + x}{\sqrt{25 - x^2}}\, dx$

7 Evaluate the following integrals, giving your answers to three significant figures:

(a) $\displaystyle\int_{0}^{2} \frac{2x + 3}{4 + x^2}\, dx$

(b) $\displaystyle\int_{0.1}^{0.4} \frac{x - 4}{\sqrt{1 - 4x^2}}\, dx$

MIXED EXERCISE

1 Use integration by parts to find $\int_0^{\frac{1}{2}} x\,e^{2x}\,dx$. [A]

2 Use integration by parts to find the value of $\int_0^1 x\cos x\,dx$, giving your answer to five decimal places. [A]

3 Show that $\int_1^e \dfrac{\ln x}{x^3}\,dx = \dfrac{e^2-3}{4e^2}$.

4 (a) By using the chain rule, or otherwise, find $\dfrac{dy}{dx}$ when $y = \ln(x^2 + 9)$.

(b) Hence show that $\int_0^3 \dfrac{x}{x^2+9}\,dx = \dfrac{1}{2}\ln 2$.

(c) Given that $\tan^{-1}1 = \dfrac{\pi}{4}$, show that

$\int_0^3 \dfrac{x+1}{x^2+9}\,dx = \dfrac{1}{2}\ln 2 + \dfrac{\pi}{12}$. [A adapted]

5 Find $\int t\,e^{2t}\,dt$.

Hence, by using the substitution $x = t^2$, evaluate $\int_0^4 e^{2\sqrt{x}}\,dx$, giving your answer in terms of e. [A]

6 Show that $\int_0^{\frac{\pi}{3}} x^2\cos 3x\,dx = -\dfrac{2\pi}{27}$.

Key point summary

1 $\int u\dfrac{dv}{dx}\,dx = uv - \int v\dfrac{du}{dx}\,dx$ is known as the integration   *p155*
by parts formula and is given, in this form, in the examination formulae booklet.

2 In the examination, the method of integration by   *p155*
parts is most commonly, although not exclusively, used to solve integrals of the form

$\int x^n \sin mx\,dx$

$\int x^n \cos mx\,dx$

$\int x^n e^{mx}\,dx$

$\int x^n \ln mx\,dx$.

3 $\int \dfrac{1}{a^2+x^2}\,dx = \dfrac{1}{a}\tan^{-1}\left(\dfrac{x}{a}\right) + c$ *p158*

$\int \dfrac{1}{\sqrt{a^2-x^2}}\,dx = \sin^{-1}\left(\dfrac{x}{a}\right) + c$

9

Test yourself	What to review

1 (a) Find $\int (5x + 1)e^{5x}\, dx$.

Section 9.2

(b) Evaluate $\int_0^1 \theta \sin \theta\, d\theta$, giving your answer to three significant figures.

2 (a) Evaluate $\int_0^{\ln 2} xe^x\, dx$.

Section 9.2

(b) Find $\int \dfrac{\ln x}{2\sqrt{x}}\, dx$.

3 Find $\int \dfrac{4x + 6}{x^2 + 36}\, dx$.

Section 9.3

4 Evaluate $\int_0^1 \dfrac{x - 4}{\sqrt{9 - 4x^2}}\, dx$ giving your answer to three significant figures.

Section 9.3

Test yourself ANSWERS

4 -1.27 (to 3 s.f.).

3 $2 \ln (x^2 + 36) + \tan^{-1}\left(\dfrac{x}{6}\right) + c$.

2 (a) $2 \ln 2 - 1$;

(b) $\sqrt{x} \ln x - 2\sqrt{x} + c$.

1 (a) $x\, e^{5x} + c$;

(b) 0.301 (to 3 s.f.).

CHAPTER 10

Volume of revolution and numerical integration

Learning objectives

After studying this chapter, you should be able to:
- evaluate volumes of revolution
- use the mid-ordinate rule to find a numerical approximation for a definite integral
- use Simpson's rule to find a numerical approximation for a definite integral.

10.1 Volumes of revolution about horizontal and vertical axes

In C1, Section 12.6, you were shown that the area of the region, R, bounded by the curve $y = \mathrm{f}(x)$, the x-axis and the lines $x = a$ and $x = b$ is given by $\int_{a}^{b} \mathrm{f}(x)\, \mathrm{d}x$. In this section you will be shown how to find the volume of the solid which is formed by rotating the same region through 2π radians ($360°$) about the x-axis.

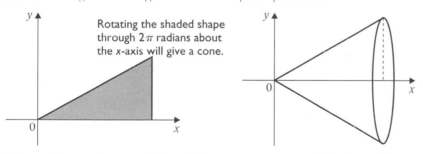

Rotating the shaded shape through 2π radians about the x-axis will give a cone.

Using this work you will be able to prove some of the formulae that you used in GCSE but were not able to prove then.

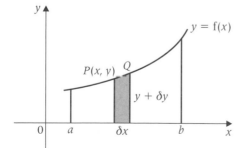

Rotating the region R through 2π radians about the x-axis gives a solid.

10

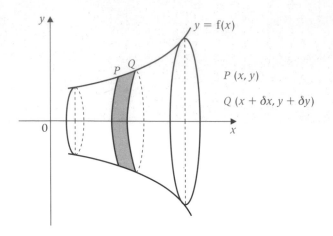

Consider the volume, δV, formed by rotating the small element, shown shaded above, and compare its volume with the volume of the two cylindrical discs.

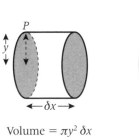

Volume $= \pi y^2 \, \delta x$

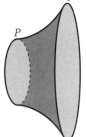

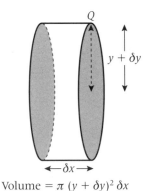

Volume $= \delta V$ Volume $= \pi (y + \delta y)^2 \, \delta x$

$$\pi y^2 \delta x < \delta V < \pi(y + \delta y)^2 \, \delta x$$

or dividing throughout by δx

$$\pi y^2 < \frac{\delta V}{\delta x} < \pi(y + \delta y)^2$$

As $\delta x \to 0$, $\delta y \to 0$ and $\dfrac{\delta V}{\delta x} \to \dfrac{\mathrm{d}V}{\mathrm{d}x}$

Hence, $\dfrac{\mathrm{d}V}{\mathrm{d}x} = \pi y^2$.

So, integrating both sides with respect to x, between the limits $x = a$ and $x = b$ covers the volume for the complete solid.

> When the region bounded by the curve $y = \mathrm{f}(x)$, the x-axis and the lines $x = a$ and $x = b$ is rotated through 2π radians about the x-axis, the volume, V, of the solid generated is called the volume of revolution and is given by
>
> $$V = \int_a^b \pi y^2 \, \mathrm{d}x = \int_a^b \pi[\mathrm{f}(x)]^2 \, \mathrm{d}x.$$

Compare with the work in C1 section 12.6.

The words 'through 2π radians' could be replaced by 'through 360°' or by 'completely'.

In a similar way you can show that:

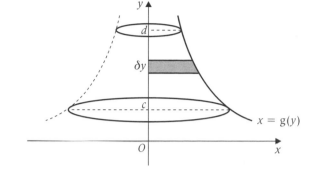

When the region bounded by the curve $x = g(y)$, the y-axis and the lines $y = c$ and $y = d$ is rotated through 2π radians about the y-axis, the volume, V, of the solid generated is given by

$$V = \int_c^d \pi x^2 \, \mathrm{d}y = \int_c^d \pi [g(y)]^2 \, \mathrm{d}y.$$

Worked example 10.1

By rotating the region bounded by the line $y = \dfrac{r}{h} x$, the x-axis and the line $x = h$ through 2π radians about the x-axis, show that the volume of a cone of height h and radius r is $\dfrac{1}{3}\pi r^2 h$.

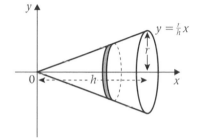

Solution

Volume of the cone = Volume of revolution of the line $y = \dfrac{r}{h} x$,

$$= \int_0^h \pi y^2 \, \mathrm{d}x = \int_0^h \pi \left[\frac{r^2}{h^2} x^2 \right] \mathrm{d}x = \frac{\pi r^2}{h^2} \int_0^h x^2 \, \mathrm{d}x$$

> The cylindrical element has radius y and height δx.

$$= \frac{\pi r^2}{h^2} \left[\frac{x^3}{3} \right]_0^h = \frac{\pi r^2 h^3}{3h^2}$$

Volume of the cone $= \dfrac{1}{3}\pi r^2 h.$

Worked example 10.2

The region bounded by the curve $y = x\sqrt{x^2 + 5}$, the x-axis and the line $x = 2$ is rotated through $360°$ about the x-axis. Find the volume of the solid generated.

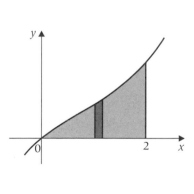

Solution

$$\text{Volume} = \int_0^2 \pi y^2 \, \mathrm{d}x = \pi \int_0^2 \left[x\sqrt{x^2 + 5} \right]^2 \mathrm{d}x = \pi \int_0^2 x^2(x^2 + 5) \, \mathrm{d}x$$

$$= \pi \int_0^2 (x^4 + 5x^2) \, \mathrm{d}x = \pi \left[\frac{x^5}{5} + \frac{5x^3}{3} \right]_0^2 = \pi \left[\left(\frac{32}{5} + \frac{40}{3} \right) - 0 \right]$$

> The cylindrical element has radius y and height δx.

Required volume $= \dfrac{296}{15}\pi.$

10

Worked example 10.3

The region bounded by the curve $y = x^3$, the y-axis and the line $y = 8$ is rotated through 2π radians about the y-axis. Find the volume of the solid generated.

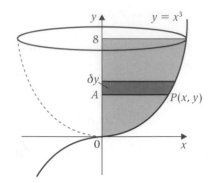

Solution

The small cylindrical element has radius $PA = |x - 0|$

$$y = x^3 \implies x = y^{\frac{1}{3}} \implies x^2 = y^{\frac{2}{3}}$$

Since the rotation is about the y-axis, the volume of the solid

$$= \int_0^8 \pi x^2 \, dy = \pi \int_0^8 y^{\frac{2}{3}} \, dy$$

$$= \pi \left[\frac{3}{5} y^{\frac{5}{3}} \right]_0^8$$

$$= \pi \left[\left(\frac{3}{5} \times 2^5 \right) - (0) \right]$$

Volume of solid $= \dfrac{96}{5} \pi.$

EXERCISE 10A

1 Find the volumes of the solids formed by rotating, through 2π radians about the x-axis, the following regions:

  (a) bounded by the curve $y = \sqrt{x}$, the x-axis and the lines $x = 1$ and $x = 2$,

  (b) bounded by the curve $y = 2x + 1$, the x-axis and the lines $x = 0$ and $x = 2$,

  (c) bounded by the curve $y = x^2 + 1$, the coordinate axes and the line $x = 1$,

  (d) bounded by the curve $y = \sec x$, the x-axis and the lines $x = 0$ and $x = \dfrac{\pi}{4}$,

  (e) bounded by the curve $y = \dfrac{2}{x}$, the x-axis and the lines $x = 2$ and $x = 4$,

  (f) bounded by the curve $y = \dfrac{3}{\sqrt{x}}$, the x-axis and the lines $x = 1$ and $x = 9$.

2 Find the volumes of the solids formed by rotating, through 2π radians about the y-axis, the following regions:

  (a) bounded by the curve $y = x^2$, the y-axis and the lines $y = 0$ and $y = 1$,

  (b) bounded by the curve $y = \sqrt{x}$, the y-axis and the lines $y = 1$ and $y = 2$,

  (c) bounded by the curve $y = x^2 + 1$, the y-axis and the line $y = 4$,

  (d) bounded by the curve $x = e^y + 1$, the y-axis and the lines $y = 0$ and $y = \ln 2$.

3 (a) Show that the curve $y = (2x - 1)^2$ only meets the x-axis at the point $(\frac{1}{2}, 0)$.

(b) The region bounded by the curve $y = (2x - 1)^2$ and the coordinate axes is rotated through $360°$ about the x-axis. Find the volume of the solid generated.

4 The region bounded by the curve $y = e^x$, the coordinate axes and the line $x - 1$ is rotated through 2π radians about the x-axis. Find the volume of the solid generated.

5 The region bounded by the curve $y = x^5$, the y-axis and the lines $y = 1$ and $y = 32$ is rotated through 2π radians about the y-axis. Find the volume of the solid generated.

6 (a) Write down the equation of the circle with centre $(0,0)$ and radius r.

(b) By rotating the region bounded by the coordinate axes and the part of the circle in the first quadrant about the x-axis, show that the volume of a hemisphere is $\frac{2}{3}\pi r^3$.

7 (a) Sketch the curve $y = |2x - 4|$, indicating the coordinates of any points where the curve meets the coordinate axes.

(b) The region enclosed by the curve $y = |2x - 4|$ and the coordinate axes is rotated completely about the y-axis to form a solid, S. Find the volume of S.

8 The points $P(3, 2)$ and $Q(0, 1)$ lie on the curve $y^2 = x + 1$.

(a) Calculate the volume of the solid generated when the region bounded by the lines $y = 0$, $x = 0$, $x = 3$ and the arc PQ of the curve is rotated completely about the x-axis. Give your answer as a multiple of π.

(b) S is the region bounded by the lines $y = 1$, $x = 3$ and the arc PQ of the curve. Show that when S is rotated completely about the x-axis the volume of the solid generated is $\dfrac{9\pi}{2}$.　　　　　　　　[A]

Hint for **(b)**: Use the answer to part **(a)** and note that the volume of a cylinder is $\pi r^2 h$.

10

9 The points $P(5, 3)$ and $Q(0, 2)$ lie on the curve $y^2 = x + 4$.

(a) Calculate the volume of the solid generated when the region bounded by the lines $y = 0$, $x = 0$, $x = 5$ and the arc PQ of the curve is rotated completely about the x-axis. Give your answer as a multiple of π.

(b) R is the region bounded by the lines $y = \dfrac{3}{5}x$, $x = 0$ and the arc PQ of the curve. Show that when R is rotated completely about the x-axis the volume of the solid generated is $\dfrac{35\pi}{2}$.

10.2 Notation for numerical integration

In all the Core modules there has been at least one chapter on integration in which you have been shown how to evaluate various definite integrals. Questions were chosen so that the integration could be done by using techniques developed in the relevant chapters. However, not all functions can be integrated by using analytical techniques, for example $\int_{-1}^{1} \sqrt{x^3 + 1}\, dx$, and so,

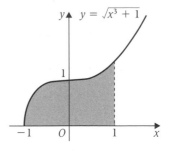

over the centuries, methods have been developed to find numerical approximations for such definite integrals. You will recall that a definite integral can represent an area so $\int_{-1}^{1} \sqrt{x^3 + 1}\, dx$, is the area of the shaded region bounded by the curve $y = \sqrt{x^3 + 1}$, the x-axis and the lines $x = -1$ and $x = 1$. In C2 you were shown the trapezium rule and learned how to apply it to find a numerical approximation for a definite integral. In the remainder of this chapter you will be introduced to two more rules for finding a numerical approximation for a definite integral.

The formulae used in these three rules appear in the formulae booklet to which you will have access in the examination. In order to apply the formulae you will need to understand the notation used.

Consider the area of the region bounded by the curve $y = f(x)$, the x-axis and the lines $x = a$ and $x = b$.

The area of the shaded region is given by $\int_{a}^{b} y\, dx$.

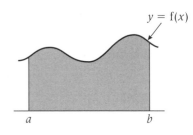

Divide the region up into n strips of equal width, h, and let the coordinates of the points of intersection with the curve be

$(x_0, y_0), (x_1, y_1), (x_2, y_2), \ldots, (x_n, y_n)$.

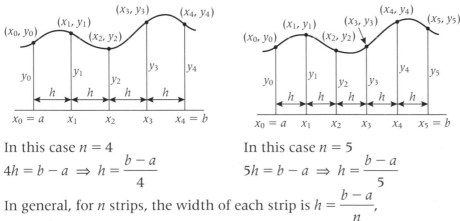

In this case $n = 4$

$$4h = b - a \implies h = \frac{b - a}{4}$$

In this case $n = 5$

$$5h = b - a \implies h = \frac{b - a}{5}$$

In general, for n strips, the width of each strip is $h = \dfrac{b - a}{n}$,

$x_0 = a$, $x_1 = a + h$, $x_2 = a + 2h$, ..., $x_k = a + kh$, ..., $x_n = b$, and the corresponding values of y, called the ordinates, are $y_0 = f(x_0)$, $y_1 = f(x_1)$, $y_2 = f(x_2)$, ..., $y_k = f(x_k)$, ..., $y_n = f(b)$.

In the left-hand diagram above we say that the region has been split into four equal strips with five ordinates.

Worked example 10.4

The region, whose area is given by $\int_0^2 (x^2 + 1)\,dx$, is divided into four strips with five ordinates. The width of each strip is h.
(a) Find the value of h.
(b) Find the value of $(y_0 + y_1 + y_2 + y_3 + y_4)$.

Solution

(a) The lower limit $a = 0$ and the upper limit $b = 2$.
The number of strips, $n = 4$.

$$h = \frac{b - a}{n} = \frac{2 - 0}{4} = 0.5$$

(b)

x	$y = x^2 + 1$
$x_0 = 0$	$y_0 = 0^2 + 1 = 1$
$x_1 = 0.5$	$y_1 = 0.5^2 + 1 = 1.25$
$x_2 = 1$	$y_2 = 1^2 + 1 = 2$
$x_3 = 1.5$	$y_3 = 1.5^2 + 1 = 3.25$
$x_4 = 2$	$y_4 = 2^2 + 1 = 5$

$$\Rightarrow y_0 + y_1 + y_2 + y_3 + y_4 = 12.5$$

EXERCISE 10B

1 The region, whose area is given by $\int_0^4 (x^2 + 1)^2\,dx$, is divided into four strips with five ordinates. The width of each strip is h.
(a) Find the value of h.
(b) Find the value of $(y_0 + y_1 + y_2 + y_3 + y_4)$.

2 The region, whose area is given by $\int_{-1}^1 e^{x^2}\,dx$, is divided into four strips with five ordinates. The width of each strip is h.
(a) Find the value of h.
(b) Find the value of $[y_0 + y_4 + 2(y_1 + y_2 + y_3)]$.

3 The region, whose area is given by $\int_0^{0.6} \sin x^2\,dx$, is divided into three strips with four ordinates. The width of each strip is h.
(a) Find the value of h.
(b) Find the value of $(y_0 + y_1 + y_2 + y_3)$.

$\sin x^2$ is $\sin(x^2)$ and **not** $\sin^2 x$.

Since calculus is involved x must be in radians so set your calculator in radian mode.

10

10.3 The mid-ordinate rule

Consider the region bounded by the curve $y = f(x)$, the x-axis and the lines $x = a$ and $x = b$ as a single strip with ordinates y_0 and y_1 which correspond to $x = x_0 = a$ and $x = x_1 = b$. The midpoint of the base joining x_0 and x_1 has x-coordinate $\dfrac{x_0 + x_1}{2}$ and is denoted by $x_{\frac{1}{2}}$. The ordinate corresponding to $x_{\frac{1}{2}}$ is denoted by $y_{\frac{1}{2}}$. We say that $y_{\frac{1}{2}}$ is the mid-ordinate of the strip with ordinates y_0 and y_1. The rectangle with height $y_{\frac{1}{2}}$ and width $h (= b - a)$, shown shaded opposite, has an area equal to $h\, y_{\frac{1}{2}}$ so, with just one strip, $\displaystyle\int_a^b y\, dx \approx h\, y_{\frac{1}{2}}$.

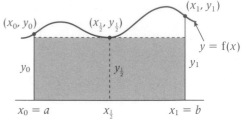

Clearly, for the shape of the curve shown in the diagram, $h\, y_{\frac{1}{2}}$ is too low an estimate for the area under the curve.

> In all the numerical methods of integration an improvement to the estimate can be obtained by increasing the number of steps (strips).

You can improve on this estimate by dividing the region into more strips of equal width h before finding the mid-ordinates. In the diagram opposite three strips have been used ($n = 3$) and so

$$\int_a^b y\, dx \approx h(y_{\frac{1}{2}} + y_{\frac{3}{2}} + y_{\frac{5}{2}}), \text{ where } h = \frac{b - a}{n}.$$

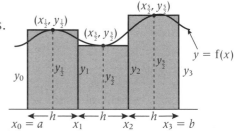

In general, if n strips are used, you have

> The mid-ordinate rule for n strips is
> $$\int_a^b y\, dx \approx h\left(y_{\frac{1}{2}} + y_{\frac{3}{2}} + \ldots + y_{n-\frac{3}{2}} + y_{n-\frac{1}{2}}\right), \text{ where } h = \frac{b - a}{n}.$$

This mid-ordinate rule is in the examination formulae booklet. The next two worked examples show you how to apply this rule to find numerical approximations for definite integrals.

In the first of the examples we use an integral that can be evaluated exactly, by using the method of integration by parts, so that a comparison of accuracies can be observed.

Worked example 10.5

By considering rectangular strips of width 0.2 use the mid-ordinate rule to obtain an approximation for $\displaystyle\int_1^{1.8} \ln x\, dx$, giving your answer to four decimal places.

Solution

The width of each rectangle is $h = 0.2$.

x-values $(x_{k+1} = x_k + h)$	Mid-x-values	Mid-ordinates $y = \ln x$
$x_0 = 1$		
	$x_{\frac{1}{2}} = 1.1$	$y_{\frac{1}{2}} = \ln 1.1 = 0.095\,31\ldots$
$x_1 = 1.2$		
	$x_{\frac{3}{2}} = 1.3$	$y_{\frac{3}{2}} = \ln 1.3 = 0.262\,36\ldots$
$x_2 = 1.4$		
	$x_{\frac{5}{2}} = 1.5$	$y_{\frac{5}{2}} = \ln 1.5 = 0.405\,46\ldots$
$x_3 = 1.6$		
	$x_{\frac{7}{2}} = 1.7$	$y_{\frac{7}{2}} = \ln 1.7 = 0.530\,62\ldots$
$x_4 = 1.8$		

Always work to a greater degree of accuracy than the accuracy required (4 d.p.) for the final answer.

$$y_{\frac{1}{2}} + y_{\frac{3}{2}} + y_{\frac{5}{2}} + y_{\frac{7}{2}} = 1.293\,76\ldots$$

$$\int_1^{1.8} \ln x \, dx \approx h\left(y_{\frac{1}{2}} + y_{\frac{3}{2}} + y_{\frac{5}{2}} + y_{\frac{7}{2}}\right) = 0.2 \times 1.293\,76\ldots = 0.258\,75\ldots$$

$$\int_1^{1.8} \ln x \, dx \approx 0.2588 \text{ to four decimal places.}$$

From Worked example 9.2, using integration by parts,

$\int \ln x \, dx = x \ln x - x$ so

$\int_1^{1.8} \ln x \, dx = 1.8 \ln 1.8 - 0.8$

$= 0.2580$ (to 4 d.p.).

Worked example 10.6

By considering two rectangular strips of equal width use the mid-ordinate rule to obtain an approximation for $\int_{-1}^{1} \sqrt{x^3 + 1} \, dx$, giving your answer to four decimal places.

Solution

The width of each rectangle is $h = \dfrac{1 - (-1)}{2} = 1$.

x-values $(x_{k+1} = x_k + h)$	Mid-x-values	Mid-ordinates $y = \ln x$
$x_0 = -1$		
	$x_{\frac{1}{2}} = -\frac{1}{2}$	$y_{\frac{1}{2}} = \sqrt{-\frac{1}{8} + 1} = 0.935\,414\ldots$
$x_1 = 0$		
	$x_{\frac{3}{2}} = \frac{1}{2}$	$y_{\frac{3}{2}} = \sqrt{\frac{1}{8} + 1} = 1.060\,660\ldots$
$x_2 = 1$		

Always work to a greater degree of accuracy than the accuracy required (4 d.p.) for the final answer.

$$y_{\frac{1}{2}} + y_{\frac{3}{2}} = 1.996\,07\ldots$$

$$\int_{-1}^{1} \sqrt{x^3 + 1} \, dx \approx h\left(y_{\frac{1}{2}} + y_{\frac{3}{2}}\right) = 1 \times 1.996\,07\ldots = 1.996\,07\ldots$$

$$\int_{-1}^{1} \sqrt{x^3 + 1} \, dx \approx 1.9961, \text{ to four decimal places.}$$

In the exam, you must always give your **final** answer to the degree of accuracy asked for in the question.

10

EXERCISE 10C

1 By considering rectangular strips of width 1 use the mid-ordinate rule to obtain an approximation for $\int_0^4 1 + \sqrt{x}\, dx$, giving your answer to three decimal places.

2 By considering four rectangular strips of equal width use the mid-ordinate rule to obtain an approximation for $\int_0^2 \sqrt{1 + e^x}\, dx$, giving your answer to four decimal places.

3 By considering rectangular strips of width 0.25 use the mid-ordinate rule to obtain an approximation for $\int_0^1 \sqrt{x(2 - x)}\, dx$, giving your answer to four decimal places.

4 By considering two rectangular strips of equal width use the mid-ordinate rule to obtain an approximation for $\int_0^{\frac{1}{2}} \cos(e^x)\, dx$, giving your answer to four significant figures.

5 **(a)** By considering rectangular strips of width 0.5 use the mid-ordinate rule to obtain an approximation for $\int_0^2 \ln(1 + \sqrt{x})\, dx$, giving your answer to three decimal places.

   (b) Hence write down an approximation for $\int_0^2 \ln(1 + \sqrt{x})^2\, dx$ giving your answer to two significant figures.

   (c) By considering one rectangular strip use the mid-ordinate rule to obtain an approximation for $\int_0^2 \ln(1 + \sqrt{x})^2\, dx$, giving your answer to two significant figures.

   (d) To what degree of accuracy are your answers to parts **(b)** and **(c)** the same?

6 By considering four rectangular strips of equal width use the mid-ordinate rule to obtain an approximation for the following three integrals, giving your answers to five decimal places:

   (a) $\int_0^1 \sin \sqrt{x}\, dx$

   (b) $\int_0^1 \cos \sqrt{x}\, dx$

   (c) $\int_0^1 \sin(1 + \sqrt{x})\, dx$

10.4 Simpson's rule

In C2 you saw that the trapezium rule approximates the area under the curve to the sum of areas of trapezia by assuming that the curve is a series of line segments. Simpson's rule approximates the area under the curve to a sum of areas under parabolas by assuming that the curve is a series of parabolic arcs.

English mathematician Thomas Simpson (1710–1761) published this rule in 1743.

For most curves this assumption is more accurate and so Simpson's rule will generally give a better approximation to a definite integral than the trapezium rule.

Consider a general curve with equation $y = f(x)$. Split the region bounded by the curve, the x-axis and the lines $x = a$ and $x = b$ into six strips each of width h. With the usual notation, the points of intersection with the curve are $P_0(x_0, y_0)$, $P_1(x_1, y_1)$, $P_2(x_2, y_2)$, $P_3(x_3, y_3)$, $P_4(x_4, y_4)$, $P_5(x_5, y_5)$ and $P_6(x_6, y_6)$ and h is the width of each strip.

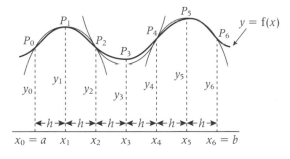

A parabola with general equation $y = Ax^2 + Bx + C$ can be drawn through any three points by choosing appropriate values for the constants A, B and C. (If the points are collinear, the 'parabola' will reduce to a straight line by equating A and B to 0.)

Firstly consider the parabola through the points P_0, P_1 and P_2. For convenience draw the y-axis to pass through the point P_1 so you have $P_0(-h, y_0)$, $P_1(0, y_1)$ and $P_2(h, y_2)$.

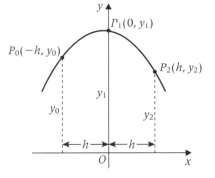

Since the three points lie on the parabola they must satisfy its equation so

$$y_0 = Ah^2 - Bh + C$$

$$y_1 = C$$

$$y_2 = Ah^2 + Bh + C.$$

The area bounded by the parabola, the x-axis and the lines $x = -h$ and $x = h$ is given by

$$\int_{-h}^{h} Ax^2 + Bx + C \, dx = \left[\frac{Ax^3}{3} + \frac{Bx^2}{2} + Cx \right]_{-h}^{h}$$

$$= \frac{2}{3}Ah^3 + 2Ch = \frac{h}{3}(2Ah^2 + 6C).$$

10

You can now write this area in terms of the known ordinates by using the above three equations

$$y_0 + y_2 = 2Ah^2 + 2C \implies 2Ah^2 + 6C = y_0 + y_2 + 4C = y_0 + y_2 + 4y_1,$$

so the area under the parabola $P_0 P_1 P_2 = \dfrac{h}{3}(y_0 + 4y_1 + y_2)$.

Similarly, the area under the parabola $P_2 P_3 P_4 = \dfrac{h}{3}(y_2 + 4y_3 + y_4)$

and the area under the parabola $P_4 P_5 P_6 = \dfrac{h}{3}(y_4 + 4y_5 + y_6)$, so, for

six strips, the area under the curve $y = f(x)$ leads to

> The area under the parabola using two equally spaced strips is given in terms of h and the three ordinates. This same result can be applied to any two equally spaced strips. If the number of strips is odd the process breaks down.

$$\int_a^b y\,dx \approx \frac{h}{3}(y_0 + 4y_1 + y_2) + \frac{h}{3}(y_2 + 4y_3 + y_4) + \frac{h}{3}(y_4 + 4y_5 + y_6)$$

$$\approx \frac{1}{3}h\left[y_0 + y_6 + 4(y_1 + y_3 + y_5) + 2(y_2 + y_4)\right],$$

where $h = \dfrac{b-a}{6}$.

Splitting the region into n strips, provided n is **even**, you can obtain the general rule as

---

Simpson's rule for n strips and $(n+1)$ ordinates, where n is **even**, is

$$\int_a^b y\,dx \approx \frac{1}{3}h\left[(y_0 + y_n) + 4(y_1 + y_3 + \ldots + y_{n-1})\right.$$
$$\left. + 2(y_2 + y_4 + \ldots + y_{n-2})\right],$$

where $h = \dfrac{b-a}{n}$.

> Simpson's rule **cannot** be used if n is odd.

Simpson's rule is in the examination formulae booklet. You may wish to remember this rule in words as

$$\frac{1}{3}h\,(\text{ends} + 4 \times \text{odds} + 2 \times \text{evens}).$$

The next two worked examples show you how to apply Simpson's rule to find numerical approximations for definite integrals.

Worked example 10.7

Use Simpson's rule with five ordinates to find an approximation to $\displaystyle\int_1^{1.8} \ln x\,dx$, giving your answer to four decimal places.

Solution

Five ordinates $\Rightarrow$ four strips (n is even so Simpson's rule can be

used with $n = 4$) $\Rightarrow h = \dfrac{1.8 - 1}{4} = 0.2$.

x-values	$y = \ln x$		
$x_0 = 1$	$y_0 = \ln 1 = 0$		
$x_1 = 1.2$	$y_1 = \ln 1.2 =$	0.182 32...	
$x_2 = 1.4$	$y_2 = \ln 1.4 =$		0.336 47...
$x_3 = 1.6$	$y_3 = \ln 1.6 =$	0.470 00...	
$x_4 = 1.8$	$y_4 = \ln 1.8 = 0.587\,78...$		

$$y_0 + y_4 = 0.587\,78...$$
$$y_1 + y_3 = 0.652\,32...$$
$$y_2 = 0.336\,47...$$

$$\int_1^{1.8} \ln x \, dx \approx \frac{1}{3} h \left[(y_0 + y_4) + 4(y_1 + y_3) + 2(y_2) \right]$$

$$\approx \frac{1}{3} \times 0.2 \times \left[0 + \ln 1.8 + 4(\ln 1.2 + \ln 1.6) + 2(\ln 1.4) \right]$$

$$\approx \frac{1}{3} \times 0.2 \times \left[0.587\,78... + 4(0.652\,32...) + 2(0.336\,47...) \right]$$

$$\approx \frac{1}{3} \times 0.2 \times (3.870\,03...) = 0.258\,002...$$

$$\int_1^{1.8} \ln x \, dx \approx 0.2580 \text{ to four decimal places.}$$

Integrating exactly, using parts, you get
$$\int_1^{1.8} \ln x \, dx = 1.8 \ln 1.8 - 0.8$$
$$= 0.258\,015\,9..., \text{ so}$$
Simpson's rule, with five ordinates, gives a correct answer to four decimal places. To obtain the same degree of accuracy using the trapezium rule, 20 ordinates (19 strips) would be required.

10

Worked example 10.8 _____

(a) Use Simpson's rule with nine ordinates to find an

approximation to $\displaystyle\int_0^1 \frac{1}{1 + x^2} \, dx$, giving your answer to eight

decimal places.

(b) Given that $\displaystyle\int_0^1 \frac{1}{1 + x^2} \, dx = \frac{\pi}{4}$ use your answer to part (a) to

estimate the value of π to seven decimal places.

Solution

(a) Nine ordinates $\Rightarrow$ eight strips (n is even so Simpson's rule can be used with $n = 8$) $\Rightarrow h = \dfrac{1 - 0}{8} = 0.125$.

x-values	$y = \dfrac{1}{1 + x^2}$	
$x_0 = 0$	$y_0 = 1$	
$x_1 = 0.125$	$y_1 =$	0.984 615 384...
$x_2 = 0.25$	$y_2 =$	0.941 176 470...
$x_3 = 0.375$	$y_3 =$	0.876 712 328...
$x_4 = 0.5$	$y_4 =$	0.8
$x_5 = 0.625$	$y_5 =$	0.719 101 123...
$x_6 = 0.75$	$y_6 =$	0.64
$x_7 = 0.875$	$y_7 =$	0.566 371 681...
$x_8 = 1$	$y_8 = 0.5$	

$$y_0 + y_8 = 1.5$$
$$y_1 + y_3 + y_5 + y_7 = 3.146\,800\,518...$$
$$y_2 + y_4 + y_6 = 2.381\,176\,471...$$

$$\int_0^1 \frac{1}{1 + x^2}\,dx \approx \frac{1}{3}h\left[(y_0 + y_8) + 4(y_1 + y_3 + y_5 + y_7) + 2(y_2 + y_4 + y_6)\right]$$

$$\approx \frac{1}{3} \times 0.125 \times \left[\begin{array}{l} 1.5 + 4(3.146\,800\,518...) \\ \qquad + 2(2.381\,176\,471...) \end{array}\right]$$

$$\approx \frac{1}{3} \times 0.125 \times (18.849\,555\,01...) = 0.785\,398\,125...$$

$$\int_0^1 \frac{1}{1 + x^2}\,dx \approx 0.785\,398\,13 \text{ to eight decimal places.}$$

(b) $\dfrac{\pi}{4} \approx 0.785\,398\,13 \Rightarrow \pi \approx 4 \times 0.785\,398\,13$

$$\pi = 3.141\,592\,5 \text{ to 7 decimal places}$$

EXERCISE 10D

1 (a) Use Simpson's rule with five ordinates (four strips) to find an approximation to $\int_0^4 1 + \sqrt{x}\,dx$, giving your answer to four decimal places.

(b) Comment on how you could obtain a better approximation to the value of the integral using Simpson's rule.

2 Use Simpson's rule with seven ordinates (six strips) to find an approximation to $\int_{-1}^1 \sqrt{x^3 + 1}\,dx$, giving your answer to four decimal places.

3 Find an approximation to $\int_0^2 \sqrt{1 + e^x}\, dx$, by using Simpson's rule with:

(a) five ordinates (four strips),

(b) seven ordinates (six strips),

(c) nine ordinates (eight strips).

Give your answers to four decimal places.

4 Find an approximation to $\int_0^1 \sqrt{x(2 - x)}\, dx$, by using Simpson's rule with:

(a) five ordinates,

(b) seven ordinates,

(c) nine ordinates.

Give your answers to four decimal places.

5 Find an approximation to $\int_0^{\frac{1}{2}} \cos(e^x)\, dx$, by using Simpson's rule with:

(a) five ordinates,

(b) seven ordinates,

(c) nine ordinates.

Give your answers to six decimal places.

6 Find an approximation to $\int_1^2 \ln(1 + \sqrt{x})\, dx$, by using Simpson's rule with:

(a) five ordinates,

(b) seven ordinates,

(c) nine ordinates.

Give your answers to seven decimal places.

7 Find an approximation to $\int_1^2 x^x\, dx$, by using Simpson's rule with:

(a) five ordinates,

(b) seven ordinates.

Give your answers to four decimal places.

8 Find approximations to:

(a) $\int_0^1 \cos\sqrt{x}\, dx$ and (b) $\int_0^1 \sin(1 + \sqrt{x})\, dx$

by using Simpson's rule with:

(i) five ordinates,

(ii) seven ordinates,

(iii) nine ordinates.

Give your answers to four decimal places.

10

9 (a) Use Simpson's rule with five ordinates (four strips) to find an approximation to $\int_0^2 \ln(1+x)\,dx$, giving your answer to four decimal places.

   (b) By using integration by parts to find the exact value of $\int_0^2 \ln(1+x)\,dx$ determine if this exact value is greater than or less than your answer to part **(a)**.

10 (a) Use Simpson's rule with nine ordinates (eight strips) to find an approximation to $\int_0^{0.5} \dfrac{1}{\sqrt{1-x^2}}\,dx$, giving your answer to eight decimal places.

   (b) Given that $\int_0^{0.5} \dfrac{1}{\sqrt{1-x^2}}\,dx = \dfrac{\pi}{6}$, use your answer to part **(a)** to estimate the value of π to seven decimal places.

> You may wish to design a spreadsheet to find out how many ordinates are required when applying Simpson's rule to this definite integral in order for the value of π to be the same as that displayed on your calculator.

MIXED EXERCISE

1 The graph opposite shows the region R enclosed by the curve $y = x - \dfrac{1}{x}$, the x-axis and the line $x = 2$. Find the exact volume of the solid formed when the region R is rotated through 2π radians about the x-axis.　　　　[A]

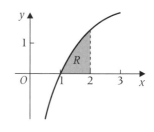

2 Use the mid-ordinate rule, with five strips of equal width, to estimate the value of $\int_0^1 \sin^{-1} x\,dx$. Give your answer to three decimal places.　　　　[A]

3 Use Simpson's rule with five ordinates (four strips) to find an approximation to $\int_0^1 x \cos x\,dx$, giving your answer to five decimal places.　　　　[A]

4 The curve with equation $y = x\sqrt{x^2+3}$ is sketched opposite. The region R, shaded on the diagram, is bounded by the curve, the x-axis and the line $x = 1$. The region R is rotated through 2π radians about the x-axis. Find the volume of the solid generated.　　　　[A]

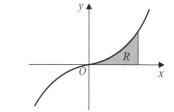

5 By considering rectangular strips of width 0.2, use the mid-ordinate rule to obtain an approximation for $\int_0^{0.6} \sqrt{(1+x^2)}\,dx$, giving your answer to four decimal places.　　　　[A]

6 The region bounded by the curve $y = \sec\!\left(2x - \dfrac{\pi}{4}\right)$, the x-axis and the lines $x = 0$ and $x = \dfrac{\pi}{4}$ is rotated through $360°$ about the x-axis. Find the volume of the solid generated.

7 Use Simpson's rule with five ordinates (four equal strips) to find an approximation to the integral $\int_0^2 \ln(x^2 + 1)\,dx$, giving your answer to three decimal places. [A]

8 The region R is bounded by the curve $y = 4 + 3\sec 2x$, the x-axis and the lines $x = -\dfrac{\pi}{6}$ and $x = \dfrac{\pi}{6}$. Find the volume of the solid formed by rotating R through 2π radians about the x-axis. Give your final answer to three decimal places.

[A adapted]

Key point summary

1 When the region bounded by the curve $y = \mathrm{f}(x)$, the   *p164*
x-axis and the lines $x = a$ and $x = b$ is rotated through 2π radians about the x-axis, the volume, V, of the solid generated is called the volume of revolution and is given by

$$V = \int_a^b \pi y^2\,dx = \int_a^b \pi[\mathrm{f}(x)]^2\,dx.$$

2 When the region bounded by the curve $x = \mathrm{g}(y)$, the   *p165*
y-axis and the lines $y = c$ and $y = d$ is rotated through 2π radians about the y-axis, the volume, V, of the solid generated is given by

$$V = \int_c^d \pi x^2\,dy = \int_c^d \pi[\mathrm{g}(y)]^2\,dy.$$

3 In all the numerical methods of integration an   *p170*
improvement to the estimate can be obtained by increasing the number of steps (strips).

4 The mid-ordinate rule for n strips is   *p170*

$$\int_a^b y\,dx \approx h(y_{\frac{1}{2}} + y_{\frac{3}{2}} + \ldots + y_{n-\frac{3}{2}} + y_{n-\frac{1}{2}}),$$

where $h = \dfrac{b - a}{n}$.

5 Simpson's rule for n strips and $(n + 1)$ ordinates,   *p174*
where n is **even**, is

$$\int_a^b y\,dx \approx \frac{1}{3}h[(y_0 + y_n) + 4(y_1 + y_3 + \ldots + y_{n-1})$$
$$+\, 2(y_2 + y_4 + \ldots + y_{n-2})],$$

where $h = \dfrac{b - a}{n}$.

10

Test yourself	What to review

1 The region bounded by the curve $y = x + \dfrac{1}{x}$, the *x*-axis and the lines $x = 1$ and $x = 3$ is rotated through 2π radians about the *x*-axis. Find the volume of the solid generated.

Section 10.1

2 The region bounded by the curve $y = \ln x$, the *y*-axis and the lines $y = 0$ and $y = 1$ is rotated through 2π radians about the *y*-axis.

Show that the volume of the solid generated is $\dfrac{\pi}{2}(e^2 - 1)$.

Section 10.1

3 By considering rectangular strips of width 1 use the mid-ordinate rule to obtain an approximation

for $\displaystyle\int_0^4 \ln(3 + \sqrt{x})\, dx$, giving your answer to three decimal places.

Section 10.3

4 (a) Use Simpson's rule with seven ordinates to find an approximation to $\displaystyle\int_0^3 \dfrac{2}{9 + x^2}\, dx$, giving your answer to eight decimal places.

(b) Given that $\displaystyle\int_0^3 \dfrac{2}{9 + x^2}\, dx = \dfrac{\pi}{6}$, use your answer to part **(a)** to estimate the value of π to seven decimal places.

Section 10.4

Test yourself ANSWERS

4 (a) 0.523 598 63;   **(b)** 3.141 591 8.

3 5.856.

1 $13\frac{1}{3}\pi\,$m.

C3: Exam style practice paper

Time allowed 1 hour 30 minutes

Answer **all** questions

1 Find $\dfrac{dy}{dx}$ when:

   (a) $y = x\tan x$ (3 marks)

   (b) $y = \ln(1 + \sin x)$ (3 marks)

2 The functions f and g are defined for real values of x by

$$f(x) = x + 2$$
$$g(x) = \frac{1}{x^2 + 1}.$$

   (a) Find the range of g.   (2 marks)

   (b) The curve with equation $y = g(x)$ has a single stationary point A. Find the coordinates of A.   (4 marks)

   (c) Solve the equation $gf(x) = \dfrac{1}{2}$.   (4 marks)

3 **(a)** Use integration by parts to find $\displaystyle\int_0^{\frac{\pi}{4}} x\sin 2x\,dx$.   (5 marks)

   (b) Use Simpson's rule with five ordinates (four strips) to find an approximation for $\displaystyle\int_0^4 \cos(e^{0.5x})\,dx$, giving your answer to four significant figures.   (4 marks)

4 A curve C has equation $y = \cos 3x + 2$.

   (a) Describe a sequence of geometrical transformations that maps the graph of $y = \cos x$ onto the graph of $y = \cos 3x + 2$.   (4 marks)

   (b) The point P, whose x-coordinate is $\dfrac{\pi}{6}$, lies on the curve C. Show that the gradient of the curve at P is -3.   (3 marks)

   (c) The line $y = 2x$ intersects the curve C at the point A, whose x-coordinate is α.

      (i) Show that $0.7 < \alpha < 0.8$.   (2 marks)

      (ii) Show that the equation $\cos 3x - 2x + 2 = 0$ can be rearranged in the form $x = \dfrac{1}{3}\cos^{-1}[2(x - 1)]$.   (1 mark)

(iii) Use the iterative formula $x_{n+1} = \dfrac{1}{3}\cos^{-1}[2(x_n - 1)]$

with $x_1 = 0.7$ to find x_4 giving your answer to three significant figures. (3 marks)

5 Use the substitution $u = 2x - 1$ to find $\int x\sqrt{2x - 1}\, dx$.

(4 marks)

6 A curve has equation $y = e^x + 1$.

(a) Sketch the curve. (3 marks)

(b) The tangent to the curve at the point P is parallel to the line $y = 3x + 1$. Find, in an exact form, the coordinates of P. (5 marks)

(c) The finite region bounded by the curve $y = e^x + 1$, the y-axis and the line $x = \ln 2$ is rotated through $360°$ about the x-axis. Show that the volume of the solid formed is $\dfrac{\pi}{2}(7 + \ln 4)$. (6 marks)

(d) The curve $y = e^x + 1$ is reflected in the line $y = x$ to form the curve with equation $y = g(x)$. Find an expression for $g(x)$. (2 marks)

7 (a) Sketch on the same diagram the graphs of $y = |2x - 3|$ and $y = 5 - x^2$ stating the coordinates of any points where the graphs meet the coordinate axes. (4 marks)

(b) Solve the equation $|2x - 3| = 5 - x^2$. (4 marks)

(c) Solve the inequality $|2x - 3| > 5 - x^2$. (2 marks)

8 Solve the equation $2\cot^2 x + 7\operatorname{cosec} x - 13 = 0$ giving all values of x to the nearest degree in the interval $0° \leqslant x \leqslant 360°$.

(7 marks)

CHAPTER 1

C4: Binomial series expansion

Learning objectives

After studying this chapter, you should be able to:

- use geometric series to expand $(1 - r)^{-1}$ for $|r| < 1$
- expand $(1 + x)^n$ in ascending powers of x, where n is rational and $|x| < 1$
- extend the method to expand $(1 + bx)^n$ and $(a + x)^n$ in ascending powers of x, where a and b are constants
- understand that the binomial series expansion of $(1 + bx)^n$ can be found when $|x| < \dfrac{1}{|b|}$
- understand that the binomial series expansion of $(a + x)^n$, where $a > 0$, can be found when $|x| < a$
- use the binomial series expansion to find approximations.

1.1 Geometric series

You have already met the geometric series

$$1 + r + r^2 + r^3 + \ldots$$

which converges when $|r| < 1$ to $\dfrac{1}{1 - r} = (1 - r)^{-1}$.

See section 12.4 of C2.

> Conversely, we could say that the series expansion of $(1 - x)^{-1}$ when $|x| < 1$ is $1 + x + x^2 + x^3 + \ldots$.

Worked example 1.1

Find the first four terms in ascending powers of x of the series expansions of:

(a) $(1 - 2x)^{-1}$, **(b)** $\left(1 + \dfrac{x}{3}\right)^{-1}$.

Solution

(a) Since $(1 - r)^{-1} = 1 + r + r^2 + r^3 + \ldots$

you can substitute $r = 2x$ to obtain

$$(1 - 2x)^{-1} = 1 + (2x) + (2x)^2 + (2x)^3 + \ldots$$
$$= 1 + 2x + 4x^2 + 8x^3 + \ldots$$

This is valid when $|r| < 1$.

The series expansion is valid when $|2x| < 1$ or when $|x| < \tfrac{1}{2}$.

(b) As before, $(1 - r)^{-1} = 1 + r + r^2 + r^3 + \dots$

Substituting $r = -\dfrac{x}{3}$ gives

The series expansion is valid when $\left|\dfrac{-x}{3}\right| < 1$ or when $|x| < 3$.

$$\left(1 + \frac{x}{3}\right)^{-1} = 1 + \left(\frac{-x}{3}\right) + \left(\frac{-x}{3}\right)^2 + \left(\frac{-x}{3}\right)^3 + \dots$$

$$= 1 - \frac{x}{3} + \frac{x^2}{9} - \frac{x^3}{27} + \dots$$

EXERCISE 1A

1 Given that $|x| < \dfrac{1}{3}$, find the first four terms in ascending powers of x of the series expansion for $(1 - 3x)^{-1}$.

2 Given that $|x| < \dfrac{1}{4}$, find the first four terms in ascending powers of x of the series expansion for $(1 + 4x)^{-1}$.

3 Find the first four terms in ascending powers of x of the series expansion for each of the following and state the values of x for which your expansion is valid:

  (a) $(1 - x)^{-1}$ **(b)** $(1 - 4x)^{-1}$

  (c) $(1 + 2x)^{-1}$ **(d)** $(1 + 3x)^{-1}$

4 Given that $|x| < 2$, find the first four terms in ascending powers of x of the series expansion of $\left(1 - \dfrac{x}{2}\right)^{-1}$.

5 Find the first four terms in ascending powers of x of the series expansion for each of the following and state the values of x for which your expansion is valid:

  (a) $\left(1 - \dfrac{x}{5}\right)^{-1}$ **(b)** $\left(1 + \dfrac{x}{4}\right)^{-1}$

  (c) $\left(1 - \dfrac{2x}{3}\right)^{-1}$ **(d)** $\left(1 + \dfrac{3x}{4}\right)^{-1}$

It is often necessary to find series expansions of $(1 + x)^n$ for values other than $n = -1$ and so a special formula has to be derived.

1.2 General binomial expansion

Previously you have used the binomial expansion

$$(a + b)^n = a^n + \binom{n}{1}a^{n-1}b + \binom{n}{2}a^{n-2}b^2 + \binom{n}{3}a^{n-3}b^3 + \dots + b^n$$

where n is a positive integer.

In the case when $a = 1$ and $b = x$,

$$(1 + x)^n = 1 + \binom{n}{1}x + \binom{n}{2}x^2 + \binom{n}{3}x^3 + \dots .$$

The notation $\binom{n}{r}$ is used when n and r are integers.

However, $\binom{n}{r} = \dfrac{n!}{(n-r)!1!}$ and therefore

$$\binom{n}{1} = \frac{n!}{(n-r)!1!} = \frac{n}{1},$$

$$\binom{n}{2} = \frac{n!}{(n-2)!2!} = \frac{n \times (n-1)}{1 \times 2},$$

$$\binom{n}{3} = \frac{n!}{(n-3)!3!} = \frac{n \times (n-1) \times (n-2)}{1 \times 2 \times 3}, \text{ etc.}$$

Consequently $\binom{n}{5}$ will have five terms as a product in the numerator, each number being one less than the previous, and the denominator will be $1 \times 2 \times 3 \times 4 \times 5$.

This allows you to obtain a series expression for $(1+x)^n$ even when n is **not** a positive integer.

> When $|x| < 1$, the binomial series expansion formula is
>
> $$(1+x)^n = 1 + nx + \frac{n \times (n-1)}{1 \times 2}x^2 + \frac{n \times (n-1) \times (n-2)}{1 \times 2 \times 3}x^3 + \dots$$

This is the key formula in this chapter.
It appears in the formulae booklet to be used in the examination.

You could experiment on a spreadsheet, or using a graphics calculator, with an expression such as $(1+x)^{-3}$, evaluating both sides of the formula for different values of x and you would find that, like the geometric series, the series is only convergent when $|x| < 1$.

The partial sums are given by
$$S_1 = 1, \ S_r = \frac{(n-r+1)x}{r}S_{r-1}.$$
For the series $(1+x)^{-3}$ you would use $n = -3$.

Worked example 1.2

Given that $|x| < 1$, use the binomial series expansion formula to obtain the series expansion of $(1+x)^{-3}$ in ascending powers of x up to and including the term in x^3.

Solution

Here $n = -3$ and therefore

$$(1+x)^{-3} = 1 + (-3)x + \frac{(-3) \times (-3-1)}{1 \times 2}x^2$$

$$+ \frac{(-3) \times (-3-1) \times (-3-2)}{1 \times 2 \times 3}x^3 + \dots$$

$$= 1 - 3x + \frac{12}{2}x^2 - \frac{60}{6}x^3 + \dots$$

$$= 1 - 3x + 6x^2 - 10x^3 + \dots .$$

> There will be an infinite number of terms when n is **not** a positive integer.

However, you should be able to see that when $n = 7$, for example, the coefficient of x^8 and all higher powers will be zero, since the numerator will then contain the factor $(n - 7)$.

> Hence, when n is a positive integer, the binomial series expansion will have a finite number of terms with the highest power being x^n.

Worked example 1.3

Use the formula

$$(1 + x)^n = 1 + nx + \frac{n \times (n - 1)}{1 \times 2}x^2 + \frac{n \times (n - 1) \times (n - 2)}{1 \times 2 \times 3}x^3 + \dots$$

to obtain the full series expansion of $(1 + x)^4$ in ascending powers of x.

Solution

Here $n = 4$ and therefore

$$(1 + x)^4 = 1 + 4x + \frac{4 \times (4 - 1)}{1 \times 2}x^2 + \frac{(4) \times (4 - 1) \times (4 - 2)}{1 \times 2 \times 3}x^3$$

$$+ \frac{(4) \times (4 - 1) \times (4 - 2) \times (4 - 3)}{1 \times 2 \times 3 \times 4}x^4$$

$$+ \frac{(4) \times (4 - 1) \times (4 - 2) \times (4 - 3) \times (4 - 4)}{1 \times 2 \times 3 \times 4 \times 5}x^5 + \dots$$

The coefficient of x^5 and all higher powers will be zero, since the numerator will then contain the factor $(4 - 4) = 0$.

Hence $(1 + x)^4 = 1 + 4x + 6x^2 + 4x^3 + x^4$.

Worked example 1.4

Given that $|x| < 1$, use the formula

$$(1 + x)^n = 1 + nx + \frac{n \times (n - 1)}{1 \times 2}x^2 + \frac{n \times (n - 1) \times (n - 2)}{1 \times 2 \times 3}x^3 + \dots$$

to obtain the series expansion of $(1 + x)^{\frac{1}{2}}$ in ascending powers of x up to and including the term in x^3.

Verify your result by squaring both sides of your solution, ignoring terms involving x^4 and higher powers of x.

Solution

Here $n = \dfrac{1}{2}$ and therefore

$$(1 + x)^{\frac{1}{2}} = 1 + \frac{1}{2}x + \frac{\left(\frac{1}{2}\right) \times \left(\frac{1}{2} - 1\right)}{1 \times 2}x^2 + \frac{\left(\frac{1}{2}\right) \times \left(\frac{1}{2} - 1\right) \times \left(\frac{1}{2} - 2\right)}{1 \times 2 \times 3}x^3 + \dots$$

$$= 1 + \frac{1}{2}x - \frac{1}{8}x^2 + \frac{1}{16}x^3 + \dots.$$

In order to square the expansion, you can write the brackets next to each other and multiply out:

$$\left(1 + \frac{1}{2}x - \frac{1}{8}x^2 + \frac{1}{16}x^3\right)\left(1 + \frac{1}{2}x - \frac{1}{8}x^2 + \frac{1}{16}x^3\right)$$

$$= 1 \times \left(1 + \frac{1}{2}x - \frac{1}{8}x^2 + \frac{1}{16}x^3\right) + \frac{1}{2}x\left(1 + \frac{1}{2}x - \frac{1}{8}x^2 + \frac{1}{16}x^3\right)$$

$$- \frac{1}{8}x^2\left(1 + \frac{1}{2}x - \frac{1}{8}x^2 + \frac{1}{16}x^3\right) + \frac{1}{16}x^3\left(1 + \frac{1}{2}x - \frac{1}{8}x^2 + \frac{1}{16}x^3\right)$$

$$= \left(1 + \frac{1}{2}x - \frac{1}{8}x^2 + \frac{1}{16}x^3\right) + \left(\frac{1}{2}x + \frac{1}{4}x^2 - \frac{1}{16}x^3 + \dots\right)$$

$$+ \left(-\frac{1}{8}x^2 - \frac{1}{16}x^3 + \dots\right) + \left(\frac{1}{16}x^3 + \dots\right)$$

$$= 1 + \left(\frac{1}{2} + \frac{1}{2}\right)x + \left(-\frac{1}{8} + \frac{1}{4} - \frac{1}{8}\right)x^2 + \left(\frac{1}{16} - \frac{1}{16} - \frac{1}{16} + \frac{1}{16}\right)x^3 + \dots$$

$= 1 + x$, ignoring terms involving x^4 and higher powers.

But $(1 + x)^{\frac{1}{2}} \times (1 + x)^{\frac{1}{2}} = 1 + x$, and so the expansion is verified.

EXERCISE 1B

Given that $|x| < 1$, find the binomial series expansion for each of the following in ascending powers of x up to and including the term in x^3:

1 $(1 + x)^7$ **2** $(1 + x)^{10}$ **3** $(1 + x)^{-4}$

4 $(1 + x)^{-2}$ **4** $(1 + x)^{-5}$ **6** $(1 + x)^{\frac{1}{3}}$

7 $(1 + x)^{-\frac{1}{2}}$ **8** $(1 + x)^{-\frac{3}{4}}$

1.3 Expansions of the form $(1 + bx)^n$

Here, the constant b may be positive or negative. The expansion of $(1 + y)^n$ is found for $|y| < 1$ and then by substituting $y = bx$.

> The expansion of $(1 + bx)^n$ is valid when $|bx| < 1$ or when $|x| < \dfrac{1}{|b|}$.

Worked example 1.5 ——————————————

Find the first four terms in the binomial expansions of:

(a) $(1 + 2x)^{-3}$ **(b)** $(1 - 8x)^{-\frac{3}{2}}$

in ascending values of x.

State the values of x for which each of the expansions is valid.

Solution

(a) $(1 + y)^{-3} = 1 - 3y + \dfrac{-3 \times -4}{1 \times 2}y^2 + \dfrac{-3 \times -4 \times -5}{1 \times 2 \times 3}y^3 + \dots$

$\qquad = 1 - 3y + 6y^2 - 10y^3 + \dots$ and is valid for $|y| < 1$.

Substituting $y = 2x$ gives

$(1 + 2x)^{-3} = 1 - 6x + 6(2x)^2 - 10(2x)^3 + \dots$

$\qquad = 1 - 6x + 24x^2 - 80x^3 + \dots.$

The expansion is valid for $|2x| < 1$ or $|x| < \dfrac{1}{2}$.

(b) $(1 + y)^{-\frac{3}{2}} = 1 - \dfrac{3}{2}y + \dfrac{-\dfrac{3}{2} \times -\dfrac{5}{2}}{1 \times 2}y^2 + \dfrac{-\dfrac{3}{2} \times -\dfrac{5}{2} \times -\dfrac{7}{2}}{1 \times 2 \times 3}y^3 + \dots$

$\qquad = 1 - \dfrac{3}{2}y + \dfrac{15}{8}y^2 - \dfrac{35}{16}y^3 + \dots.$

Substituting $y = -8x$ gives

$(1 - 8x)^{-\frac{3}{2}} = 1 - \dfrac{3}{2}(-8x) + \dfrac{15}{8}(-8x)^2 - \dfrac{35}{16}(-8x)^3 + \dots$

$\qquad = 1 + 12x + 120x^2 + 1120x^3 + \dots.$

The expansion is valid for $|8x| < 1$ or $|x| < \dfrac{1}{8}$.

EXERCISE 1C ——————————————

In questions **1** to **10**, find the first four terms in the binomial expansion in ascending values of x. State the values of x for which each of the expansions is valid.

1 $(1 - x)^{-1}$ **2** $(1 + 3x)^{-2}$

3 $(1 + 2x)^{-4}$ **4** $(1 - x)^{-2}$

5 $(1 - 2x)^{-5}$ **6** $(1 + \frac{1}{3}x)^{-3}$

7 $(1 + 4x)^{-\frac{3}{2}}$ **8** $(1 - 3x)^{-\frac{4}{3}}$

9 $(1 + 4x)^{-\frac{5}{4}}$ **10** $(1 + \frac{1}{2}x)^{-4}$

11 Given that $|x| < 3$, write down the binomial expansion of $\left(1 + \dfrac{x}{3}\right)^{\frac{1}{2}}$ in ascending powers of x up to and including the term in x^3.

                                                     [A]

12 The binomial expansion of $(1 + 3x)^{-4}$ is $1 + ax + bx^2 + cx^3 + \ldots$, where a, b, c are constants.

   (a) Determine the values of a, b and c.

   (b) State the range of values of x for which the expansion is valid. [A]

13 (a) Obtain the binomial expansion of $(1 + 2x)^{-2}$ in ascending powers of x up to and including the term in x^3.

   (b) State the range of values of x for which the full expansion is valid. [A]

1.4 Expansions of the form $(a + x)^n$

This section makes use of the fact that $(a + x) = a\left(1 + \dfrac{x}{a}\right)$ and hence that $(a + x)^n = a^n\left(1 + \dfrac{x}{a}\right)^n$.

> The expansion of $(a + x)^n$ is valid when $|x| < a$, where a is a positive constant.

Worked example 1.6

Find the binomial expansion of $(2 + x)^{-3}$ in ascending powers of x up to and including the term in x^3.

For what values of x is the expansion valid?

Solution

The important first step is to write $(2 + x)^{-3} = 2^{-3}\left(1 + \dfrac{x}{2}\right)^{-3}$.

Since $(1 + y)^{-3} = 1 - 3y + \dfrac{-3 \times -4}{1 \times 2}y^2 + \dfrac{-3 \times -4 \times -5}{1 \times 2 \times 3}y^3 + \ldots$

$= 1 - 3y + 6y^2 - 10y^3 + \ldots$,

then $\left(1 + \dfrac{x}{2}\right)^{-3} = 1 - 3\left(\dfrac{x}{2}\right) + 6\left(\dfrac{x}{2}\right)^2 - 10\left(\dfrac{x}{2}\right)^3 + \ldots$

$= 1 - \dfrac{3}{2}x + \dfrac{3}{2}x^2 - \dfrac{5}{4}x^3 + \ldots$.

| Putting $y = \dfrac{x}{2}$. |

Hence $(2 + x)^{-3} = 2^{-3}\left(1 + \dfrac{x}{2}\right)^{-3} = \dfrac{1}{8}\left(1 - \dfrac{3}{2}x + \dfrac{3}{2}x^2 - \dfrac{5}{4}x^3 + \ldots\right)$.

The expansion of $\left(1 + \dfrac{x}{2}\right)^{-3}$ is valid for $\left|\dfrac{x}{2}\right| < 1$ or when $|x| < 2$.

> The expansion of $(a + x)^n$ can be found by writing
> $(a + x)^n = a^n\left(1 + \dfrac{x}{a}\right)^n$.

1.5 An approach using differentiation

An alternative to using the formula in the booklet for the binomial expansion is to derive a series expansion from first principles.

If you assume the expansion is of the form $a + bx + cx^2 + dx^3 + \dots$, you can use differentiation in order to find the values of the constants a, b, c, etc.

> This section is included for completeness to show you how the terms of the binomial expansion can be obtained.
>
> It is not the recommended approach when asked to find a binomial expansion since it is easy to make slips in the differentiation and any error in finding one term is propagated to the next, etc.

Worked example 1.7

Given that the series expansion of $(8 - x)^{\frac{1}{3}}$ is of the form

$$a + bx + cx^2 + dx^3 + \dots,$$

find a, b and c by successively substituting $x = 0$ and differentiating with respect to x.

Solution

Assume $(8 - x)^{\frac{1}{3}} = a + bx + cx^2 + dx^3 + \dots$.

Substituting $x = 0$ gives $a = 8^{\frac{1}{3}} = 2$.

Differentiating both sides with respect to x gives

$$-1 \times \frac{1}{3}(8 - x)^{-\frac{2}{3}} = b + 2cx + 3dx^2 + \dots$$

Substituting $x = 0$ gives

$$-1 \times \frac{1}{3} \times 8^{-\frac{2}{3}} = -\frac{1}{3} \times \frac{1}{4} = -\frac{1}{12} = b.$$

Differentiating both sides once more gives

$$\frac{1}{3} \times \left(\frac{-2}{3}\right)(8 - x)^{-\frac{5}{3}} = 2c + 6dx + \dots$$

Substituting $x = 0$ gives

$$\frac{1}{3} \times \left(\frac{-2}{3}\right)8^{-\frac{5}{3}} = -\frac{2}{9} \times \frac{1}{32} = -\frac{1}{144} = 2c.$$

> You can continue this process to find the value of d, etc.

Hence $c = -\dfrac{1}{288}$.

Therefore $(8 - x)^{\frac{1}{3}} = 2 - \dfrac{1}{12}x - \dfrac{1}{288}x^2 + \dots$.

EXERCISE 1D

For questions **1** to **8**, find the first four terms in the binomial expansions in ascending values of x and state the values of x for which each of the expansions is valid:

1 $(3 + x)^{-1}$

2 $(2 + x)^{-2}$

3 $(3 - x)^{-4}$

4 $(5 - x)^{-2}$

5 $(2 + x)^{-4}$

6 $(4 + 3x)^{-2}$

7 $(3 - 2x)^{-1}$

8 $(4 + x)^{-\frac{3}{2}}$

9 Assume that the expansion of $(1 + x)^n$ is of the form $a + bx + cx^2 + dx^3 + \ldots$ and use the approach of Worked example 1.7 to find the values of a, b, c and d and hence show that

$$(1 + x)^n = 1 + nx + \frac{n \times (n - 1)}{1 \times 2}x^2 + \frac{n \times (n - 1) \times (n - 2)}{1 \times 2 \times 3}x^3 + \ldots .$$

10 Given that $|x| < 1$ and $(1 + x)^{-1} = 1 - x + x^2 - x^3 + \ldots$, differentiate both sides to obtain series expansions of:

(a) $(1 + x)^{-2}$

(b) $(1 + x)^{-3}$

1.6 The use of binomial expansions for approximations

Your calculator uses algorithms based on series expansions to work out things like logarithms and trigonometric functions. The binomial series expansions can be used to find various approximations.

Worked example 1.8

Given that x is small, find the binomial expansion of $(4 - x)^{\frac{1}{2}}$ in ascending powers of x up to and including the term in x^3. Hence find the value of $\sqrt{3.9999}$ giving your answer to 12 decimal places.

Solution

Again, the first step is to write $(4 - x)^{\frac{1}{2}} = 4^{\frac{1}{2}}\left(1 - \frac{x}{4}\right)^{\frac{1}{2}}$.

$$\text{Since} \quad (1 + y)^{\frac{1}{2}} = 1 + \frac{1}{2}y + \frac{\frac{1}{2} \times -\frac{1}{2}}{1 \times 2}y^2 + \frac{\frac{1}{2} \times -\frac{1}{2} \times -\frac{3}{2}}{1 \times 2 \times 3}y^3 + \ldots$$

$$= 1 + \frac{1}{2}y - \frac{1}{8}y^2 + \frac{1}{16}y^3 + \ldots,$$

you can substitute $y = \frac{x}{4}$ to give

$$\left(1 - \frac{x}{4}\right)^{\frac{1}{2}} = 1 + \frac{1}{2}\left(-\frac{x}{4}\right) - \frac{1}{8}\left(-\frac{x}{4}\right)^2 + \frac{1}{16}\left(-\frac{x}{4}\right)^3 + \ldots$$

$$= 1 - \frac{x}{8} - \frac{x^2}{128} - \frac{x^3}{1024} + \ldots .$$

Hence $(4 - x)^{\frac{1}{2}} = 4^{\frac{1}{2}}\left(1 - \frac{x}{4}\right)^{\frac{1}{2}} = 2\left(1 - \frac{x}{8} - \frac{x^2}{128} - \frac{x^3}{1024}\right)$

$$= 2 - \frac{x}{4} - \frac{x^2}{64} - \frac{x^3}{512} + \ldots .$$

Substituting $x = 0.0001$ into the binomial expansion will allow you to find $\sqrt{3.9999}$.

The expansion gives $2 - \dfrac{10^{-4}}{4} - \dfrac{10^{-8}}{64} - \dfrac{10^{-12}}{512} + \ldots$

$$= 2 - 0.000\,025$$
$$- 0.000\,000\,000\,156\,25$$
$$- 0.000\,000\,000\,000\,001\,953\,125$$

Hence $\sqrt{3.9999} \approx 1.999\,974\,999\,844$ (to 12 decimal places).

> The next term is too small to make a contribution to the first 12 decimal places. Very few calculators could give you an answer to this level of accuracy.

Worked example 1.9

(a) Given that $|x| < 2$, find the binomial expansion of $(2 - x)^{-1}$ in ascending powers of x up to the term in x^4.

(b) Find $\displaystyle\int_0^1 (2 - x)^{-1}\,dx$.

(c) By integrating the series expansion term by term between the limits $x = 0$ and $x = 1$ find a two significant figure approximation for $\ln 2$.

Solution

(a) $(2 - x)^{-1} = 2^{-1}\left(1 - \dfrac{x}{2}\right)^{-1} = \dfrac{1}{2}\left[1 + \dfrac{x}{2} + \left(\dfrac{x}{2}\right)^2 + \left(\dfrac{x}{2}\right)^3 + \left(\dfrac{x}{2}\right)^4 + \ldots\right]$

$$= \dfrac{1}{2} + \dfrac{x}{4} + \dfrac{x^2}{8} + \dfrac{x^3}{16} + \dfrac{x^4}{32} + \ldots.$$

> Since
> $(1 - r)^{-1} = 1 + r + r^2 + \ldots.$

(b) Since $\dfrac{d}{dx}\ln(2 - x) = \dfrac{-1}{(2 - x)} = -(2 - x)^{-1}$,

$$\int_0^1 (2 - x)^{-1}\,dx = \left[-\ln(2 - x)\right]_0^1 = -\ln 1 - (-\ln 2) = \ln 2.$$

(c) Integrating the series from (a) term by term gives

$$\int_0^1 \left(\dfrac{1}{2} + \dfrac{x}{4} + \dfrac{x^2}{8} + \dfrac{x^3}{16} + \dfrac{x^4}{32} + \ldots\right) dx$$

$$= \left[\dfrac{x}{2} + \dfrac{x^2}{8} + \dfrac{x^3}{24} + \dfrac{x^4}{64} + \dfrac{x^5}{160} + \ldots\right]_0^1$$

$$\approx 0.5 + 0.125 + 0.0417 + 0.0156 + 0.0063$$

$$\approx 0.69 \text{ (to 2 significant figures)}.$$

EXERCISE 1E

1 (a) Given that $|y| < 1$, find the expansion of $(1 + y)^{-1}$ in ascending powers of y up to and including the term in y^4. Hence find the expansion of $\dfrac{1}{1 + x^2}$ up to and including the term in x^8.

(b) By integrating the series expansion in (a), find an approximation for $\displaystyle\int_0^{0.2} \dfrac{1}{1 + x^2}\,dx$, giving your answer to six decimal places.

(c) Explain why the binomial expansion could not be used to find an approximation for $\displaystyle\int_0^5 \dfrac{1}{1 + x^2}\,dx$.

2 (a) Given that $|x| < 1$, find the binomial expansion of $(1 - x)^{\frac{1}{3}}$ in ascending powers of x up to and including the term in x^4.

(b) By substituting $x = 0.1$ in your expansion to **(a)**, find a five significant figure approximation for
(i) $\sqrt[3]{0.9}$, **(ii)** $\sqrt[3]{900}$.

3 (a) Given that $|x| < \dfrac{1}{2}$, show that the expansion of $(1 - 2x)^{\frac{1}{2}}$ in ascending powers of x is $1 - x - \dfrac{1}{2}x^2 - \dfrac{1}{2}x^3 + \dots$.

(b) (i) Show that $\sqrt{0.98} = \dfrac{7}{10}\sqrt{2}$.

(ii) By substituting $x = 0.01$ in the expansion of **(a)**, find a seven significant figure approximation for $\sqrt{2}$.

4 (a) Given that x is small in magnitude, find the expansion of $(1 - 4x)^{\frac{1}{2}}$ in ascending powers of x to the term involving x^3.

(b) By using a suitable value of x, find an approximation for $\sqrt{0.96}$.

(c) Use the result from **(b)** to deduce approximations for
(i) $\sqrt{96}$, **(ii)** $\sqrt{6}$.

1.7 Further techniques

Worked examination question

(a) Find the binomial expansion of $(1 + 4x)^{-2}$ in ascending powers of x up to the term in x^3. State the set of values for which the expansion is valid.

(b) Hence find the coefficient of x^3 in the expansion of $\dfrac{(1 - 5x)}{(1 + 4x)^2}$.

Solution

(a) $(1 + y)^{-2} = 1 - 2y + \dfrac{-2 \times -3}{1 \times 2}y^2 + \dfrac{-2 \times -3 \times -4}{1 \times 2 \times 3}y^3 + \dots$

$\qquad = 1 - 2y + 3y^2 - 4y^3 + \dots$.

Hence $(1 + 4x)^{-2} = 1 - 2(4x) + 3(4x)^2 + 4(4x)^3$
$\qquad\qquad\qquad = 1 - 8x + 48x^2 + 256x^3$.

The expansion is valid for $|4x| < 1$ or $|x| < \dfrac{1}{4}$.

(b) The expression is the same as $(1 - 5x)(1 + 4x)^{-2}$ and so you multiply the expansion from **(a)** by $(1 - 5x)$.

The only contributions to the term in x^3 are from multiplying 1 by $256x^3$ and $-5x$ by $48x^2$.

Hence the coefficient of x^3 is $256 - 5 \times 48 =$ ~~key~~ -496

MIXED EXERCISE

1 The function f is defined by $f(x) = \dfrac{2x}{(1 + 2x)^3}$, $\quad x \neq -\dfrac{1}{2}$.

Obtain the first three terms of the binomial expansion of $f(x)$ in ascending powers of x. [A]

2 The binomial expansion of $(1 + ax)^n$ begins $1 + 2x + \dfrac{3}{2}x^2$.

Find the value of the constants a and n.

3 (a) Find the binomial expansion of:

(i) $(1 - x)^{-1}$, (ii) $(1 - 2x)^{-1}$

in ascending powers of x up to and including the term in x^3.

(b) Show that $\dfrac{2}{(1 - 2x)} - \dfrac{1}{(1 - x)} \equiv \dfrac{1}{(1 - x)(1 - 2x)}$.

(c) Hence, For otherwise, find the binomial expansion of

$f(x) = \dfrac{1}{(1 - x)(1 - 2x)}$ in ascending powers of x up to

and including the term in x^3.

(d) Determine the coefficient of x^n in the series expansion of $f(x)$.

4 (a) Determine the binomial expansion of $\left(1 - \dfrac{x}{10}\right)^{-3}$ in

ascending powers of x, up to and including the term in x^3.

(b) Show that the coefficient of x^n in this expansion is

$K(n + 1)(n + 2) \times \dfrac{1}{10^n}$ for a rational number K whose

value is to be determined.

(c) Determine the value of $\left(\dfrac{1}{0.999}\right)^3$ correct to 14 decimal

places. [A]

5 Use the binomial expansion to show that

$$\dfrac{x^2}{\sqrt{4 - x^2}} = \dfrac{1}{2}x^2 + \dfrac{1}{16}x^4 + kx^6 + \dots \quad (|x| < 2)$$

for some constant k, and state its value.

Hence show, by integrating the first three terms of the series

that $I = \displaystyle\int_0^1 \dfrac{x^2}{\sqrt{4 - x^2}} \, dx$ is approximately 0.1808. [A]

6 Given that $|x| < \dfrac{1}{6}$, write down the binomial expansion of $(1 - 6x)^{-\frac{1}{2}}$ in ascending powers of x up to and including the term in x^3.

Hence obtain the coefficient of x^3 in the expansion of

$$\dfrac{2 + 3x}{\sqrt{1 - 6x}}.$$ [A]

7 (a) Given that $|x| < \dfrac{1}{2}$, write down the binomial expansions of:

(i) $(1 + 2x)^{-\frac{1}{2}}$, **(ii)** $\dfrac{1 + 5x}{\sqrt{1 + 2x}}$

in ascending powers of x up to and including the term in x^3.

(b) By putting $x = 0.04$ in the expansion for **(a) (ii)** deduce an approximate value of $\dfrac{1}{\sqrt{3}}$, giving your answer to three decimal places.   [A]

8 (a) Write down and simplify binomial series in ascending powers of x up to and including the terms in x^3 for

(i) $(1 + x)^{-1}$, **(ii)** $(1 - 2x)^{-1}$, **(iii)** $(1 - 2x)^{-2}$.

(b) (i) Using your series from **(a)**, expand

$$f(x) = \dfrac{3}{(1 + x)} + \dfrac{1}{(1 - 2x)} + \dfrac{2}{(1 - 2x)^2}$$ in ascending

powers of x up to and including the term in x^3.

(ii) State the range of values of x for which the expansion is valid.

9 (a) Obtain the binomial expansion of $(1 + x)^{\frac{1}{2}}$ in ascending powers of x up to and including the term in x^2.

(b) (i) Hence, or otherwise, find the series expansion of $(4 + 2x)^{\frac{1}{2}}$ in ascending powers of x up to and including the term in x^2.

(ii) State the range of values of x for which the expansion is valid.

10 (a) Obtain the expansion of $(16 + y)^{\frac{1}{2}}$ in ascending powers of y up to and including the term in y^2.

(b) State the set of values of y for which the expansion is valid.

(c) Hence show that if k^3 and higher powers of k are neglected

$$\sqrt{16 + 4k + k^2} = 4 + \dfrac{k}{2} + \dfrac{3k^2}{32}.$$ [A]

Key point summary

1 Using the formula for the sum of an infinite   *p183*
geometric series, the expansion of $(1 - x)^{-1}$ when
$|x| < 1$ is $1 + x + x^2 + x^3 + \dots + x^n + \dots$.

2 When $|x| < 1$, the binomial series expansion formula is   *p185*

$$(1 + x)^n = 1 + nx + \frac{n \times (n - 1)}{1 \times 2} x^2$$
$$+ \frac{n \times (n - 1) \times (n - 2)}{1 \times 2 \times 3} x^3 + \dots$$

3 When n is a positive integer, the series will have a finite   *p186*
number of terms with the highest power being x^n.

4 When n is real but **not** a positive integer, the series will   *p186*
have an infinite number of terms.

5 The expansion of $(1 + bx)^n$, where b is a constant, is   *p187*
valid for $|bx| < 1$ or when $|x| < \frac{1}{|b|}$.

6 The expansion of $(a + x)^n$ can be found by writing   *p189*
$$(a + x)^n = a^n \left(1 + \frac{x}{a}\right)^n.$$

7 The expansion of $(a + x)^n$, where a is a positive   *p189*
constant, is valid for $|x| < a$.

Test yourself	What to review
1 Given that $|x| < 1$, use the formula for the infinite geometric series to find the first five terms in ascending powers of x in the expansion of $(1 + 2x)^{-1}$.	*Section 1.1*
2 Given that $|x| < 1$, use the binomial series expansion formula to find the first four terms in ascending powers of x in the expansion of $(1 + x)^{-6}$.	*Section 1.2*
3 Given that $|x| < \frac{1}{3}$, find the expansion of $(1 + 3x)^{\frac{3}{2}}$ in ascending powers of x up to and including the term in x^3.	*Section 1.3*
4 Given that $|x| < 8$, write down the binomial expansion of $(8 + x)^{\frac{1}{3}}$ in ascending powers of x up to and including the term in x^2.	*Section 1.4*

Test yourself (continued)	What to review

5 Given that $|x| < 1$, use the binomial series expansion formula to find the first three non-zero terms in ascending powers of x in the expansion of $(1 + x^2)^{-3}$.
Use these three terms to find an approximation for
$$\int_0^{0.1} \frac{1}{(1 + x^2)^3} \, dx.$$

Section 1.6

6 Given that $|x| < \frac{1}{4}$, write down the binomial expansion of $(1 - 4x)^{-\frac{1}{2}}$ in ascending powers of x up to and including the term in x^3.
Hence obtain the coefficient of x^3 in the expansion of
$$\frac{1 - 3x}{\sqrt{1 - 4x}}.$$

Section 1.7

Test yourself **ANSWERS**

1 $1 - 2x + 4x^2 - 8x^3 + 16x^4$

2 $1 - 6x + 21x^2 - 56x^3$

3 $1 + \frac{9}{2}x + \frac{27}{8}x^2 - \frac{27}{16}x^3$

4 $2 + \frac{12}{x} - \frac{288}{x^2}$

5 $1 - 3x^2 + 6x^4$, 0.099 012.

6 $1 + 2x + 6x^2 + 20x^3$, 2.

CHAPTER 2

C4: Rational functions and division of polynomials

Learning objectives

After studying this chapter, you should be able to:

- simplify rational expressions
- multiply and divide rational expressions
- add and subtract rational expressions
- divide a polynomial by a linear expression of the form $(ax + b)$
- recall and use the factor theorem for divisors $(ax + b)$
- recall and use the remainder theorem for divisors $(ax + b)$.

2.1 Rational expressions

A **rational expression** is an algebraic fraction.

Just like in ordinary fractions, the term at the top of the fraction is called the **numerator** and the term at the bottom of the fraction is called the **denominator**.

You will recall from C1 section 5.1 that an expression of the form $ax^n + bx^{n-1} + \ldots + px^2 + qx + r$ (where $a, b, \ldots, p, q, r$ are constants and n is a non-negative integer) is called a **polynomial** in x. The **degree** of a polynomial is given by the highest power of the variable.

An algebraic fraction in which the degree of the numerator is less than the degree of the denominator is called a **proper fraction**.

If the algebraic fraction is not a proper fraction it is called an **improper fraction**.

$\dfrac{x - 7}{3(x^2 - 49)}$ is an example of a rational expression.
Its numerator is $x - 7$ and its denominator is $3(x^2 - 49)$.
The degree of $x - 7$ is 1 and the degree of $3(x^2 - 49)$ is 2.
$\dfrac{x - 7}{3(x^2 - 49)}$ is a proper fraction.

2.2 Simplifying rational expressions

To simplify rational expressions:
- factorise all algebraic expressions
- cancel any factors that are common to the numerator and denominator.

Worked example 2.1

Simplify $\dfrac{2x^2 - 8}{x^2 + 3x + 2}$.

Solution

$$\frac{2x^2 - 8}{x^2 + 3x + 2} = \frac{2(x^2 - 4)}{(x + 2)(x + 1)}$$

| Factorised numerator and denominator. |

$$= \frac{2(x + 2)(x - 2)}{(x + 2)(x + 1)}$$

| Factorised fully using the difference of two squares. |

$$= \frac{2(x - 2)}{(x + 1)}$$

| Cancelled the common factor $(x + 2)$. |

Worked example 2.2

Simplify $\dfrac{36 - 4x^2}{x^2 + x - 12}$.

Solution

$$\frac{36 - 4x^2}{x^2 + x - 12} = \frac{4(9 - x^2)}{(x - 3)(x + 4)}$$

| Factorised numerator and denominator. |

$$= \frac{4(3 - x)(3 + x)}{(x - 3)(x + 4)}$$

| Factorised fully using the difference of two squares. |

$$= \frac{-4(x - 3)(3 + x)}{(x - 3)(x + 4)}$$

| $(3 - x) = -(x - 3)$. |

$$= \frac{-4(3 + x)}{(x + 4)}$$

| Cancelled the common factor $(x - 3)$. |

EXERCISE 2A

Simplify the following rational expressions:

1 $\dfrac{2x^2 - 8x}{x^2 - 16}$

2 $\dfrac{x^2 + x - 6}{3x^2 - 12}$

3 $\dfrac{25 - x^2}{9x^2 - 49x + 20}$

4 $\dfrac{50 + 20x - 16x^2}{8x^2 - 50}$

5 $\dfrac{12x^2 - 11x - 36}{6x^2 - 19x - 36}$

6 $\dfrac{(x^2 - x - 6)(x^2 + 4x + 3)}{(x + 1)(x^2 - 9)}$

7 $\dfrac{(6x^2 + 15x)(2x^2 + x - 3)}{(2x^2 + 5x + 3)(4x^3 - 25x)}$

2.3 Multiplying and dividing rational expressions

> To multiply rational expressions:
> - factorise all algebraic expressions
> - write as a single fraction
> - cancel any factors that are common to the numerator and denominator.

Worked example 2.3

Simplify $\dfrac{x^2 - 2x}{x^2 - 3x + 2} \times \dfrac{x^2 - 1}{x^2}$.

Solution

$$\frac{x^2 - 2x}{x^2 - 3x + 2} \times \frac{x^2 - 1}{x^2} = \frac{x(x-2)}{(x-2)(x-1)} \times \frac{(x-1)(x+1)}{x^2}$$

Factorised numerators and denominators.

$$= \frac{x(x-2)(x-1)(x+1)}{x^2(x-2)(x-1)}$$

Written as a single fraction.

$$= \frac{x(x+1)}{x^2}$$

Cancelled the common factor $(x-2)$ $(x-1)$.

$$= \frac{(x+1)}{x}$$

Cancelled the common factor x.

$$= \frac{x+1}{x}$$

You cannot cancel the xs here because x is not a common factor.

> To divide by a rational expression:
> - change the division to a multiplication of the reciprocal
> - factorise all algebraic expressions
> - write as a single fraction
> - cancel any factors that are common to the numerator and denominator.

To divide by $\dfrac{a}{b}$

change $\div \dfrac{a}{b}$ to $\times \dfrac{b}{a}$.

$$\frac{6}{25} \div \frac{3}{5} = \frac{6}{25} \times \frac{5}{3} = \frac{2}{5}$$

Worked example 2.4

Simplify $\dfrac{2x + 6}{4x} \div \dfrac{x^2 - 9}{x + 1}$.

Solution

$$\frac{2x + 6}{4x} \div \frac{x^2 - 9}{x + 1} = \frac{2x + 6}{4x} \times \frac{x + 1}{x^2 - 9}$$

Dividing by a fraction is same as multiplying by its reciprocal.

$$= \frac{2(x + 3)}{4x} \times \frac{x + 1}{(x + 3)(x - 3)}$$

Factorised numerators and denominators.

$$= \frac{2(x + 3)(x + 1)}{4x(x + 3)(x - 3)}$$

Written as a single fraction.

$$= \frac{2(x + 1)}{4x(x - 3)}$$

Cancelled the common factor $(x + 3)$.

$$= \frac{(x + 1)}{2x(x - 3)}$$

Cancelled the common factor 2.

EXERCISE 2B

Simplify:

1 $\dfrac{x^2 + 5x - 6}{x^2 - 1} \times \dfrac{x + 1}{2x}$

2 $\dfrac{1}{x^2 - 9} \div \dfrac{1}{x - 3}$

3 $\dfrac{x^2 + 3x + 2}{5x^2 + 10x} \times \dfrac{15x}{x^2 - 1}$

4 $\dfrac{2x + 3}{x^2 - x} \div \dfrac{4x^2 - 9}{(x - 1)^2}$

5 $\dfrac{x^2 + 5x + 6}{x^2 - 9} \times \dfrac{3 - x}{6}$

6 $\dfrac{2x + 3}{x^2 + x - 2} \div \dfrac{4x^2 + 12x + 9}{(x - 1)^2}$

2.4 Adding and subtracting rational expressions

To add/subtract rational expressions:
- factorise all algebraic expressions
- write each rational expression with the same common denominator
- add/subtract to get a single rational expression
- simplify the numerator
- cancel any factors that are common to the numerator and denominator.

Worked example 2.5

Express $\dfrac{1}{x+1} - \dfrac{x-3}{x^2+3x+2}$ as a single fraction.

Solution

$$\frac{1}{x+1} - \frac{x-3}{x^2+3x+2} = \frac{1}{x+1} - \frac{x-3}{(x+1)(x+2)}$$

Factorised the denominator.

$$= \frac{x+2}{(x+1)(x+2)} - \frac{x-3}{(x+1)(x+2)}$$

Lowest common denominator is $(x+1)(x+2)$.

$$= \frac{x+2-(x-3)}{(x+1)(x+2)}$$

Written as a single fraction.

$$= \frac{5}{(x+1)(x+2)}$$

Simplified the numerator.

Worked example 2.6

Given that $x > 0$, show that $\dfrac{x}{x+4} + \dfrac{5x+4}{x^2+4x} > 1$.

Solution

$$\frac{x}{x+4} + \frac{5x+4}{x^2+4x} = \frac{x}{x+4} + \frac{5x+4}{x(x+4)}$$

Factorised the denominator.

$$= \frac{x^2}{x(x+4)} + \frac{5x+4}{x(x+4)}$$

Lowest common denominator is $x(x+4)$.

$$= \frac{x^2+5x+4}{x(x+4)}$$

Written as a single fraction.

$$= \frac{(x+1)(x+4)}{x(x+4)}$$

Simplified the numerator.

$$= \frac{x+1}{x}$$

Cancelled the common factor $(x+4)$.

$$= 1 + \frac{1}{x}$$

Now, $x > 0 \Rightarrow \dfrac{1}{x} > 0 \Rightarrow 1 + \dfrac{1}{x} > 1 \Rightarrow \dfrac{x}{x+4} + \dfrac{5x+4}{x^2+4x} > 1$

EXERCISE 2C _____

1 Write each of the following as a single fraction in its simplest form:

(a) $\dfrac{1}{x-2} - \dfrac{1}{x+2}$　　　　　(b) $\dfrac{x+1}{x^2-4} - \dfrac{5}{x+2}$

(c) $\dfrac{3}{x+3} + \dfrac{4}{x+2}$　　　　　(d) $\dfrac{x+1}{x^2-x-6} + \dfrac{1}{x+2}$

(e) $\dfrac{4}{x^2-1} + \dfrac{2}{x+1}$　　　　　(f) $\dfrac{x-20}{x^2-5x-6} - \dfrac{3}{x+1}$

(g) $\dfrac{4}{x^2-9} - \dfrac{2}{x^2+3x}$　　　　　(h) $\dfrac{3}{x^2-x} - \dfrac{4}{x^2-1}$

2 Given that $x > 0$, show that $\dfrac{x}{x+2} + \dfrac{(x+1)(x+6)}{x^2+2x} > 2$.

3 Write as a single fraction in its simplest form:

(a) $\dfrac{1}{x+2} - \dfrac{3}{(x+2)^2}$

(b) $\dfrac{1}{x-1} + \dfrac{2}{(x-1)^2} + \dfrac{1}{(x-1)^3}$

2.5 Dividing a polynomial by a linear expression

You will recall that in C1 section 6.5 you divided certain types of polynomials by a linear expression of the form $(x - a)$:

$$\frac{x+b}{x-a},\ \frac{x^2+bx+c}{x-a},\ \frac{x^3+bx^2+cx+d}{x-a},\ \text{where } a, b, c \text{ and } d$$

are integers.

In this section we extend this work to dividing any polynomial by any linear expression. The method we use is called **algebraic long division**.

Worked example 2.7 _____

Divide $2x^3 - 7x^2 - x + 12$ by $(x - 3)$.

Solution

$$
\begin{array}{r}
2x^2 \\
x - 3\overline{)2x^3 - 7x^2 - x + 12} \\
\underline{2x^3 - 6x^2 } \\
-x^2 - x + 12
\end{array}
$$

> Dividing $2x^3$ by x gives $2x^2$. (We can think of this as 'what do we need to multiply x by to get $2x^3$?') Now multiply the answer ($2x^2$) by $x - 3$ to give $2x^3 - 6x^2$. We then subtract it from $2x^3 - 7x^2 - x + 12$ to get the remainder $-x^2 - x + 12$. The process is repeated until the remainder is at least one degree less than the divisor $(x - 3)$.

$$\begin{array}{r} 2x^2 - x - 4 \\ x - 3\overline{)2x^3 - 7x^2 - x + 12} \\ \underline{2x^3 - 6x^2} \\ -x^2 - x + 12 \\ \underline{-x^2 + 3x} \\ -4x + 12 \\ \underline{-4x + 12} \\ 0 \end{array}$$

> Dividing $-x^2$ by x gives $-x$. Now multiply the answer $(-x)$ by $x - 3$ to give $-x^2 + 3x$. We then subtract it from $-x^2 - x + 12$ to get the remainder $-4x + 12$. The degrees of $-4x + 12$ is not less than the degree of $(x - 3)$ so the process is repeated.

> Dividing $-4x$ by x gives -4.

so $(2x^3 - 7x^2 - x + 12) \div (x - 3) = (2x^2 - x - 4)$ remainder 0,

which can be written as

$$(2x^3 - 7x^2 - x + 12) = (x - 3)(2x^2 - x - 4).$$

> The expression we are dividing by is called the **divisor**. Here the divisor is $(x - 3)$.

Since the remainder is 0 we can deduce that $(x - 3)$ is a factor of $2x^3 - 7x^2 - x + 12$.

> The answer we get is called the **quotient**. Here the quotient is $(2x^2 - x - 4)$.

In most cases the remainder will not be zero. However, the remainder will always be at least one degree less than the divisor.

> When a polynomial is divided by a linear expression the remainder will always be a constant and the quotient will always be one degree less than the polynomial.

Worked example 2.8

Divide $2x^3 - 3x^2 - 4x + 23$ by $(x + 2)$.

Solution

$$\begin{array}{r} 2x^2 - 7x + 10 \\ x + 2\overline{)2x^3 - 3x^2 - 4x + 23} \\ \underline{2x^3 + 4x^2} \\ -7x^2 - 4x + 23 \\ \underline{-7x^2 - 14x} \\ 10x + 23 \\ \underline{10x + 20} \\ 3 \end{array}$$

> $2x^3$ divided by x is $2x^2$.

> $2x^2$ multiplied by $x + 2$ gives $2x^3 + 4x^2$.

> Subtracted.

> $-7x$ multiplied by $x + 2$ gives $-7x^2 - 14x$.

> Subtracted.

> 10 multiplied by $x + 2$ gives $10x + 20$.

so $(2x^3 - 3x^2 - 4x + 23) \div (x + 2) \equiv (2x^2 - 7x + 10)$ remainder 3,

which can be written as

$$\frac{2x^3 - 3x^2 - 4x + 23}{x + 2} \equiv 2x^2 - 7x + 10 + \frac{3}{x + 2}$$

or $(2x^3 - 3x^2 - 4x + 23) = (x + 2)(2x^2 - 7x + 10) + 3.$

> Similar to numerical fractions, e.g. $23 \div 5 = 4$ remainder 3.
> $$\frac{23}{5} = 4 + \frac{3}{5}.$$

> Polynomial $\equiv$ divisor $\times$ quotient $+$ remainder

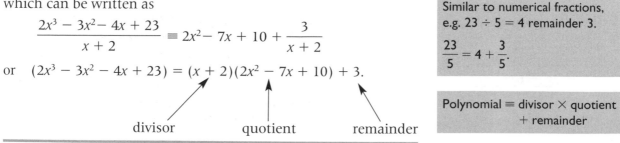

$\qquad\qquad$ divisor $\qquad\qquad$ quotient $\qquad\qquad$ remainder

In the next example the linear divisor has a coefficient of x that is not 1. Also the polynomial has some terms in x missing.

> Before starting to use algebraic long division write the polynomial and divisor in descending powers of x and include all powers of x in the polynomial, inserting zero coefficients if necessary.

Worked example 2.9

Divide $1 - 3x^2 + 2x^3$ by $(2x - 1)$.

Solution

Rewrite the polynomial in descending powers of x and, since there is no term in x, write $0x$.

$$
\begin{array}{r}
x^2 - x - \dfrac{1}{2} \\
2x - 1 \overline{\smash{)}\ 2x^3 - 3x^2 + 0x + 1} \\
\underline{2x^3 - x^2} \\
-2x^2 + 0x + 1 \\
-2x^2 + x \\
\underline{} \\
-x + 1 \\
-x + \dfrac{1}{2} \\
\underline{} \\
\dfrac{1}{2}
\end{array}
$$

$2x^3$ divided by $2x$ is x^2.

x^2 multiplied by $2x - 1$ gives $2x^3 - x^2$.

Subtracted.

$-x$ multiplied by $2x - 1$ gives $-2x^2 + x$.

Subtracted.

$-\dfrac{1}{2}$ multiplied by $2x - 1$ gives $-x + \dfrac{1}{2}$.

So $(1 - 3x^2 + 2x^3) \div (2x - 1) \equiv \left(x^2 - x - \dfrac{1}{2}\right)$ remainder $\dfrac{1}{2}$,

which can be written as

$$
\frac{1 - 3x^2 + 2x^3}{2x - 1} \equiv x^2 - x - \frac{1}{2} + \frac{\frac{1}{2}}{2x - 1} \equiv x^2 - x - \frac{1}{2} + \frac{1}{2(2x - 1)}
$$

or

$$
(1 - 3x^2 + 2x^3) \equiv (2x - 1)\left(x^2 - x - \frac{1}{2}\right) + \frac{1}{2}.
$$

Polynomial $\equiv$ divisor $\times$ quotient + remainder

EXERCISE 2D

1 Divide:

   (a) $3x + 7$ by x,

   (b) $3x + 7$ by $x + 2$,

   (c) $3x + 7$ by $2x - 4$,

   (d) x^2 by $x + 1$,

   (e) $4x^3 - 5x^2 - 2x - 1$ by $x - 1$,

   (f) $2x^3 + 7x^2 - 4x + 5$ by $x - 2$,

   (g) $3x^3 - 4x^2 + 2x - 1$ by $3x + 1$,

   (h) $6x^3 + 2x^2 - 7x + 3$ by $2x - 1$,

   (i) $5x^3 - 2x^2 + 3x - 1$ by $5x + 2$,

   (j) $5x^3 - 2x^2 + 3x - 1$ by $5x - 2$,

   (k) $1 - 7x - x^2 + 4x^3$ by $2x + 1$,

   (l) $9x^3 - 1$ by $3x - 2$.

2 Find **(i)** the quotient and **(ii)** the remainder when the polynomial $P(x)$ is divided by $(2x + 1)$:

   (a) $P(x) = 4x^2 - 3$,   **(b)** $P(x) = 6x^3 - 2x + 7$,

   (c) $P(x) = 2x^4 - 1$,   **(d)** $P(x) = 2x^3 + 8x^2 - 9$.

3 When the polynomial $8x^3 - 2x^2 + x + k$, where k is a constant, is divided by $2x - 1$ the remainder is 3. Find the value of k and the quotient.

2.6 The factor theorem for divisors of the form $(ax + b)$

In C1 section 6.2 you were shown the factor theorem in a simplified form:

   $(x - a)$ is a factor of the polynomial $P(x) \Leftrightarrow P(a) = 0$

You will now consider a more general form of this result.

Suppose $P(x) = (ax + b)Q(x)$, where $Q(x)$ is a polynomial and a and b are integers with no factor in common,

then $P\left(-\dfrac{b}{a}\right) = 0$.

The converse of this statement is also true.

> Choose the value of x to make $(ax + b) = 0$.

   A more general form of the **factor theorem** is

   $(ax + b)$ is a factor of the polynomial $P(x) \Leftrightarrow P\left(-\dfrac{b}{a}\right) = 0$.

Worked example 2.10

Given that $P(x) = 4x^3 - 6x^2 + 1$, find $P(\frac{1}{2})$ and hence find a linear factor of $P(x)$.

Solution

$P(\frac{1}{2}) = 4(\frac{1}{2})^3 - 6(\frac{1}{2})^2 + 1 = \frac{1}{2} - \frac{3}{2} + 1 = 0$

Hence $(2x-1)$ is a factor.

Note: It would not be wrong to say that $(x - \frac{1}{2})$ is a factor, but it is usual to find factors where the coefficients are integers.

Dividing $P(x)$ by $(2x - 1)$ shows that
$P(x) = (2x - 1)(2x^2 - 2x - 1)$.

However this could also be written as
$P(x) = (x - \frac{1}{2})(4x^2 - 4x - 2)$.

Although this second form is still correct, it is often requested that the factors have integer coefficients.

Worked example 2.11

(a) Show that $(3x + 2)$ is a factor of $P(x) = 3x^3 + 8x^2 + x - 2$.

(b) Find the quadratic factor $Q(x)$, where $P(x) = (3x + 2)\,Q(x)$.

(c) Find the exact values of the roots of the equation $P(x) = 0$.

Solution

(a) $P(-\frac{2}{3}) = 3(-\frac{2}{3})^3 + 8(-\frac{2}{3})^2 + (-\frac{2}{3}) - 2$
$= -\frac{8}{9} + \frac{32}{9} - \frac{2}{3} - 2 = 0.$

Hence $(3x + 2)$ is a factor.

(b) Let $Q(x) = ax^2 + bx + c$.
$(3x + 2)(ax^2 + bx + c) \equiv 3x^3 + 8x^2 + x - 2.$

Comparing the coefficient of x^3 gives $3a = 3 \Rightarrow a = 1$.

Comparing constant terms: $2c = -2 \Rightarrow c = -1$.

$(3x + 2)(x^2 + bx - 1) \equiv 3x^3 + 8x^2 + x - 2.$

Comparing coefficients of x: $-3 + 2b = 1 \Rightarrow b = 2$.

Hence $Q(x) = x^2 + 2x - 1$.

(c) $P(x) = 0 \Rightarrow (3x + 2)(x^2 + 2x - 1) = 0.$

Either $(3x + 2) = 0 \Rightarrow x = -\frac{2}{3}$.

Or $x^2 + 2x - 1 = 0$. The quadratic cannot be factorised and so the quadratic equation formula must be used.
Alternatively, completing the square gives

$$(x + 1)^2 - 1 - 1 = 0 \Rightarrow (x + 1)^2 = 2 \Rightarrow (x + 1) = \pm\sqrt{2}$$
$$\Rightarrow x = -1 \pm \sqrt{2}$$

The three exact values are $x = -\frac{2}{3}, -1 + \sqrt{2}, -1 - \sqrt{2}$.

> You could use algebraic long division to find $Q(x)$:
>
> $$\begin{array}{r} x^2 + 2x - 1 \\ 3x + 2 \overline{)\ 3x^3 + 8x^2 + x - 2} \\ \underline{3x^3 + 2x^2} \\ 6x^2 + x - 2 \\ \underline{6x^2 + 4x} \\ -3x - 2 \\ \underline{-3x - 2} \\ 0 \end{array}$$

> Notice that the term 'exact values' is used to warn you to leave your answers as surds and not to find an approximation from your calculator.

Worked example 2.12

Express $P(x) = 6x^3 - 37x^2 + 5x + 6$ as a product of linear factors with integer coefficients.

> A little thought will tell you that one of the factors must be of the form $x \pm d$. Can you see why?

Solution

Because the first term is $6x^3$, the linear factors must be of the form $(6x \pm a)$, $(3x \pm b)$, $(2x \pm c)$ or $(x \pm d)$.
Furthermore, any value of d must be 1, 2, 3 or 6.

Try $P(1) = 6 - 37 + 5 + 6 = -20$.
Next try $P(2) = 6 \times 8 - 37 \times 4 + 5 \times 2 + 6 = 48 - 148 + 10 + 6$
$$= -84.$$

$P(3) = 6 \times 27 - 37 \times 9 + 5 \times 3 + 6 = 162 - 333 + 15 + 6 = -150$

$P(6) = 6 \times 216 - 37 \times 36 + 5 \times 6 + 6 = 1296 - 1332 + 30 + 6 = 0.$

Therefore $x - 6$ is a factor.

Let $P(x) = (x - 6)(ax^2 + bx + c) \equiv 6x^3 - 37x^2 + 5x + 6$.

Comparing coefficients of x^3: $a = 6$.
Comparing constant terms: $-6c = 6 \Rightarrow c = -1$.
Therefore $(x - 6)(6x^2 + bx - 1) \equiv 6x^3 - 37x^2 + 5x + 6$.

Comparing coefficients of x^2: $b - 36 = -37 \Rightarrow b = -1$.

The quadratic factor is $6x^2 - x - 1$, which can be factorised since the discriminant is $(-1)^2 - 4 \times 6 \times (-1) = 25$ which is a perfect square.

$$6x^2 - x - 1 = (3x + 1)(2x - 1)$$

Hence $P(x) = (x - 6)(3x + 1)(2x - 1)$.

> You could have searched for other factors by trying appropriate fractions. You should verify that the factorisation is correct by showing that $P(\frac{1}{2}) = 0$ and that $P(-\frac{1}{3}) = 0$.

EXERCISE 2E

1 Use the factor theorem to show that $(3x + 1)$ is a factor of $Q(x) = 18x^2 - 9x - 5$. Hence factorise $Q(x)$.

2 Show that $(2x + 5)$ is a factor of $f(x) = 15 - 14x - 8x^2$.
Hence factorise $f(x)$.

3 Given that $P(x) = 6x^3 - 5x^2 - 2x + 1$, find $P(\frac{1}{2})$ and $P(\frac{1}{3})$.
Hence write down a factor of $P(x)$.
Show that $(2x + 1)$ is also a factor of $P(x)$.

4 Show that $(4x - 3)$ and $(3x - 2)$ are factors of
$$P(x) = 36x^3 - 27x^2 - 16x + 12.$$

Express $P(x)$ as the product of linear factors with integer coefficients.

5 Find the values of the constants a and b for which $(2x + 1)$ and $(3x - 2)$ are factors of the polynomial

$$P(x) = 12x^3 + ax^2 - x + b.$$

Hence find a third factor of $P(x)$.

6 Given that $(3x + 1)$ is a factor of

$$f(x) = 6x^3 + kx^2 + x + 1$$

find the value of the constant k.
Factorise $f(x)$, and hence show that the equation $f(x) = 0$ has only one real root.

7 Show that $(2x + 1)$ is a factor of $P(x) = 2x^3 - 3x^2 - 4x - 1$.
Find the quadratic factor $Q(x)$, where $P(x) = (2x + 1) Q(x)$.
Find the exact values of the roots of the equation $P(x) = 0$.

8 It is given that $f(x) = 4x^3 - 17x^2 + 16x - 3$.

(a) Use the factor theorem to show that $(x - 3)$ and $(4x - 1)$ are factors of $f(x)$.

(b) Express $f(x)$ as a product of three linear factors.   [A]

9 Factorise the following polynomials completely:

(a) $2x^3 + 9x^2 + 7x - 6$

(b) $4x^3 - 8x^2 - 29x - 12$

(c) $6x^3 - 25x^2 - 29x + 20$

(d) $9x^3 - 9x^2 - 22x + 8$

10 Explain why any linear factor of the polynomial

$$p(x) = 2x^3 - 5x^2 + x + 1$$

must be one of the following $(x + 1)$, $(x - 1)$, $(2x + 1)$, $(2x - 1)$.
Hence prove that $p(x)$ has no linear factors with integer coefficients.

2.7 The remainder theorem for divisors of the form $(ax + b)$

In C1 section 6.6 you were shown the remainder theorem for a divisor $(x - a)$. In this section we apply the same principles to find the remainder theorem for a more general divisor $(ax + b)$.

In section 2.5 we noted that dividing a polynomial by a linear expression led to the identity

Polynomial $\equiv$ Divisor $\times$ Quotient $+$ Remainder.

When the divisor is a linear expression (degree 1), the remainder is a constant (degree 0)
so

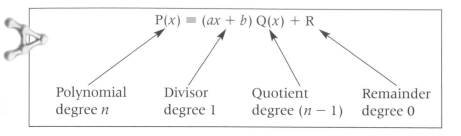

$$P(x) \equiv (ax + b)\, Q(x) + R$$

| Polynomial | Divisor | Quotient | Remainder |
| degree n | degree 1 | degree $(n-1)$ | degree 0 |

Since this identity is true for all values of x, consider the case when $x = -\dfrac{b}{a}$,

Choose the value of x so that $(ax + b) = 0$.

so
$$P\left(-\frac{b}{a}\right) = (-b + b)\, Q\left(-\frac{b}{a}\right) + R$$

$$\Rightarrow \quad P\left(-\frac{b}{a}\right) = 0 + R$$

$$\Rightarrow \quad R = P\left(-\frac{b}{a}\right)$$

This is called the **remainder theorem**.

If a polynomial $P(x)$ is divided by $(ax + b)$, the remainder is $P\left(-\dfrac{b}{a}\right)$.

If the remainder is 0 then $(ax + b)$ is a factor of $P(x)$ so $P\left(-\dfrac{b}{a}\right) = 0 \Leftrightarrow (ax + b)$ is a factor of $P(x)$. This result is called the **factor theorem** and was established in the previous section.

Worked example 2.13

Find the remainder when the polynomial $1 - 3x^2 + 2x^3$ is divided by $(2x - 1)$.

Solution

Let $P(x) = 1 - 3x^2 + 2x^3$.

$$P(x) \equiv (2x - 1)\, Q(x) + R$$

$$P\left(\frac{1}{2}\right) = 0 + R$$

$$R = P\left(\frac{1}{2}\right) = 1 - 3\left(\frac{1}{2}\right)^2 + 2\left(\frac{1}{2}\right)^3 = 1 - \frac{3}{4} + \frac{1}{4} = \frac{1}{2}$$

When $1 - 3x^2 + 2x^3$ is divided by $(2x - 1)$ the remainder is $\dfrac{1}{2}$.

See Worked example 2.9.

Worked example 2.14

When $4x^3 - 3x^2 + kx + 3$ is divided by $2x + 1$ the remainder is $\dfrac{1}{4}$.
Find the value of the constant k.

Solution

Let $P(x) = 4x^3 - 3x^2 + kx + 3$

$P(x) \equiv (2x + 1)\, Q(x) + R$

$$\Rightarrow \quad P\!\left(-\frac{1}{2}\right) = 0 + R$$

$$\Rightarrow \quad R = P\!\left(-\frac{1}{2}\right) = 4\left(-\frac{1}{2}\right)^3 - 3\left(-\frac{1}{2}\right)^2 + k\left(-\frac{1}{2}\right) + 3$$

$$\Rightarrow \quad \frac{1}{4} = -\frac{1}{2} - \frac{3}{4} - \frac{1}{2}k + 3$$

$$\Rightarrow \quad \frac{1}{2}k = 3 - \frac{1}{4} - \frac{5}{4} = \frac{3}{2}$$

$$\Rightarrow \quad k = 3$$

EXERCISE 2F

1 Find the remainder when the polynomial $2x^3 + 8x^2 - 5$ is
divided by:

(a) $2x - 1$

(b) $2x - 3$

(c) $2x + 1$

2 Find the remainder when the polynomial $P(x)$ is divided by
the linear expression $f(x)$:

(a) $P(x) = 8x^3 - 3$ $f(x) = 2x - 1$

(b) $P(x) = -2x^3 + 2x^2 - 3x - 1$ $f(x) = 2x + 1$

(c) $P(x) = 16x^4 - 3x^2 + 1$ $f(x) = 4x - 1$

(d) $P(x) = 5 - 3x + 9x^2 - 9x^3$ $f(x) = 2 - 3x$

(e) $P(x) = 3 - 54x^3$ $f(x) = 1 + 3x$

3 The polynomial $P(x)$ is defined by $P(x) = 4x^3 - 6x^2 + 4x + k$,
where k is a constant. When $P(x)$ is divided by $2x - 1$ the
remainder is 3. Show that $k = 2$.

4 When $4x^3 - kx^2 + 6x + 5$ is divided by $2x + 1$ the remainder
is -7. Find the value of the constant k.

5 The polynomial $P(x) = x^3 + ax^2 - x + 12$ leaves a remainder of 9 when divided by $x - 3$. Find the remainder when $P(x)$ is divided by $3x - 2$.

6 The polynomial $P(x) = x^3 + 3x^2 - 2x + k$ has a factor $x + 1$. Find the remainder when $P(x)$ is divided by $2x + 1$.

7 Given that $P(x)$, where $P(x) = 8x^3 + kx + 4$ and k is a constant, is such that the remainder on dividing $P(x)$ by $(x + 1)$ is twice the remainder on dividing $P(x)$ by $(2x - 1)$, find the value of k.

8 When divided by $(2x - 1)$ the polynomial $P(x) = 8x^3 - ax^2 + bx - 6$ leaves a remainder of 2. When divided by $(x - 2)$, $P(x)$ leaves a remainder of 14.

   (a) Find the values of the constants a and b.

   (b) Show that $(x - 1)$ is **not** a factor of $P(x)$.

9 The polynomial $P(x) = 3x^3 - 11x^2 + 8x + k$ leaves a remainder of -18 when divided by $x + 1$.

   (a) Find the value of the constant k.

   (b) Show that $(3x + 1)$ is a factor of $P(x)$.

   (c) Hence find the roots of the equation $P(x) = 0$.

10 The polynomials $f(x)$ and $g(x)$ are defined by $f(x) = (2x + 3)(px^2 + x + 2)$, $g(x) = 4x^3 + px + 12$, where p is a constant. When $f(x)$ and $g(x)$ are divided by $2x - 1$, the remainder is R in each case. Find the values of p and R.

11 The polynomial $p(x)$ is given by $p(x) = 2x^3 - 3x^2 + 7x - 6$.

   (a) Find the remainder when $p(x)$ is divided by $(x + 3)$.

   (b) (i) Show that $x^2 + 2x - 3$ and $p(x)$ have a common linear factor.

       (ii) Hence write $\dfrac{x^2 + 2x - 3}{p(x)}$ as a simplified algebraic fraction. [A]

Key point summary

1 To simplify rational expressions: *p198*
   - factorise all algebraic expressions
   - cancel any factors that are common to the numerator and denominator.

2 To multiply rational expressions: *p200*
- factorise all algebraic expressions
- write as a single fraction
- cancel any factors that are common to the numerator and denominator.

3 To divide by a rational expression: *p200*
- change the division to a multiplication of the reciprocal
- factorise all algebraic expressions
- write as a single fraction
- cancel any factors that are common to the numerator and denominator.

4 To add/subtract rational expressions: *p201*
- factorise all algebraic expressions
- write each rational expression with the same common denominator
- add/subtract to get a single rational expression
- simplify the numerator
- cancel any factors that are common to the numerator and denominator.

5 Polynomial $\equiv$ divisor $\times$ quotient + remainder   *p203*

6 When a polynomial is divided by a linear expression   *p204*
the remainder will always be a constant and the quotient will always be one degree less than the polynomial.

7 Before starting to use algebraic long division write   *p205*
the polynomial and divisor in descending powers of x and include all powers of x in the polynomial, inserting zero coefficients if necessary.

8 A more general form of the **factor theorem** is   *p206*
$(ax + b)$ is a factor of the polynomial
$$P(x) \Leftrightarrow P\left(-\frac{b}{a}\right) = 0.$$

9 **The remainder theorem**:   *p210*
If a polynomial $P(x)$ is divided by $(ax + b)$, the remainder is $P\left(-\frac{b}{a}\right)$.

2

Test yourself	What to review
1 Simplify $\dfrac{18x - 2x^3}{x^2 + 2x - 15}$.	*Section 2.2*
2 Simplify $\left(\dfrac{2x^2 + x - 1}{x^2 - 1} \times \dfrac{2x^2 - 5x + 3}{2x^2 - 7x + 3}\right) \div \left(\dfrac{6x^2 + x - 2}{3x^2 - 7x - 6}\right)$.	*Section 2.3*
3 Express $\dfrac{4}{x + 5} + \dfrac{1}{x - 2} - \dfrac{7}{x^2 + 3x - 10}$ in its simplest form.	*Section 2.4*
4 The polynomial p(x) is defined by p(x) = $4x^3 - 31x + 15$.	
(a) Use the factor theorem to show that $2x - 1$ is a factor of p(x).	*Section 2.6*
(b) Write p(x) as a product of three linear factors.	*Section 2.5*
5 (a) Find the remainder when the polynomial $6 + x - 9x^3$ is divided by $(3x + 5)$.	*Section 2.7*
(b) Hence, or otherwise, write $\dfrac{6 + x - 9x^3}{3x + 5}$ as $ax^2 + bx + c + \dfrac{d}{3x + 5}$, where a, b, c and d are integers to be found.	*Section 2.5*

Test yourself ANSWERS

5 (a) 46;   **(b)** $a = -3$, $b = 5$, $c = -8$, $d = 46$.

4 (b) $(2x - 1)(x + 3)(2x - 5)$.

3 $\dfrac{5}{x + 5}$.

2 $\dfrac{2x - 3}{2x - 1}$.

1 $\dfrac{-2x(x + 3)}{x + 5}$.

CHAPTER 3

C4: Partial fractions and applications

3

Learning objectives

After studying this chapter, you should be able to:
- split a rational expression into its partial fractions
- use partial fractions to write rational functions as a series expansion
- use partial fractions to find and evaluate integrals.

3.1 Partial fractions

In section 2.4 you were shown how to add and subtract rational expressions to produce a single fraction. Sometimes it is more useful to split a single fraction into the sums or differences of other rational expressions. This process is called **writing a rational expression in terms of its partial fractions**.

> Only proper fractions can be expressed in terms of partial fractions. If the given rational expression is improper you must first carry out a long division to obtain a proper fraction.

There are two types of rational expressions for which you may be expected to find the corresponding partial fractions.

> A proper fraction which has up to three linear factors in the denominator,
> $$\frac{p(x)}{(x-a)(x-b)(x-c)},$$ has three partial fractions of
> the form $\dfrac{A}{(x-a)} + \dfrac{B}{(x-b)} + \dfrac{C}{(x-c)}$.

> A proper fraction which has a repeated linear factor in the denominator,
> $$\frac{q(x)}{(x-a)^3},$$ has partial fractions of the form
> $$\frac{A}{(x-a)} + \frac{B}{(x-a)^2} + \frac{C}{(x-a)^3}.$$

$$\frac{2}{2x+1} - \frac{1}{x+2} \equiv \frac{3}{(2x+1)(x+2)}$$

Reversing the process we would say 'Express $\dfrac{3}{(2x+1)(x+2)}$ in partial fractions'.

You used algebraic long division in section 2.5.

In the examination for this module you will not be asked to write a rational expression that has a quadratic factor (which cannot be factorised) in the denominator, for example,
$$\frac{2}{(x-1)(x^2+1)}.$$

See Q3, Ex2C

In each case A, B and C are constants which need to be determined.

Worked example 3.1

Express $\dfrac{x + 4}{(x + 1)(x - 2)}$ in partial fractions.

Solution

$\dfrac{x + 4}{(x + 1)(x - 2)} \equiv \dfrac{A}{x + 1} + \dfrac{B}{x - 2}$, where A and B are constants.

Only linear factors in the denominator.

$\dfrac{x + 4}{(x + 1)(x - 2)} \equiv \dfrac{A(x - 2) + B(x + 1)}{(x + 1)(x - 2)}$

Writing both sides with the same denominators.

Since the denominators are the same we can equate the numerators.

$x + 4 \equiv A(x - 2) + B(x + 1)$

Let $x = 2 \Rightarrow 6 = 3B \Rightarrow B = 2$

Let $x = -1 \Rightarrow 3 = -3A \Rightarrow A = -1$

So $\dfrac{x + 4}{(x + 1)(x - 2)} \equiv \dfrac{-1}{x + 1} + \dfrac{2}{x - 2}$.

When the denominator has only linear factors the **cover up** method can be used to find A and B. For example, the denominator for B is $x - 2$. The value of x for this to be zero is 2. **Cover up** the $(x - 2)$ factor in the denominator of

$\dfrac{x + 4}{(x + 1)(x - 2)}$ and evaluate

$\dfrac{x + 4}{(x + 1)}$ when $x = 2$ to find B.

This gives $B = \dfrac{6}{3}$. Similarly

$A = \dfrac{3}{-3}$.

In past examination questions partial fractions have often been tested in the first part and the result is then required to, for example, evaluate an integral.

Because of this it is well worthwhile spending time checking your values for the constants. In the above example this could be done by comparing coefficients of x: $\Rightarrow 1 = A + B$.
Since $1 = -1 + 2$, you can be confident that the values are correct to use in the next part of the question.

You will be shown how to evaluate integrals using partial fractions later in this chapter.

In some questions you will be told the form of the partial fractions. The next worked example is such a question.

Worked example 3.2

Express $\dfrac{1}{x^2(x - 1)}$ in the form $\dfrac{A}{x} + \dfrac{B}{x^2} + \dfrac{C}{x - 1}$, and state the values of the constants A, B and C.

Note the denominator has a repeated linear factor, x, and a single linear factor $x - 1$, hence the form of the three partial fractions.

Solution

$\dfrac{1}{x^2(x - 1)} \equiv \dfrac{A}{x} + \dfrac{B}{x^2} + \dfrac{C}{x - 1}$

$\dfrac{1}{x^2(x - 1)} \equiv \dfrac{Ax(x - 1) + B(x - 1) + Cx^2}{x^2(x - 1)}$

Writing both sides with the same denominators. Be careful with repeated factors. A **common error** is to write the numerator as $Ax^2(x - 1) + Bx(x - 1) + Cx^3$.

Since the denominators are the same we can equate the numerators.

$$1 \equiv Ax(x - 1) + B(x - 1) + Cx^2$$

Let $x = 1 \Rightarrow 1 = C \Rightarrow C = 1$

Let $x = 0 \Rightarrow 1 = -B \Rightarrow B = -1$

Equate coefficients of $x^2 \Rightarrow 0 = A + C \Rightarrow A = -1$

So $\dfrac{1}{x^2(x - 1)} \equiv \dfrac{-1}{x} + \dfrac{-1}{x^2} + \dfrac{1}{x - 1}.$

To check you must use a method that involves all the constants:
Let $x = 2$
$\Rightarrow 1 = 2A + B + 4C$
$1 = -2 - 1 + 4$;
this is true.

The next worked example considers the case when the denominator of the given rational expression is not factorised.

Worked example 3.3

Express $\dfrac{x - 2}{x^2 - 1}$ as a sum of two partial fractions.

Note the denominator is not factorised.

Solution

The first step is to fully factorise the denominator:
$x^2 - 1 \equiv (x + 1)(x - 1)$

$$\frac{x - 2}{x^2 - 1} \equiv \frac{x - 2}{(x + 1)(x - 1)} \equiv \frac{A}{x + 1} + \frac{B}{x - 1}$$

$$\frac{x - 2}{(x + 1)(x - 1)} \equiv \frac{A(x - 1) + B(x + 1)}{(x + 1)(x - 1)}$$

The denominator has two single linear factors.

Since the denominators are the same we can equate the numerators:

$$x - 2 \equiv A(x - 1) + B(x + 1)$$

Let $x = 1 \Rightarrow -1 = 2B \Rightarrow B = \dfrac{-1}{2}$

Let $x = -1 \Rightarrow -3 = -2A \Rightarrow A = \dfrac{3}{2}$

Check: Let $x = 0$:
$-2 = A(-1) + B(1)$
$-2 = -1.5 - 0.5$ which is true.

So $\dfrac{x - 2}{x^2 - 1} \equiv \dfrac{\frac{3}{2}}{x + 1} + \dfrac{-\frac{1}{2}}{x - 1}$,

which can be written as

$$\frac{x - 2}{x^2 - 1} \equiv \frac{1}{2}\left(\frac{3}{x + 1} - \frac{1}{x - 1}\right).$$

The next Worked example looks at a rational expression that is improper. As we have stated earlier, before starting to form partial fractions the rational expression must be written as a proper fraction. The method of algebraic long division is extended to division by non-linear expressions to cope with this case.

Worked example 3.4

Express $\dfrac{x^2 - x + 1}{(x - 2)(x + 1)}$ in partial fractions.

> Since the degrees of the numerator and denominator are the same, both 2, the fraction is improper so we need to divide out.

Solution

The first step involves a long division

$$x^2 - x - 2 \overline{\smash{\big)}\,x^2 - x + 1} \quad \overset{\displaystyle 1}{}$$
$$\underline{x^2 - x - 2}$$
$$3$$

> $(x - 2)(x + 1) = x^2 - x - 2$

$$\frac{x^2 - x + 1}{(x - 2)(x + 1)} \equiv 1 + \frac{3}{x^2 - x - 2} \equiv 1 + \frac{3}{(x - 2)(x + 1)}$$

Consider $\dfrac{3}{(x - 2)(x + 1)} \equiv \dfrac{A}{x - 2} + \dfrac{B}{x + 1}$

$\Rightarrow 3 \equiv A(x + 1) + B(x - 2)$

Let $x = 2 \Rightarrow A = 1$

Let $x = -1 \Rightarrow B = -1$

So $\dfrac{x^2 - x + 1}{(x - 2)(x + 1)} \equiv 1 + \dfrac{1}{x - 2} + \dfrac{-1}{x + 1}$

or $\dfrac{x^2 - x + 1}{(x - 2)(x + 1)} \equiv 1 + \dfrac{1}{x - 2} - \dfrac{1}{x + 1}$.

EXERCISE 3A

1 Express $\dfrac{x - 1}{(x + 2)(x + 1)}$ in the form $\dfrac{A}{x + 2} + \dfrac{B}{x + 1}$, and state the values of the constants A and B.

2 Express $\dfrac{3x + 7}{(x - 2)(x - 1)}$ in the form $\dfrac{A}{x - 2} + \dfrac{B}{x - 1}$, and state the values of the constants A and B.

3 Express $\dfrac{x}{(1 - x)(2 + x)}$ in the form $\dfrac{A}{1 - x} + \dfrac{B}{2 + x}$, and state the values of the constants A and B.

4 Express $\dfrac{4}{x^2 - 4}$ in the form $\dfrac{A}{x + 2} + \dfrac{B}{x - 2}$, and state the values of the constants A and B.

5 Express $\dfrac{2}{(x-1)^2(x+1)}$ in the form $\dfrac{A}{(x-1)^2} + \dfrac{B}{x-1} + \dfrac{C}{x+1}$
   and state the values of the constants A, B and C.

6 Express $\dfrac{x^2}{x^2-1}$ in the form $A + \dfrac{B}{x-1} + \dfrac{C}{x+1}$, and
   state the values of the constants A, B and C.

7 Express $\dfrac{3x}{(x+2)(x+1)}$ in partial fractions.

8 Express $\dfrac{3x-5}{x^2-1}$ in partial fractions.

9 Express $\dfrac{x^2+1}{x(x+1)^2}$ in partial fractions.

10 Express $\dfrac{18}{(x+2)(x-1)^2}$ in partial fractions.

11 Express $\dfrac{x^2}{4-x^2}$ in partial fractions.

12 Express $\dfrac{3x}{(x+2)(x+1)^2}$ in partial fractions.

13 Express $\dfrac{x^2+2}{(x-1)^3}$ in partial fractions.

14 Express $\dfrac{x+4}{(x+3)(x+2)(x+1)}$ in partial fractions.

3.2 Integration using partial fractions

In this section you will apply your knowledge of partial fractions
to integration.

> To solve integration problems using partial fractions, the
> following two integrals, where a and b are constants, are
> often needed:
>
> $$\int \frac{1}{ax+b}\,dx = \frac{1}{a}\ln|ax+b| + c$$
>
> $$\int \frac{1}{(ax+b)^2}\,dx = -\frac{1}{a(ax+b)} + c$$

$\ln p + \ln q = \ln(pq).$

$\ln p - \ln q = \ln\left(\dfrac{p}{q}\right).$

You may also need to apply the laws of logarithms to obtain
printed answers in examination questions.

$\ln p^n = n\ln p.$

Worked example 3.5

Show that $\displaystyle\int_1^2 \left(\frac{2x+2}{2x+1}\right) dx = 1 + \frac{1}{2} \ln \frac{5}{3}$.

Solution

Firstly, write $\dfrac{2x+2}{2x+1}$ as a proper fraction.

> See section 3.1.

$$\frac{2x+2}{2x+1} = \frac{2x+1+1}{2x+1} = \frac{2x+1}{2x+1} + \frac{1}{2x+1} = 1 + \frac{1}{2x+1}$$

So $\displaystyle\int_1^2 \left(\frac{2x+2}{2x+1}\right) dx = \int_1^2 1 + \frac{1}{2x+1}\, dx$

> $\displaystyle\int\left(\frac{1}{2x+1}\right) dx = \frac{1}{2}\int\left(\frac{2}{2x+1}\right) dx$

$$= \left[x + \frac{1}{2} \ln |2x+1| \right]_1^2$$

$$= \left(2 + \frac{1}{2} \ln 5 \right) - \left(1 + \frac{1}{2} \ln 3 \right)$$

$$= 1 + \frac{1}{2}(\ln 5 - \ln 3)$$

$$= 1 + \frac{1}{2} \ln \frac{5}{3}.$$

Worked example 3.6

By writing $\dfrac{x+4}{(x+1)(x-2)}$ in partial fractions, find the value of

$\displaystyle\int_3^4 \frac{x+4}{(x+1)(x-2)}\, dx$, leaving your answer in the form

$p \ln 4 - q \ln 5$, where p and q are integers to be found.

Solution

$\dfrac{x+4}{(x+1)(x-2)} \equiv \dfrac{A}{x+1} + \dfrac{B}{x-2}$, where A and B are constants.

Applying the cover-up rule gives $A = \dfrac{-1+4}{-1-2} = -1$

> See Worked example 3.1.

$$\text{and } B = \frac{2+4}{2+1} = 2$$

$$\int_3^4 \frac{x+4}{(x+1)(x-2)}\, dx = \int_3^4 \frac{-1}{x+1} + \frac{2}{x-2}\, dx$$

$$= \left[-\ln|x+1| + 2\ln|x-2| \right]_3^4$$

$$= (-\ln 5 + 2\ln 2) - (-\ln 4 + 2\ln 1)$$

$$= -\ln 5 + \ln 2^2 + \ln 4 - 0$$

$$= 2\ln 4 - \ln 5$$

Worked examination question

(a) Express $\dfrac{2x}{(3 + x)^2}$ in the form $\dfrac{A}{3 + x} + \dfrac{B}{(3 + x)^2}$, where A and B are constants to be determined.

(b) Show that $\displaystyle\int_0^3 \dfrac{2x}{(3 + x)^2}\,dx = p + \ln q$, where p and q are integers to be determined. [A]

Solution

(a) $\dfrac{2x}{(3 + x)^2} \equiv \dfrac{A}{3 + x} + \dfrac{B}{(3 + x)^2} \equiv \dfrac{A(3 + x) + B}{(3 + x)^2}$

$\Rightarrow 2x \equiv A(3 + x) + B$

Comparing coefficients of x gives $2 = A$.

Putting $x = -3$ gives $-6 = B$.

$\dfrac{2x}{(3 + x)^2} \equiv \dfrac{2}{3 + x} - \dfrac{6}{(3 + x)^2}$

Check: Put $x = 0$,
$0 = 3A + B$
$0 = 6 - 6$ is true.

(b) $\displaystyle\int_0^3 \dfrac{2x}{(3 + x)^2}\,dx = \int_0^3 \dfrac{2}{3 + x}\,dx - \int_0^3 \dfrac{6}{(3 + x)^2}\,dx,$

$= \left[2 \ln|3 + x| - \dfrac{6(3 + x)^{-1}}{-1}\right]_0^3$

$= \left[2 \ln|3 + x| + \dfrac{6}{3 + x}\right]_0^3$

$= 2 \ln 6 + 1 - (2 \ln 3 + 2)$

$= -1 + 2 \ln \dfrac{6}{3}$

$= -1 + 2 \ln 2$

$\displaystyle\int_0^3 \dfrac{2x}{(3 + x)^2}\,dx = -1 + \ln 4, \ (p = -1, q = 4)$

EXERCISE 3B

1 Find:

(a) $\displaystyle\int_0^1 \left(\dfrac{x + 2}{x + 1}\right) dx$ (b) $\displaystyle\int_{\frac{1}{2}}^1 \left(\dfrac{2x - 3}{2x + 1}\right) dx$ (c) $\displaystyle\int_1^2 \left(\dfrac{x^2 + 2x}{x + 1}\right) dx$

2 Find:

(a) $\displaystyle\int_1^2 \dfrac{2}{x(x + 1)}\,dx$ (b) $\displaystyle\int_2^3 \dfrac{4}{(x - 1)(x + 1)}\,dx$

(c) $\displaystyle\int_3^4 \dfrac{1}{(x - 1)(x - 2)}\,dx$ (d) $\displaystyle\int_2^3 \dfrac{5}{(2x - 3)(x + 1)}\,dx$

(e) $\displaystyle\int_1^2 \dfrac{18}{x^2(x + 3)}\,dx$ (f) $\displaystyle\int_2^3 \dfrac{1}{(x - 1)(x + 1)^2}\,dx$

3 (a) Express $\dfrac{2 - 5x^2}{(2 - x)(1 + x)^2}$ in terms of three partial fractions.

  (b) Hence show that $\displaystyle\int_0^1 \dfrac{2 - 5x^2}{(2 - x)(1 + x)^2}\,dx = \ln 2 - \dfrac{1}{2}$.

4 Find:

  (a) $\displaystyle\int \dfrac{x + 1}{x(2x + 1)}\,dx$ **(b)** $\displaystyle\int \dfrac{x(2x + 1)}{x + 1}\,dx$

5 (a) Express $\dfrac{2x^2 + 3x + 12}{(2x - 1)(x + 3)}$ in the form $A + \dfrac{B}{2x - 1} + \dfrac{C}{x + 3}$,

  where A, B and C are constants to be determined.

  (b) Hence find $\displaystyle\int_1^5 \dfrac{2x^2 + 3x + 12}{(2x - 1)(x + 3)}\,dx$, giving your answer in

  the form $p + q \ln 2 + r \ln 3$, where p, q and r are integers
  to be found.

6 A curve C has equation $y = \dfrac{2x - 1}{(x - 2)(5 - x)}$.

  (a) Express $\dfrac{2x - 1}{(x - 2)(5 - x)}$ in partial fractions.

  (b) Find the area bounded by the curve C, the x-axis and the
  lines $x = 3$ and $x = 4$.

7 (a) Express $\dfrac{2x}{(3 + x)^2}$ in the form $\dfrac{A}{3 + x} + \dfrac{B}{(3 + x)^2}$, and state

  the value of the constants A and B.

  (b) Show that $\displaystyle\int_0^3 \dfrac{2x}{(3 + x)^2}\,dx = p + \ln q$, where p and q are

  constants to be determined.   [A]

3.3 Series expansions using partial fractions

In section 1.2 you were shown the binomial expansion for real n.

$$(1 + x)^n = 1 + nx + \frac{n(n - 1)}{1.2}x^2 + \ldots + \frac{n(n - 1)\ldots(n - r + 1)}{1.2\ldots r}x^r + \ldots,$$

which is valid for $|x| < 1$. In this section you will apply your
knowledge of partial fractions and the binomial expansion to
find series expansions for rational functions.

> You will frequently need to use the following two applications
> (for $n = -1$ and $n = -2$) of the binomial expansion:
>
> $(1 + y)^{-1} = 1 - y + y^2 - y^3 + y^4 - \ldots$ valid for $|y| < 1$
>
> $(1 + y)^{-2} = 1 - 2y + 3y^2 - 4y^3 + \ldots$ valid for $|y| < 1$.

Recall the sum to infinity of the
geometric series.

Worked example 3.7

Find the binomial expansion of $\dfrac{1}{x+2}$ in ascending powers of x up to and including the term in x^3 and state the values of x for which the series is valid.

Solution

$$\frac{1}{x+2} = \frac{1}{2+x} = \frac{1}{2\left(1+\dfrac{x}{2}\right)} = \frac{1}{2} \times \left(1 + \frac{x}{2}\right)^{-1}$$

Firstly write it in the form $k(1+y)^{-1}$.

Using the expansion of $(1+y)^{-1}$ with $y = \dfrac{x}{2}$, you get

$$\frac{1}{x+2} = \frac{1}{2} \times \left[1 - \frac{x}{2} + \left(\frac{x}{2}\right)^2 - \left(\frac{x}{2}\right)^3 + \ldots\right]$$

$$= \frac{1}{2} - \frac{x}{4} + \frac{x^2}{8} - \frac{x^3}{16} + \ldots \text{ valid for } \left|\frac{x}{2}\right| < 1 \Rightarrow |x| < 2.$$

A common error is not to square or cube the 2. Using brackets carefully usually avoids this error.

3

Worked example 3.8

Find the binomial expansion of $\dfrac{4}{(2+3x)^2}$ in ascending powers of x up to and including the term in x^3 and state the values of x for which the series is valid.

Solution

$$\frac{4}{(2+3x)^2} = \frac{4}{\left[2\left(1+\dfrac{3x}{2}\right)\right]^2} = \frac{4}{4\left(1+\dfrac{3x}{2}\right)^2} = \frac{1}{\left(1+\dfrac{3x}{2}\right)^2} = \left(1 + \frac{3x}{2}\right)^{-2}$$

Firstly write it in the form $k(1+y)^{-2}$.

Using the expansion of $(1+y)^{-2}$ with $y = \dfrac{3x}{2}$, you get

$$\frac{4}{(2+3x)^2} = \left[1 - 2\left(\frac{3x}{2}\right) + 3\left(\frac{3x}{2}\right)^2 - 4\left(\frac{3x}{2}\right)^3 + \ldots\right]$$

$$= 1 - 3x + \frac{27x^2}{4} - \frac{27x^3}{2} + \ldots \text{ valid for}$$

$$\left|\frac{3x}{2}\right| < 1 \Rightarrow |3x| < 2 \Rightarrow |x| < \frac{2}{3}.$$

A common error is not to square or cube the $\dfrac{3}{2}$. Using brackets carefully usually avoids this error.

The next worked example uses partial fractions.

Worked example 3.9

By writing $\dfrac{3}{(1+x)(2-x)}$ in partial fractions, find the binomial expansion of $\dfrac{3}{(1+x)(2-x)}$ in ascending powers of x up to and including the term in x^3 and state the values of x for which the series is valid.

Solution

$$\frac{3}{(1+x)(2-x)} \equiv \frac{A}{1+x} + \frac{B}{2-x}$$

$$\Rightarrow \quad 3 \equiv A(2-x) + B(1+x)$$

Putting $x = 2$ gives $3 = 3B \Rightarrow B = 1$.

Putting $x = -1$ gives $3 = 3A \Rightarrow A = 1$.

> Check: coefficients of x
> $\Rightarrow 0 = -A + B$ which is true
> when $A = 1$ and $B = 1$.

$$\frac{3}{(1+x)(2-x)} = \frac{1}{1+x} + \frac{1}{2-x} = \frac{1}{1+x} + \frac{1}{2\left(1 - \dfrac{x}{2}\right)}$$

$$= (1+x)^{-1} + \frac{1}{2}\left(1 - \frac{x}{2}\right)^{-1}$$

$$= (1 - x + x^2 - x^3 + \ldots)$$

$$+ \frac{1}{2} \times \left[1 - \left(-\frac{x}{2}\right) + \left(-\frac{x}{2}\right)^2 - \left(-\frac{x}{2}\right)^3 + \ldots\right]$$

$$= (1 - x + x^2 - x^3 + \ldots)$$

$$+ \frac{1}{2} \times \left(1 + \frac{x}{2} + \frac{x^2}{4} + \frac{x^3}{8} + \ldots\right)$$

$$= \frac{3}{2} - \frac{3}{4}x + \frac{9}{8}x^2 - \frac{15}{16}x^3 + \ldots$$

The expansion for $(1+x)^{-1}$ is valid for $|x| < 1$.

The expansion for $\left(1 - \dfrac{x}{2}\right)^{-1}$ is valid for $\left|-\dfrac{x}{2}\right| < 1 \Rightarrow |x| < 2$.

Since both expansions have been used the final series is only valid for those values of x which satisfy the stricter of the two inequalities $|x| < 1$ and $|x| < 2$, that is $|x| < 1$.
The full expansion is valid for $|x| < 1$ so

> A useful partial check is to check that both sides are the same when $x = 0$; here both sides
> equal $\dfrac{3}{2}$.

$$\frac{3}{(1+x)(2-x)} = \frac{3}{2} - \frac{3}{4}x + \frac{9}{8}x^2 - \frac{15}{16}x^3 \ldots, \text{ valid for } |x| < 1.$$

The next worked example has a repeated linear factor in the denominator.

Worked example 3.10

By writing $\dfrac{4}{(1-x)^2(1+x)}$ in partial fractions, find the binomial

expansion of $\dfrac{4}{(1-x)^2(1+x)}$ in ascending powers of x up to and

including the term in x^2.

Solution

$$\frac{4}{(1-x)^2(1+x)} \equiv \frac{A}{1-x} + \frac{B}{(1-x)^2} + \frac{C}{1+x}$$

$$\equiv \frac{A(1-x)(1+x) + B(1+x) + C(1-x)^2}{(1-x)^2(1+x)}$$

$$\Rightarrow 4 \equiv A(1-x)(1+x) + B(1+x) + C(1-x)^2$$

Putting $x = 1$ gives $4 = 2B \Rightarrow B = 2$.

Putting $x = -1$ gives $4 = 4C \Rightarrow C = 1$.

Coefficients of x^2 gives $0 = -A + C \Rightarrow A = 1$.

Check: putting $x = 0$
$\Rightarrow 4 = A + B + C$ which is true
when $A = 1$, $B = 2$ and $C = 1$.

3

$$\frac{4}{(1-x)^2(1+x)} = \frac{1}{1-x} + \frac{2}{(1-x)^2} + \frac{1}{1+x}$$

$$\frac{4}{(1-x)^2(1+x)} = (1-x)^{-1} + 2(1-x)^{-2} + (1+x)^{-1}$$

$$= \left[1 - (-x) + (-x)^2 - \ldots\right] + 2\left[1 - 2(-x) + 3(-x)^2 - \ldots\right] + \left[1 - x + x^2 - \ldots\right]$$

$$= (1 + x + x^2 + \ldots) + 2(1 + 2x + 3x^2 + \ldots) + (1 - x + x^2 - \ldots)$$

$$= 4 + 4x + 8x^2 + \ldots$$

$$\frac{4}{(1-x)^2(1+x)} = 4 + 4x + 8x^2 + \ldots .$$

Check that both sides are the
same when $x = 0$; here both
sides equal 4.

EXERCISE 3C

1 (a) Write down the binomial expansion in ascending powers of x up to and including the term in x^3 for:

  (i) $(1 + x)^{-1}$ **(ii)** $(1 + 2x)^{-1}$

(b) Express $\dfrac{1}{(1 + 2x)(1 + x)}$ in the form $\dfrac{A}{1 + 2x} + \dfrac{B}{1 + x}$, and

  state the values of the constants A and B.

(c) Hence find the binomial expansion of $\dfrac{1}{(1 + 2x)(1 + x)}$ in

  ascending powers of x up to and including the term in x^3
  and state the values of x for which the full series is valid.

2 (a) Write down the binomial expansion in ascending powers of x up to and including the term in x^2 for:

  (i) $(1 + x)^{-2}$ **(ii)** $(2 + x)^{-1}$

(b) Express $\dfrac{1}{(1 + x)^2(2 + x)}$ in the form $\dfrac{A}{1 + x} + \dfrac{B}{(1 + x)^2} + \dfrac{C}{2 + x}$,

  and state the values of the constants A, B and C.

(c) Hence find the binomial expansion of $\dfrac{1}{(1 + x)^2(2 + x)}$ in

  ascending powers of x up to and including the term in x^2
  and state the values of x for which the full series is valid.

3 (a) Write down the binomial expansion in ascending powers of x up to and including the term in x^3 for:
  (i) $(1-x)^{-1}$ **(ii)** $(1-2x)^{-1}$

(b) Hence find the binomial expansion of $\dfrac{3x}{(1-x)(1-2x)}$ in ascending powers of x up to and including the term in x^3.

4 (a) Express $\dfrac{1-x}{(1+2x)^2(1+x)}$ in the form
  $$\dfrac{A}{1+2x}+\dfrac{B}{(1+2x)^2}+\dfrac{C}{1+x},$$
  and state the values of the constants A, B and C.

(b) Hence find the binomial expansion of $\dfrac{1-x}{(1+2x)^2(1+x)}$ in ascending powers of x up to and including the term in x^2 and state the values of x for which the full series is valid.

5 Find the binomial expansion of $\dfrac{3}{(1-x)(1+2x)}$ in ascending powers of x up to and including the term in x^3. State the values of x for which the full series is valid.

6 Find the binomial expansion of $\dfrac{2x+3}{(1+x)(2+x)}$ in ascending powers of x up to and including the term in x^3. State the values of x for which the full series is valid.

7 Find the binomial expansion of $\dfrac{1-4x}{(1-2x)(1-3x)}$ in ascending powers of x up to and including the term in x^3. State the values of x for which the full series is valid.

8 Find the binomial expansion of $\dfrac{x}{(2+x)(1+2x)^2}$ in ascending powers of x up to and including the term in x^3. State the values of x for which the full series is valid.

9 Find the binomial expansion of $\dfrac{9(1+x^2)}{(3+x)(1+x)(1-x)}$ in ascending powers of x up to and including the term in x^2. State the values of x for which the full series is valid.

10 Find the binomial expansion of $\dfrac{3+2x^2}{(2x+1)(x-3)^2}$ in ascending powers of x up to and including the term in x^2. State the values of x for which the full series is valid.

MIXED EXERCISE

1 (a) Express $\dfrac{4-x}{(1-x)(2+x)}$ in the form $\dfrac{A}{1-x}+\dfrac{B}{2+x}$.

  (b) (i) Show that the first **three** terms in the expansion of $\dfrac{1}{2+x}$ in ascending powers of x are $\dfrac{1}{2}-\dfrac{x}{4}+\dfrac{x^2}{8}$.

    (ii) Obtain also the first **three** terms in the expansion of $\dfrac{1}{1-x}$ in ascending powers of x.

  (c) Hence, or otherwise, obtain the first **three** terms in the expansion of $\dfrac{4-x}{(1-x)(2+x)}$ in ascending powers of x. [A]

2 (a) Given that $\dfrac{25x+1}{(2x-1)(x+1)^2}\equiv\dfrac{A}{2x-1}+\dfrac{B}{x+1}+\dfrac{C}{(x+1)^2}$,

  (i) show that
$$25x+1\equiv A(x+1)^2+B(x+1)(2x-1)+C(2x-1),$$
  (ii) find the values of A, B and C.

  (b) Hence, find $\displaystyle\int_1^2\dfrac{25x+1}{(2x-1)(x+1)^2}\,dx$ leaving your answer in the form $p+q\ln 2$. [A]

3 (a) Express $\dfrac{x+13}{(x-2)(x+3)}$ in partial fractions.

  (b) A curve has equation $y=\dfrac{x+13}{(x-2)(x+3)}$.

    (i) Find the value of $\dfrac{dy}{dx}$ when $x=-1$.

    (ii) Hence find the equation of the tangent to the curve at the point where $x=-1$. [A]

4 (a) Express $\dfrac{1}{(1-x)(2-x)}$ as the sum of two partial fractions.

  (b) Hence expand $\dfrac{1}{(1-x)(2-x)}$ in a series of ascending powers of x up to and including the term in x^2.

  (c) State the values of x for which the full expansion in part **(b)** is valid. [A]

5 (a) Express $\dfrac{5x-2}{(x-1)(x+2)}$ in partial fractions.

  (b) Hence find the value of $\displaystyle\int_2^3\dfrac{5x-2}{(x-1)(x+2)}\,dx$, leaving your answer in the form $p\ln 5+q\ln 2$, where p and q are integers to be found. [A]

6 (a) Express $f(x)=\dfrac{10+19x+6x^2}{(2-x)(1+x)^2}$ as the sum of three partial fractions in the form $\dfrac{A}{2-x}+\dfrac{B}{1+x}+\dfrac{C}{(1+x)^2}$.

(b) Hence expand f(x) in ascending powers of x up to and including the term in x^3. [A]

7 (a) Express $\dfrac{2x^2 - x + 11}{(2x - 3)(x + 2)}$ in the form $A + \dfrac{B}{2x - 3} + \dfrac{C}{x + 2}$.

(b) Hence find $\displaystyle\int_2^6 \dfrac{2x^2 - x + 11}{(2x - 3)(x + 2)}\,dx$, giving your answer in the form $p + q \ln 2 + r \ln 3$, where p, q and r are integers. [A]

8 The function f is given by $f(x) = \dfrac{9}{(1 + 2x)(4 - x)}$.

(a) Express f(x) in partial fractions.

(b) (i) Show that the first three terms in the expansion of $\dfrac{1}{4 - x}$ in ascending powers of x are $\dfrac{1}{4} + \dfrac{x}{16} + \dfrac{x^2}{64}$.

(ii) Obtain a similar expansion for $\dfrac{1}{1 + 2x}$.

(iii) Hence, or otherwise, obtain the first three terms in the expansion of f(x) in ascending powers of x.

(iv) Find the range of values of x for which the expansion of f(x) in ascending powers of x is valid.

(c) (i) Find $\displaystyle\int f(x)\,dx$.

(ii) Hence find, to two significant figures, the error in using the expansion of f(x) up to and including the term in x^2 to evaluate $\displaystyle\int_0^{0.25} f(x)\,dx$. [A]

Key point summary

1 Only proper fractions can be expressed in terms of partial fractions. If the given rational expression is improper you must first carry out a long division to obtain a proper fraction.   *p215*

2 A proper fraction which has up to three linear factors   *p215* in the denominator, $\dfrac{p(x)}{(x - a)(x - b)(x - c)}$, has three partial fractions of the form $\dfrac{A}{(x - a)} + \dfrac{B}{(x - b)} + \dfrac{C}{(x - c)}$.

3 A proper fraction which has a repeated linear factor in *p215* the denominator, $\dfrac{q(x)}{(x - a)^3}$, has partial fractions of the form $\dfrac{A}{(x - a)} + \dfrac{B}{(x - a)^2} + \dfrac{C}{(x - a)^3}$.

4 To solve integration problems using partial fractions, *p219*
the following two integrals, where a and b are constants,
are often needed:

$$\int \frac{1}{ax + b}\, dx = \frac{1}{a} \ln |ax + b| + c$$

$$\int \frac{1}{(ax + b)^2}\, dx = -\frac{1}{a(ax + b)} + c$$

5 You will frequently need to use the following two *p222*
applications (for $n = -1$ and $n = -2$) of the binomial
expansion:

$$(1 + y)^{-1} = 1 - y + y^2 - y^3 + y^4 - \ldots \text{ valid for } |y| < 1$$

$$(1 + y)^{-2} = 1 - 2y + 3y^2 - 4y^3 + \ldots \text{ valid for } |y| < 1.$$

3

Test yourself	**What to review**
1 Express $\dfrac{18x - 10x^2}{(1 - x)^2(3 - x)}$ as the sum of three partial fractions.	*Section 3.1*
2 Show that $\displaystyle\int_1^2 \frac{1}{x(x + 1)}\, dx = \ln \frac{4}{3}$.	*Section 3.2*
3 (a) Express $\dfrac{x^2}{(x + 3)(x - 1)}$ in the form $A + \dfrac{B}{x + 3} + \dfrac{C}{x - 1}$	*Section 3.1*
(b) Hence show that: $\displaystyle\int_2^3 \frac{x^2}{(x + 3)(x - 1)}\, dx = 1 + \frac{1}{4} \ln 2 - \frac{9}{4} \ln \left(\frac{6}{5}\right).$	*Section 3.2*
4 Find the binomial expansion of $\dfrac{1}{(1 - x)(1 + 2x)}$ in ascending powers of x up to and including the term in x^3. State the values of x for which the full series is valid.	*Section 3.3*
5 Find the binomial expansion of $\dfrac{13 + 54x}{(1 + 5x)^2(2 - x)}$ in ascending powers of x up to and including the term in x^2. State the values of x for which the full series is valid.	*Section 3.3*

Test yourself ANSWERS

1 $\dfrac{4}{(1 - x)^2} - \dfrac{1}{(1 - x)} - \dfrac{6}{(3 - x)}$

2 $1 - x + 3x^2 - 5x^3, \; |x| > \dfrac{1}{2}.$

3 $A = 1, \; B = -\dfrac{9}{4}, \; C = \dfrac{4}{4}.$

5 $6\dfrac{2}{1} - 34\dfrac{3}{4}x + 200\dfrac{8}{1}x^2, \; |x| > \dfrac{1}{5}.$

CHAPTER 4

C4: Implicit differentiation and applications

Learning objectives

After studying this chapter, you should be able to:
- differentiate functions defined implicitly
- find equations of tangents and normals to curves specified implicitly

4.1 Implicit functions

When a curve is defined by an equation of the form $y = f(x)$ we say that y is an **explicit function** of x.

Most of the mathematics you have studied so far involves explicit functions; you have been able to rearrange the equation of a curve into the form $y = f(x)$.

Sometimes this is not the case, for example it is not possible to rearrange $y^2 + x \ln y = x + \sin y$ to write it in the form $y = f(x)$. In such cases we say that y is an **implicit function** of x.

> Since $xy + y = 3$ can be rearranged in the form $y = \dfrac{3}{x + 1}$, y is an explicit function of x.

> $\sin yx = y + \ln x$ and $xy + y^3 = e^x$ are two other examples of y as an implicit function of x.

> y is an **implicit function** of x if y cannot be written in the form $y = f(x)$.

4.2 Differentiation of functions defined implicitly

When a curve is defined explicitly by an equation of the form $y = f(x)$ you can find $\dfrac{dy}{dx}$ by applying your earlier knowledge.

For example $y = x + \sin x \Rightarrow \dfrac{dy}{dx} = 1 + \cos x.$

In this section you will be shown how to differentiate functions defined implicitly to find expressions for $\dfrac{dy}{dx}$.

We will start by considering some of the standard results which you will need when differentiating implicit functions.

> In the exam you will not be required to find $\dfrac{d^2y}{dx^2}$ for curves defined implicitly.

To differentiate y^2 with respect to x you use the chain rule.

See C3 section 5.3.

Let $z = y^2$.

The chain rule $\dfrac{dz}{dx} = \dfrac{dz}{dy} \times \dfrac{dy}{dx}$

gives $\dfrac{d}{dx}(z) = 2y \times \dfrac{dy}{dx} \Rightarrow \dfrac{d}{dx}(y^2) = 2y\dfrac{dy}{dx}.$

This can be generalised to give

$$\frac{d}{dx}(y^n) = ny^{n-1}\frac{dy}{dx}$$

Apply the chain rule with $z = y^n$.

Similarly,

$$\frac{d}{dx}(\ln y) = \frac{1}{y}\frac{dy}{dx}$$

Apply the chain rule with $z = \ln y$.

and

$$\frac{d}{dx}(\sin y) = \cos y \frac{dy}{dx}$$

Apply the chain rule with $z = \sin y$.

and for $f(y)$, a general function of y,

$$\frac{d}{dx}\left[f(y)\right] = f'(y)\frac{dy}{dx}, \text{ where } f'(y) = \frac{df}{dy}.$$

Apply the chain rule with $z = f(y)$.

Worked example 4.1

Use the chain rule with $z = \cos 2y$ to show that
$\dfrac{d}{dx}(\cos 2y) = -2\sin 2y \dfrac{dy}{dx}.$

Solution

Let $z = \cos 2y$

$\Rightarrow \dfrac{dz}{dy} = -2\sin 2y.$

The chain rule $\dfrac{dz}{dx} = \dfrac{dz}{dy} \times \dfrac{dy}{dx}$

gives $\dfrac{dz}{dx} = -2\sin 2y \times \dfrac{dy}{dx} \Rightarrow \dfrac{d}{dx}(\cos 2y) = -2\sin 2y\dfrac{dy}{dx}.$

4

Worked example 4.2

A curve is defined by the equation $y^2 + \dfrac{1}{y} = x^2 + x$.

Find the value of $\dfrac{dy}{dx}$ at the point $(-2, 1)$.

Solution

$$\frac{d}{dx}(y^2 + y^{-1}) = \frac{d}{dx}(x^2 + x)$$

$$\frac{d}{dx}(y^2) + \frac{d}{dx}(y^{-1}) = 2x + 1$$

$$2y\frac{dy}{dx} + (-1)y^{-2}\frac{dy}{dx} = 2x + 1$$

At the point $(-2, 1)$, $\quad 2(1)\dfrac{dy}{dx} + (-1)(1)^{-2}\dfrac{dy}{dx} = 2(-2) + 1$

$$\Rightarrow \quad \frac{dy}{dx} = -3$$

Worked example 4.3

Differentiate $y^3 + \ln y + e^{3y} + 7x^2$ with respect to x.

Solution

$$\frac{d}{dx}(y^3 + \ln y + e^{3y} + 7x^2) = \frac{d}{dx}(y^3) + \frac{d}{dx}(\ln y) + \frac{d}{dx}(e^{3y}) + \frac{d}{dx}(7x^2)$$

$$= 3y^2\frac{dy}{dx} + \frac{1}{y}\frac{dy}{dx} + 3e^{3y}\frac{dy}{dx} + 14x$$

In examination questions you may have to apply the product rule and/or the quotient rule when differentiating implicit functions. The next two worked examples illustrate this.

> C3 chapter 6.

Worked example 4.4

Given that $xy - y^3 = x^2 + 3$, find an expression for $\dfrac{dy}{dx}$.

Solution

$$\frac{d}{dx}(xy - y^3) = \frac{d}{dx}(x^2 + 3)$$

$$\Rightarrow \quad \frac{d}{dx}(xy) - \frac{d}{dx}(y^3) = 2x$$

$$\Rightarrow \quad \left[x\frac{dy}{dx} + y(1)\right] - 3y^2\frac{dy}{dx} = 2x$$

$$\Rightarrow \quad \frac{dy}{dx}(x - 3y^2) = 2x - y$$

$$\Rightarrow \quad \frac{dy}{dx} = \frac{2x - y}{x - 3y^2}$$

> xy is a product so use the product rule:
> $$\frac{d}{dx}(uv) = u\frac{dv}{dx} + v\frac{du}{dx}.$$

Worked example 4.5

(a) Differentiate $\dfrac{\ln y}{x + 1}$ with respect to x.

(b) A curve is given by the equation $x^2 - xy^2 = \dfrac{\ln y}{x + 1}$.

Find the gradient of the curve at the point $(1, 1)$.

Solution

(a) $\dfrac{d}{dx}\left(\dfrac{\ln y}{x + 1}\right) = \dfrac{(x + 1)\dfrac{d}{dx}(\ln y) - (\ln y)\dfrac{d}{dx}(x + 1)}{(x + 1)^2}$

$= \dfrac{(x + 1)\dfrac{1}{y}\dfrac{dy}{dx} - (\ln y)(1)}{(x + 1)^2}$

Use the quotient rule:

$\dfrac{d}{dx}\left(\dfrac{u}{v}\right) = \dfrac{v\dfrac{du}{dx} - u\dfrac{dv}{dx}}{v^2}$.

(b) $\dfrac{d}{dx}(x^2 - xy^2) = \dfrac{d}{dx}\left(\dfrac{\ln y}{x + 1}\right)$

$\Rightarrow 2x - \dfrac{d}{dx}(xy^2) = \dfrac{(x + 1)\dfrac{1}{y}\dfrac{dy}{dx} - (\ln y)(1)}{(x + 1)^2}$

Need to find the value of $\dfrac{dy}{dx}$ at $(1, 1)$. Also use (a).

Now xy^2 is a product so you use the product rule to differentiate

$\dfrac{d}{dx}(uv) = u\dfrac{dv}{dx} + v\dfrac{du}{dx}$

$\Rightarrow 2x - \left[x\left(2y\dfrac{dy}{dx}\right) + y^2(1)\right] = \dfrac{(x + 1)\dfrac{1}{y}\dfrac{dy}{dx} - (\ln y)(1)}{(x + 1)^2}$

Be careful if a negative sign is in front of the product; use brackets.

At $(1, 1)$, $\quad 2 - \left(2\dfrac{dy}{dx} + 1\right) = \dfrac{2\dfrac{dy}{dx} - 0}{4} \quad \Rightarrow \quad 1 = 2\dfrac{1}{2}\dfrac{dy}{dx}$.

Substitute values for x and y at the first opportunity.

So, at the point $(1, 1)$ the gradient of the curve is $\dfrac{2}{5}$.

EXERCISE 4A

1 Use the chain rule with $z = \ln y$ to show that
$\dfrac{d}{dx}(\ln y) = \dfrac{1}{y}\dfrac{dy}{dx}$.

2 Use the chain rule with $z = \tan 2y$ to show that
$\dfrac{d}{dx}(\tan 2y) = 2\sec^2 2y\,\dfrac{dy}{dx}$.

3 Differentiate with respect to x:

(a) $2y$ (b) y^6 (c) $\dfrac{1}{y^2}$

(d) $x^2 + y^2$ (e) $\tan y$ (f) $\ln y$

(g) $y^2 + 3x^2 + 4y$ (h) $\cos 3y$ (i) $e^{2y} + 2y + 7$

(j) $(2y - 1)^4$

4 Use the product rule to differentiate with respect to x:

(a) xy (b) $x^2 y^2$

(c) xe^y (d) $y \sin x$

(e) $e^{2x + y}$ (f) $y \sin y$

(g) $x \ln y$ (h) $x^2 e^{2y}$

5 Use the quotient rule to differentiate with respect to x:

(a) $\dfrac{y}{x}$ (b) $\dfrac{x + 1}{y}$

(c) $\dfrac{y}{\sin x}$ (d) $\dfrac{2x}{x + y}$

(e) $\dfrac{\sin x}{x + \cos y}$ (f) $\dfrac{e^{2y}}{x + e^y}$

6 A curve has equation $3x^2 - 4y^2 = 4x$. Show that the gradient of the curve at the point $(2, 1)$ is 1.

7 A curve has equation $x^2 - 4xy = y - 7x$. Find the gradient of the curve at the point $(2, 2)$.

8 A curve is given by the equation $x^2 y + 2x = 3y + 2 \ln y$. Find the value of $\dfrac{dy}{dx}$ at the point $(1, 1)$.

9 A curve is defined by the equation $y^2 + e^x = 4y + 1$.
Show that $2(y - 2) \dfrac{dy}{dx} = -e^x$.

10 A curve is defined by the equation $x^3 + y^3 = 3xy$.
Show that $\dfrac{dy}{dx} = \dfrac{y - x^2}{y^2 - x}$.

4.3 Tangents and normals to curves whose equations are given implicitly

In C1 chapter 10 you learned how to find the equations of tangents and normals to curves, where the equations of the curves were in the form of a polynomial. This was then extended in C2 and C3 to more complicated equations of curves as your knowledge of calculus was extended. In this section you will learn how to find the equations of tangents and normals to curves whose equations are given implicitly.

In C1 you found the equations of tangents to circles by using the fact that a radius is perpendicular to a tangent. The next worked example shows you how to use implicit differentiation to find the equation of the tangent.

Worked example 4.6

Find the equation of the tangent to the circle
$x^2 + y^2 - 4x + 10y + 16 = 0$ at the point $P(4, -2)$.

C1 worked example 8.16.

Solution

Differentiating $x^2 + y^2 - 4x + 10y + 16 = 0$ with respect to x
gives

$$2x + 2y\frac{dy}{dx} - 4 + 10\frac{dy}{dx} = 0$$

At $P(4, -2)$, $\quad 8 - 4\frac{dy}{dx} - 4 + 10\frac{dy}{dx} = 0$

$$\Rightarrow \quad 6\frac{dy}{dx} = -4 \Rightarrow \frac{dy}{dx} = -\frac{2}{3}$$

Equation of the tangent at $P(4, -2)$ is $y - (-2) = -\dfrac{2}{3}(x - 4)$

$$\text{or } 3y + 2x = 2.$$

Worked example 4.7

A curve C has equation $3x^2y - 2y = y^2 + 2x$.
Find an equation of the normal to C at the point $P(-1, 2)$.

Solution

Differentiating $3x^2y - 2y = y^2 + 2x$ with respect to x gives

$$\left(6xy + 3x^2\frac{dy}{dx}\right) - 2\frac{dy}{dx} = 2y\frac{dy}{dx} + 2.$$

Using the product rule.

At $P(-1, 2)$, $\quad \left(-12 + 3\frac{dy}{dx}\right) - 2\frac{dy}{dx} = 4\frac{dy}{dx} + 2.$

$$\Rightarrow \qquad\qquad -3\frac{dy}{dx} = 14 \Rightarrow \frac{dy}{dx} = -\frac{14}{3}$$

$\Rightarrow \quad$ gradient of normal at $P(-1, 2)$ is $\dfrac{3}{14}$

Using $m_1 \times m_2 = -1$.

$\Rightarrow \quad$ equation of normal at $P(-1, 2)$ is $y - 2 = \dfrac{3}{14}(x + 1)$.

Worked example 4.8

A curve C has implicit equation $xy + y^2 = x^2 + 5$.

(a) Find the equation of the tangent to C at the point $P(1, -3)$.

(b) (i) Find the equation of the normal to C at the point P.
    (ii) This normal intersects C again at the point A. Find the
coordinates of A.

Solution

(a) Differentiating both sides of $xy + y^2 = x^2 + 5$ with respect to x, applying the product rule for xy, gives

$$\left[x\frac{dy}{dx} + y(1) \right] + 2y\frac{dy}{dx} = 2x$$

At $P(1, -3)$, $\left(\dfrac{dy}{dx} - 3 \right) - 6\dfrac{dy}{dx} = 2 \implies \dfrac{dy}{dx} = -1.$

The equation of the tangent to C at the point P is
$y - (-3) = -1(x - 1)$ or $y + x + 2 = 0$.

(b) (i) The gradient of the normal to C at P is 1.
The equation of the normal to C at $P(1, -3)$ is
$y - (-3) = 1(x - 1)$ or $y = x - 4$.

Using $m_1 \times m_2 = -1$.

(ii) To find the points where the normal $y = x - 4$ intersects the curve $xy + y^2 = x^2 + 5$, you solve the equations simultaneously.

$\implies \quad x(x - 4) + (x - 4)^2 = x^2 + 5$

$\implies \quad x^2 - 4x + x^2 - 8x + 16 = x^2 + 5$

$\implies \quad x^2 - 12x + 11 = 0$

$\implies \quad (x - 11)(x - 1) = 0 \implies x = 1$ (at P) and $x = 11$.

When $x = 11$, $y = 11 - 4 = 7$.

The normal intersects the curve again at the point $A(11, 7)$.

EXERCISE 4B

1 Find the equation of the tangent to the parabola $y^2 = 4(x - 1)$ at the point $(2, 2)$.

2 Find the equation of the normal to the curve $y^3 = 4x^2 + 2y + 2x - 2$ at the point $(1, 2)$.

3 A curve C has equation $x^2 - y^2 = 9$. The point $P(5, 4)$ lies on the curve.

   (a) Show that the equation of the tangent to C at P is $4y = 5x - 9$.

   (b) Find the equation of the normal to C at P.

4 Find the equation of the normal to the curve $\dfrac{x^2}{4} + \dfrac{y^2}{9} = 2$ at the point $(2, 3)$.

5 A curve C has equation $xy - y^2 = 2$. The point $P(3, 2)$ lies on the curve.

   (a) Find the equation of the tangent to C at P.

   (b) The normal to C at P intersects C again at the point A. Find the coordinates of A.

6 A curve C has equation $2x + y \ln y = 4$. The point $P(2, 1)$ lies on the curve. Show that the normal to C at P passes through the point $(0, 0)$.

7 The equation of a curve C is $y \cos x = x + y^2$. Find the equation of the tangent to C at the point $(0, 1)$.

8 The two points on the curve $x^2 + y^2 - xy = 84$ where the gradient of the curve is $\dfrac{1}{3}$ are A and B. Find the equations of the tangents at A and B.

MIXED EXERCISE

1 Given that $y^3 + 3y = x^3$, use implicit differentiation to show that $\dfrac{dy}{dx} = \dfrac{x^2}{y^2 + 1}$. [A]

2 The equation of a curve C is $x^2 + 5y^2 = 21$.

(a) Show that the gradient of C at the point $(4, -1)$ is $\dfrac{4}{5}$.

(b) Hence find the equation of the normal to C at the point $(4, -1)$. [A]

3 The equation of a curve is $3x^2 + y^2 = 2xy + 8x - 2$. Show that the coordinates of the points on the curve at which $\dfrac{dy}{dx} = 2$ satisfy the equation $x + y = 4$. [part A]

4 A curve has implicit equation $y^3 + xy = 4x - 2$.

(a) Show that the value of $\dfrac{dy}{dx}$ at the point $(1, 1)$ is $\dfrac{3}{4}$.

(b) Find the equation of the normal to the curve at the point $(1, 1)$. [A]

5 A curve has implicit equation $xy + y^2 = x^2 + 5$.

Find the value of $\dfrac{dy}{dx}$ at the point $(1, 2)$. [part A]

6 A curve is given by the equation $3(x + 1)^2 - 9(y - 1)^2 = 32$.

(a) Find the coordinates of the two points on the curve at which $x = 3$.

(b) Find the gradient of the curve at each of these two points. [A]

7 A curve has equation $xy = 3x^2 - y^2 + 3$.
Find, in surd form, the x-coordinates of the points P and Q on the curve at which $\dfrac{dy}{dx} = 0$.

8 Find the gradient of the curve with equation $xy^2 + 12 = 5xy$ at the point $(2, 2)$. [A]

9 A curve is defined by the equation $y^2 + yx \ln x = x$, $x > 0$.

(a) Show that the line $x = 1$ intersects the curve at the point $A(1, 1)$ and at another point B. Find the coordinates of B.

(b) Use implicit differentiation to obtain an equation connecting x, y and $\dfrac{dy}{dx}$.

> Hint: Extension of the product rule:
> $(uvw)' = (uv)w' + (uw)v' + (vw)u'$

(c) Hence,

(i) verify that the curve has a stationary value at A,

(ii) find the gradient of the tangent to the curve at B and show that this tangent passes through the origin. [A]

10 A curve C is defined by the equation $2x + y = (x + y)^2$.

(a) Show that $\dfrac{dy}{dx} = \dfrac{2(x + y - 1)}{1 - 2(x + y)}$.

(b) Find the coordinates of the stationary point of C. [A]

Key point summary

1 y is an **implicit function** of x if y cannot be written in the form $y = f(x)$; for example the equation $y^2 + x \ln y = x + \sin y$.   *p230*

2 $\dfrac{d}{dx}(y^n) = ny^{n-1}\dfrac{dy}{dx}$ *p231*

$\dfrac{d}{dx}(\ln y) = \dfrac{1}{y}\dfrac{dy}{dx}$

$\dfrac{d}{dx}(\sin y) = \cos y \dfrac{dy}{dx}$

In general,

$\dfrac{d}{dx}\left[f(y)\right] = f'(y)\dfrac{dy}{dx}$ where $f'(y) = \dfrac{df}{dy}$.

Test yourself	What to review
1 Given that $\dfrac{1}{y} + \cos y = 5 - x^3$, use implicit differentiation to show that $\dfrac{dy}{dx} = \dfrac{3x^2y^2}{1 + y^2 \sin y}$.	*Section 4.2*
2 A curve C has equation $e^{2y} + x^2 + x \sin y = 5$. Find the gradient of the curve at the point $(2, 0)$.	*Section 4.2*

Test yourself (continued) | What to review

3 A curve C is defined by the equation $x^2y - 2x - 2y^2 = 1$.
The line $y = 1$ intersects C at the points $(-1, 1)$ and A.

   (a) Find the coordinates of A.

   (b) Find the equation of the tangent to C at the point $(-1, 1)$.

Section 4.3

4 A curve has equation $x^2 + y^2 = 11 - 3xy$. Show that the equation of the normal to the curve at the point $(2, 1)$ is $7y = 8x - 9$.

Section 4.3

Test yourself ANSWERS

3 (a) $(3, 1)$;   **(b)** $3y + 4x + 1 = 0$.

2 -1.

4

CHAPTER 5

C4: Parametric equations for curves and differentiation

Learning objectives

After studying this chapter, you should be able to:
- sketch graphs of curves represented in parametric form
- find gradients of curves defined parametrically, and use them to find stationary values and equations of tangents and normals
- find a Cartesian equation for a curve given in parametric form.

5.1 Parametric equations for curves

It is often very difficult to sketch a curve when it is described implicitly, such as $y^3 - x^2 + 3y^2 + 6x + 3y = 8$.

You may be lucky enough to stumble on a few points which lie on the curve such as $(2, 0)$ and $(3, -1)$.

However, the same curve is given by

$$x = t^3 + 3$$
$$y = t^2 - 1$$

This is known as a **parametric form** where t is the **parameter**.

As t varies, you obtain values for x and y simultaneously.

For instance, when $t = 0$ you find that $x = 3$ and $y = -1$, so the point $(3, -1)$ lies on the curve.

Substituting integer values of t from 1 to 3 gives you the additional points $(4, 0)$, $(11, 3)$, $(30, 8)$ and you can soon sketch the curve for $t \geq 0$.

It looks like half a parabola.

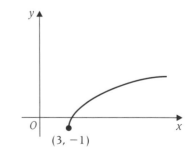

Moving into negative values of t: when $t = -1$, you obtain the point $(2, 0)$; when $t = -2$ the point produced is $(-5, 3)$; and when $t = -3$, you obtain the point $(-24, 8)$.

You may notice symmetry because the y-coordinate is the same when $t = k$ as when $t = -k$, for any value of k.

The full curve looks
something like

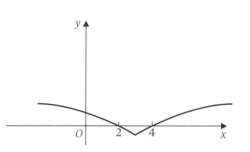

If you have a graphics calculator
you should be able to find a setting
that allows you to input equations
of curves in parametric form. Try
it for this particular curve.

The curve has a sharp point at
$(3, -1)$ which is called a cusp.

Worked example 5.1

Find the coordinates of five points on the curve defined by

$$x = t^2 + 2, \quad y = 4 - t.$$

Find the coordinates of the point where the curve crosses the
x-axis and sketch the curve.

Solution

When $t = 0$, the point $(2, 4)$ is obtained.
Putting $t = 1$ gives you the point $(3, 3)$.
Substituting $t = 2$, gives $(6, 2)$.
Trying some negative values of t:
$t = -1$ leads to $(3, 5)$ and $t = -2$ produces $(6, 6)$.

It is helpful to have some positive
and some negative values of t in
order to predict the shape of the
full curve.

You may notice symmetry in the values. This time, the
x-coordinate is the same when $t = \pm k$, for any value of k.

The curve crosses the x-axis when $y = 0$, so $t = 4$.
Therefore the x-coordinate is found by substituting $t = 4$ into
$x = t^2 + 2$, giving $x = 18$.

Hence the curve crosses
the x-axis at $(18, 0)$.

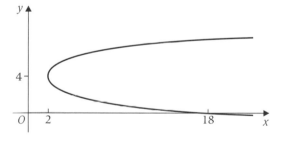

EXERCISE 5A

For each of the following curves defined parametrically:

    1 $x = 2t^2, \quad y = t + 1.$ **2** $x = 2t, \quad\quad y = t^2 - 4.$
    3 $x = t^2, \quad\quad y = 12 - 3t.$ **4** $x = t^2 - 4, \: y = t + 3.$
    5 $x = t - 5, \: y = t^2 - 9.$

(a) find the five points on the curve corresponding to

         $t = -2, -1, 0, 1$ and 2,

(b) find the coordinates of any points where the curve crosses
the coordinate axes,

(c) sketch the curve.

Check your graphs by using a
graphics calculator.

5.2 Gradients of curves defined parametrically

The chain rule was used in the C3 module. If you consider the three variables y, x and t then

$$\frac{dy}{dx} = \frac{dy}{dt} \times \frac{dt}{dx}.$$

But you also learned that $\dfrac{dt}{dx} = \dfrac{1}{\dfrac{dx}{dt}}$.

> Hence to find the gradient of a curve defined parametrically in terms of t, use
>
> $$\frac{dy}{dx} = \frac{\dfrac{dy}{dt}}{\dfrac{dx}{dt}}.$$

Worked example 5.2

A curve is defined by the parametric equations

$$x = 1 + 3t^2, \quad y = t - 2t^2.$$

(a) Find $\dfrac{dy}{dx}$ in terms of t.

(b) Hence find the equation of the normal to the curve at the point where $t = 1$.

Solution

(a) $\dfrac{dx}{dt} = 6t, \dfrac{dy}{dt} = 1 - 4t$

$$\frac{dy}{dx} = \frac{\dfrac{dy}{dt}}{\dfrac{dx}{dt}}, \text{ so } \frac{dy}{dx} = \frac{1 - 4t}{6t}$$

(b) When $t = 1$, $\dfrac{dy}{dx} = \dfrac{1 - 4}{6} = -\dfrac{1}{2}$.

Hence the normal has gradient equal to 2.
When $t = 1$, the point on the curve has coordinates $(4, -1)$.
The normal has equation $y + 1 = 2(x - 4)$ or $y = 2x - 9$.

Worked example 5.3

A curve is defined parametrically by

$$x = 2t^2 + 3, \quad y = t^2 - 4t + 5,$$

where t is a parameter.

(a) Find the gradient of the curve at the point where $t = 3$.

(b) Find the coordinates of the stationary point of the curve.

Solution

(a) $\dfrac{dx}{dt} = 4t, \dfrac{dy}{dt} = 2t - 4$

$\dfrac{dy}{dx} = \dfrac{\frac{dy}{dt}}{\frac{dx}{dt}}$, so $\dfrac{dy}{dx} = \dfrac{2t - 4}{4t} = \dfrac{t - 2}{2t}$

When $t = 3$, $\dfrac{dy}{dx} = \dfrac{3 - 2}{6} = \dfrac{1}{6}$.

The gradient has value $\dfrac{1}{6}$.

(b) Stationary points occur when $\dfrac{dy}{dx} = 0$.

Since $\dfrac{dy}{dx} = \dfrac{t - 2}{2t}$, the stationary point occurs when $t = 2$.

Substituting $t = 2$ into $x = 2t^2 + 3$, $y = t^2 - 4t + 5$, gives $(11, 1)$ as the coordinates of the stationary point.

> If you sketch the curve using a graphics calculator, you can confirm that the stationary point is at $(11, 1)$.

5

EXERCISE 5B

1 A curve is defined by the parametric equations $x = t^2 + 1$, $y = t^3$, where t is a parameter.

   (a) Find $\dfrac{dy}{dx}$ in terms of t.

   (b) Find the gradient of the curve at the point where $t = 2$.

   (c) Hence find the equation of the tangent to the curve at the point where $t = 2$.

2 A curve has parametric equations $x = 4t^2 - 36$, $y = 2t + 1$.

   (a) Find the coordinates of the points where the curve crosses the coordinate axes.

   (b) Find $\dfrac{dy}{dx}$ in terms of t.

   (6) Determine the equation of the normal to this curve at the point where $t = 3$.

3 A curve is given by the parametric equations $x = 2t + 3$, $y = \dfrac{2}{t}$.
Find the equation of the normal at the point on the curve where $t = 2$.   [A]

4 (a) Sketch the curve given by the parametric equation $x = 4t^2$, $y = 8t$ for $t \geqslant 0$.

   (b) Find $\dfrac{dy}{dx}$ in terms of t.

   (c) Hence find the equation of the tangent to the curve at the point where $t = 0.5$.   [A]

5 A curve is given by the parametric equations $x = 3t - 1$, $y = \dfrac{1}{t}$.

   (a) Find $\dfrac{dy}{dx}$ in terms of t.

   (b) Hence find the equation of the normal to the curve at the point where $t = 1$. [A]

6 A curve is given by the parametric equations $x = 1 - t^2$, $y = 2t$.

   (a) Find $\dfrac{dy}{dx}$ in terms of t.

   (b) Hence find the equation of the normal to the curve at the point where $t = 3$. [A]

7 A curve is given parametrically by

   $x = 3t^2 - t, \quad y = t^3 - 3t^2 + 5,$

   where t is a parameter.

   (a) Find $\dfrac{dy}{dx}$ in terms of t.

   (b) Find the two values of t for which $\dfrac{dy}{dx} = 0$.

   (c) Hence find the coordinates of the stationary points of the curve.

8 A curve is defined by the parametric equations

   $x = t + 5, \quad y = t - 2\sqrt{t}$, where $t \ (\geqslant 0)$ is a parameter.

   (a) Find $\dfrac{dy}{dx}$ in terms of t.

   (b) Find the coordinates of the stationary point of the curve.

   (c) Find the equation of the normal to this curve at the point where $t = 4$.

9 A curve is defined by the parametric equations

   $x = t - \dfrac{1}{t}, \quad y = t + \dfrac{1}{t}$, where t is a parameter.

   (a) Find $\dfrac{dx}{dt}$ and $\dfrac{dy}{dt}$ in terms of t. Hence show that

   $$\frac{dy}{dx} = \frac{t^2 - 1}{t^2 + 1}.$$

   (b) Hence find the coordinates of the stationary points of the curve.

   (c) Find the equation of the normal to this curve at the point where $t = 3$.

5.3 Parametric equations involving trigonometric functions

The equations of curves may involve trigonometric functions. After differentiation you must remember to use radians when substituting values of the parameters.

Worked example 5.4

A curve is defined parametrically by

$$x = t + \cos 3t, \; y = 4 + \sin 2t,$$

where t is a parameter.

(a) Find the gradient of the curve at the point where $t = 0$.

(b) Hence find an equation of the tangent to the curve at the point where $t = 0$.

Solution

(a) $\dfrac{\mathrm{d}x}{\mathrm{d}t} = 1 - 3 \sin 3t; \; \dfrac{\mathrm{d}y}{\mathrm{d}t} = 2 \cos 2t$

$\dfrac{\mathrm{d}y}{\mathrm{d}x} = \dfrac{\dfrac{\mathrm{d}y}{\mathrm{d}t}}{\dfrac{\mathrm{d}x}{\mathrm{d}t}}, \;$ so $\; \dfrac{\mathrm{d}y}{\mathrm{d}x} = \dfrac{2 \cos 2t}{1 - 3 \sin 3t}$

When $t = 0$, $\dfrac{\mathrm{d}y}{\mathrm{d}x} = \dfrac{2}{1 - 0} = 2$, so the gradient of the curve is equal to 2 at the point where $t = 0$.

(b) The tangent will have gradient 2.
The point where $t = 0$ has coordinates $(0 + \cos 0, 4 + \sin 0)$ or $(1, 4)$.
The tangent has equation $y - 4 = 2(x - 1)$ which can be written in the form $y = 2x + 2$.

Worked example 5.5

A curve is defined parametrically by $x = 3 - \cos \theta, \; y = 2 + \sec \theta$, where θ is a parameter.

(a) Show that $\dfrac{\mathrm{d}y}{\mathrm{d}x} = \sec^2 \theta$.

(b) Find an equation of the normal to the curve at the point where $\theta = \dfrac{\pi}{3}$.

> The parameter does not always have to be t.
> Often with trigonometric functions, the parameter is θ.

Solution

(a) $\dfrac{dx}{d\theta} = \sin\theta; \quad \dfrac{dy}{d\theta} = \sec\theta\tan\theta = \dfrac{\sin\theta}{\cos^2\theta}$

$$\dfrac{dy}{dx} = \dfrac{\dfrac{dy}{d\theta}}{\dfrac{dx}{d\theta}}, \text{ so } \dfrac{dy}{dx} = \dfrac{\sin\theta}{\sin\theta\cos^2\theta} = \dfrac{1}{\cos^2\theta} = \sec^2\theta$$

(b) When $\theta = \dfrac{\pi}{3}$, $\cos\theta = \dfrac{1}{2}$ and $\sec\theta = 2$.

The gradient of the curve is 4 so the normal has gradient $-\dfrac{1}{4}$.

The point on the curve has coordinates $(2\frac{1}{2}, 4)$.

The normal has equation $y - 4 = -\dfrac{1}{4}(x - 2\frac{1}{2})$ or, by multiplying both sides by 8 and simplifying, $8y + 2x = 37$.

EXERCISE 5C

1 A curve is defined by the parametric equations

$x = 3\sin t \quad$ and $\quad y = \cos t$.

(a) Show that at the point P, where $t = \dfrac{\pi}{4}$, the gradient of the curve is $-\dfrac{1}{3}$.

(b) Find the equation of the tangent to the curve at the point P, giving your answer in the form $y = mx + c$.

2 A curve is defined by the parametric equations

$x = 3 + \sin 2t \quad$ and $\quad y = 8t + 4\cos 2t$.

(a) Show that at the point Q, where $t = 0$, the gradient of the curve is 4.

(b) Find the equation of the tangent to the curve at the point Q, giving your answer in the form $y = mx + c$.

3 A curve is defined by $x = 1 + \sin 4t$, $y = 5 - \cos 2t$, where t is a parameter. The point P on the curve is where $t = \dfrac{\pi}{4}$.

(a) Find $\dfrac{dy}{dx}$ in terms of t.

(b) Find the gradient of the curve at P.

(c) Find the coordinates of the point where the normal at P cuts the x-axis.

4 A curve has parametric equations

$$x = 3 - 2 \sin \theta, \quad y = 2\theta + 4 \cos \theta,$$

where θ is a parameter.

(a) Show that $\dfrac{dy}{dx} = 2 \tan \theta - \sec \theta$.

(b) Find all values of θ in the interval $0 < \theta < 2\pi$ for which the curve has a stationary point.

(c) Find the equation of the tangent to this curve at the point where $\theta = 0$.

5 A curve is defined by the parametric equations

$$x = 3 + \sin \theta, \quad y = 2 - \sec \theta.$$

(a) Show that at the point P, where $\theta = \dfrac{\pi}{4}$, the gradient of the curve is -2.

(b) Find the equation of the tangent to the curve at the point P, giving your answer in the form $y = mx + c$.

5.4 Finding a Cartesian equation

It is often possible to eliminate the parameter and thus obtain a Cartesian equation of the curve.

Worked example 5.6

Find a Cartesian equation for each of the following curves:

(a) $x = 3p^2, y = 6p$, where p is a parameter,

(b) $x = 2q^2, y = q^3$, where q is a parameter,

(c) $x = t^3 + 3, y = t^2 - 1$, where t is a parameter.

Solution

(a) The key here is to square y so that both x and y are multiples of p^2:

$$y = 6p \implies y^2 = 36p^2$$

Since $x = 3p^2$, it follows that $12x = 36p^2$

$$\implies y^2 = 12x \text{ is a Cartesian equation for the curve.}$$

(b) By cubing x and squaring y, you will have two expressions involving q^6.

$$x = 2q^2 \implies x^3 = 8q^6 \quad \text{and} \quad y^2 = q^6 \quad \text{so } 8y^2 = 8q^6.$$

Therefore a Cartesian equation for the curve is $x^3 = 8y^2$.

(c) The equations $x = t^3 + 3$ and $y = t^2 - 1$ need to be rewritten in the form $x - 3 = t^3$ and $y + 1 = t^2$.

Then you can see that by squaring t^3 and cubing t^2 you obtain t^6. Hence a Cartesian equation for the curve is $(x - 3)^2 = (y + 1)^3$.

It is usual to leave answers in this compact form rather than multiplying out. However, if you did multiply out the brackets you would obtain the curve described in the opening section of this chapter.

5.5 Use of trigonometric identities

You have already met the following identities:

$$\cos^2 \theta + \sin^2 \theta = 1$$
$$\sec^2 \theta = 1 + \tan^2 \theta$$
$$\csc^2 \theta = 1 + \cot^2 \theta,$$

where $\sec \theta = \dfrac{1}{\cos \theta}$, $\csc \theta = \dfrac{1}{\sin \theta}$ and $\cot \theta = \dfrac{1}{\tan \theta}$.

These are often useful in finding a Cartesian equation of a curve where the parametric form involves trigonometric functions.

Worked example 5.7

A curve is defined by the parametric equations
$x = 3 \cos \theta$, $y = 2 \sin \theta$, where θ is a parameter.

Find a Cartesian equation for the curve.

Solution

Writing the equations in the form $\dfrac{x}{3} = \cos \theta$ and $\dfrac{y}{2} = \sin \theta$ you

can then use the identity $\cos^2 \theta + \sin^2 \theta = 1$ to produce the Cartesian equation

$$\left(\frac{x}{3}\right)^2 + \left(\frac{y}{2}\right)^2 = 1.$$

You may wish to square the individual terms and write
$\dfrac{x^2}{9} + \dfrac{y^2}{4} = 1$, or multiplying throughout by 36 gives

$4x^2 + 9y^2 = 36$.

> This illustrates that the Cartesian equation is not unique. However, each of these three forms of the answer is acceptable.
> That is why a question often asks for **a** Cartesian equation rather than **the** Cartesian equation.

Worked example 5.8

A curve has parametric form $x = 3 + \sec \theta$, $y = 1 - \tan \theta$.
Eliminate θ to find a Cartesian equation for the curve.

Solution

Writing the equations in the form $\sec \theta = x - 3$ and $\tan \theta = 1 - y$ prompts you to use the identity $\sec^2 \theta = 1 + \tan^2 \theta$.

Hence a Cartesian equation is $(x - 3)^2 = 1 + (1 - y)^2$.

This multiplies out to give $x^2 - 6x + 9 = 1 + 1 - 2y + y^2$ or
$x^2 - y^2 - 6x + 2y + 7 = 0$.

EXERCISE 5D

1 Find a Cartesian equation for each of the following curves:
   (a) $x = p^2$, $y = 2p$, where p is a parameter,
   (b) $x = 3q^2$, $y = 6q^3$, where q is a parameter,
   (c) $x = t^3 - 5$, $y = t^2 + 2$, where t is a parameter.

2 A curve is defined by the parametric equations $x = t^2 - 1$, $y = 3t + 2$, where t is a parameter.

    (a) Express t in terms of y.

    (b) Hence obtain a Cartesian equation for the curve.

3 A curve is defined parametrically by $x = 5 + 8t^2$, $y = 4t$, where t is a parameter. Find a Cartesian equation for the curve.

4 A curve is defined by the parametric equations $x = 2t + 3$, $y = 8t^2 + 6t + 1$, where t is a parameter. Find a Cartesian equation for the curve.

5 A curve is defined by the parametric equations $x = t^2 - 1$, $y = 2t + 3$, where t is a parameter. Find a Cartesian equation for the curve.

6 A curve is defined by the parametric equations $x = 3t$, $y = \dfrac{2}{t}$, where t is a non-zero parameter. Find a Cartesian equation for the curve.

7 A curve is defined by the parametric equations $x = 2\cos\theta$, $y = 5\sin\theta$, where θ is a parameter. Find a Cartesian equation for the curve.

8 Eliminate the parameter θ to find a Cartesian equation for the curve defined by $x = 3 + 2\csc\theta$, $y = 5 - 4\sin\theta$.

9 Eliminate the parameter θ to find a Cartesian equation for the curve defined by $x = 4 + \tan\theta$, $y = 2\sec\theta - 1$.

10 A curve is defined by the parametric equations $x = \dfrac{1}{p - 2}$, $y = \dfrac{1}{p + 4}$, where p is a parameter.

    (a) Find $\dfrac{1}{x}$ and $\dfrac{1}{y}$ in terms of p.

    (b) Hence find a Cartesian equation for the curve.

11 A curve is defined by the parametric equations $x = 3 + 2\csc\theta$, $y = 2 - 3\cot\theta$, where θ is a parameter. Find a Cartesian equation for the curve.

12 A curve is defined by the parametric equations $x = 1 + \sin\theta$, $y = 2 - \sec\theta$, where θ is a parameter. Find a Cartesian equation for the curve.

13 A curve is defined by the parametric equations $x = 2^p + 2^{-p}$, $y = 2^p - 2^{-p}$, where p is a parameter.

    (a) Find $x + y$ and $x - y$ in terms of p.

    (b) Hence find a Cartesian equation for the curve.

5.6 Linking the ideas

Examination questions may require a combination of the techniques considered in the previous sections.

Worked example 5.9

A curve is defined by the parametric equations

$$x = 1 + \frac{3}{t}, \quad y = 2t^3 - 6t^2.$$

(a) Find $\frac{dy}{dx}$, giving your answer in the form $at^3 + bt^4$.

(b) Find an equation of the tangent to the curve at the point where $t = 1$.

(c) Find a Cartesian equation of the curve expressing your answer in the form $y = f(x)$.

Solution

(a) $\dfrac{dx}{dt} = -\dfrac{3}{t^2}, \quad \dfrac{dy}{dt} = 6t^2 - 12t$

$$\frac{dy}{dx} = \frac{\dfrac{dy}{dt}}{\dfrac{dx}{dt}},$$

so $\dfrac{dy}{dx} = (6t^2 - 12t) \div \dfrac{-3}{t^2} = \dfrac{(6t^4 - 12t^3)}{-3} = 4t^3 - 2t^4$

(b) When $t = 1$, $\dfrac{dy}{dx} = 2$ so the tangent has gradient 2.

When $t = 1$, $x = 4$ and $y = -4$ so it passes through $(4, -4)$.

The tangent has equation $y + 4 = 2(x - 4)$ or $y = 2x - 12$.

(c) The equation for x can be rearranged as

$$x - 1 = \frac{3}{t} \implies t = \frac{3}{(x - 1)}.$$

> Any correct form will score full marks provided it is in the form $y = f(x)$ as requested.

Substituting into the equation for y:

$$y = 2t^3 - 6t^2 \implies y = 2\left(\frac{3}{x - 1}\right)^3 - 6\left(\frac{3}{x - 1}\right)^2$$

$$\implies y = \frac{54}{(x - 1)^3} - \frac{54}{(x - 1)^2}.$$

Clearly the parametric equations may be a little more complicated than those previously considered. Sometimes the differentiation requires the use of the product or quotient rules. In other cases, a Cartesian equation is difficult to obtain or may need more detailed consideration.

Worked example 5.10

A curve is defined by the parametric equations

$$x = \frac{t}{3 - t}, \quad y = \frac{t^2}{3 - t}, \quad (t \neq 3).$$

(a) Determine a Cartesian equation for the curve.

(b) Find $\dfrac{dy}{dx}$, in terms of t, simplifying your answer.

Solution

(a) Since the expressions for x and y have the same denominator, dividing y by x gives $\dfrac{y}{x} = t$. Substitute $t = \dfrac{y}{x}$ into the equation for x to obtain

$$x = \frac{\dfrac{y}{x}}{3 - \dfrac{y}{x}}.$$

Multiplying top and bottom of the right-hand side by x gives

$$x = \frac{y}{3x - y} \quad \Rightarrow \quad x(3x - y) = y \quad \Rightarrow \quad 3x^2 = xy + y$$

(b) Using the quotient rule:

$$\frac{dx}{dt} = \frac{[(3 - t) \times 1] - [t \times (-1)]}{(3 - t)^2} = \frac{3}{(3 - t)^2},$$

$$\frac{dy}{dt} = \frac{[(3 - t) \times 2t] - [t^2 \times (-1)]}{(3 - t)^2} = \frac{6t - t^2}{(3 - t)^2}$$

$$\frac{dy}{dx} = \frac{\dfrac{dy}{dt}}{\dfrac{dx}{dt}}, \quad \text{so} \quad \frac{dy}{dx} = \frac{6t - t^2}{3} = 2t - \frac{1}{3}t^2.$$

> You could leave your answer in either of these two forms. The main thing is to cancel the $(3 - t)^2$ terms.

Worked example 5.11

Eliminate the parameter θ to find a Cartesian equation for the curve defined by

$$x = 1 + 4 \sec \theta, \quad y = 2 - 3 \cos \theta.$$

Comment on the Cartesian equation obtained and the original curve defined parametrically.

Solution

The trigonometric identity required here is $\sec\theta = \dfrac{1}{\cos\theta}$,

which is more easily used in the form $\sec\theta \times \cos\theta = 1$.

Rearranging the equations gives

$$\sec\theta = \frac{x-1}{4} \text{ and } \cos\theta = \frac{2-y}{3}.$$

Hence a Cartesian equation is $\dfrac{x-1}{4} \times \dfrac{2-y}{3} = 1$.

This is usually written with brackets as $(x-1)(2-y) = 12$.

Note that sometimes when a curve has been defined in terms of trigonometric functions there is a restriction on the values that x and y take. This is not always obvious in the final Cartesian form. For instance, since $\sec\theta$ cannot take values between -1 and $+1$, it follows that x cannot take values between -3 and 5. Similarly, for this curve, y can only take values between -1 and 5.

The point $(0, 14)$ clearly lies on the curve with equation $(x-1)(2-y) = 12$ but it does not lie on the original curve defined parametrically.

Therefore, the original curve given in parametric form is only **part** of the Cartesian curve with equation $(x-1)(2-y) = 12$.

MIXED EXERCISE

1 A curve is defined by the parametric equations

$$x = 2t + 1, \quad y = 8t^3 - 4t^2.$$

(a) Find $\dfrac{dy}{dx}$ in terms of t.

(b) Find an equation of the normal to the curve at the point where $t = 1$.

(c) Find a Cartesian equation of the curve expressing your answer in the form $y = f(x)$.

2 A curve is defined parametrically by

$$x = t^2 + t, \quad y = t^2 - t \quad (t > 0).$$

(a) Show that $\dfrac{dy}{dx} = 1 - \dfrac{2}{2t+1}$.

(b) Determine an equation for the normal to the curve at the point where $t = 2$.

(c) Find a Cartesian equation of the curve. [A]

3 A curve is defined by the parametric equations
$x = 2 + e^p$, $y = 5 + e^{2p}$, where p is a parameter.

  (a) Find $\dfrac{dy}{dx}$ in terms of p.

  (b) Find the equation of the tangent to the curve at the point where $p = 0$.

  (c) Find a Cartesian equation for the curve.

4 A curve is defined by the parametric equations
$x = \sec \theta + \tan \theta$, $y = \sec \theta - \tan \theta$, where θ is a parameter.

  (a) **(i)** Find xy in terms of θ.

      (ii) Hence, or otherwise, find a Cartesian equation for this curve.

  (b) **(i)** Show that $\dfrac{dy}{dx} = \dfrac{\sin \theta - 1}{\sin \theta + 1}$.

      (ii) Hence, find an equation of the tangent to the curve at the point where $\theta = 0$.

5 A curve is defined by the parametric equations
$x = \dfrac{t^3}{t^2 + 1}$, $y = \dfrac{t^2}{t^2 + 1}$, where t is a parameter.

  (a) Show that a Cartesian equation for the curve is
$y^3 = x^2(1 - y)$.

  (b) Find $\dfrac{dy}{dx}$ in terms of t, giving your answer in its simplest form.

6 A curve is defined by the parametric equations
$x = \cos^3 \theta$, $y = \sin^3 \theta$, where θ is parameter.

  (a) Find $x^{\frac{2}{3}}$ in terms of θ and hence find a Cartesian equation for the curve.

  (b) Sketch the curve.

  (c) Show that $\dfrac{dy}{dx} = -\tan \theta$.

7 A curve is defined by the parametric equations
$x = 2 - \sin \theta$, $y = 1 - \tan \theta$, where θ is a parameter.

  (a) Find a Cartesian equation for the curve.

  (b) Show that $\dfrac{dy}{dx} = (\cos \theta)^n$, where n is an integer.

8 A curve is defined by the parametric equations
$$x = \frac{6t}{1 + t}, \quad y = \frac{3t^2}{1 + t} \quad (t \neq -1).$$

  (a) Show that $t = \dfrac{2y}{x}$ and hence determine a Cartesian equation for the curve.

  (b) Find an equation for the tangent to the curve at the point where $t = 2$.

9 A curve is defined by the parametric equations
$x = e^\theta \cos \theta$, $y = e^\theta \sin \theta$, where θ is a parameter.

(a) Show that $\dfrac{dy}{dx} = \dfrac{\sin \theta + \cos \theta}{\cos \theta - \sin \theta}$.

(b) Hence find an equation of the normal to the curve at the point where $\theta = 0$.

10 A curve has parametric equations $x = \dfrac{t}{1 + t}$, $y = \dfrac{1}{1 + t}$.

(a) Find $\dfrac{dy}{dx}$ in its simplest form. What can you deduce about the curve?

(b) By eliminating t, find a relationship between x and y.

(c) Find the point in your answer to **(b)** which is not included in the parametric form.

Key point summary

1 To find the gradient of a curve defined parametrically   *p242*

in terms of t, use $\dfrac{dy}{dx} = \dfrac{\dfrac{dy}{dt}}{\dfrac{dx}{dt}}$.

2 Parametric equations can be converted to Cartesian   *p247*
equations by eliminating the parameter using algebraic
rearrangements or trigonometric identities.

Test yourself

What to review

1 A curve is defined by the parametric equations   *Section 5.1*

$x = 3 - t$, $y = t^2 - 1$.

(a) Find the points on the curve corresponding to
$t = 0$ and $t = -2$.

(b) Find the coordinates of the points where the curve
crosses the coordinate axes.

(c) Sketch the curve.

2 A curve is defined by the parametric equations   *Section 5.2*

$x = 6 - 2t$, $y = 3t^2 - 4t - 1$.

(a) Find $\dfrac{dy}{dx}$ in terms of t.

(b) Find the equation of the normal to the curve at the
point where $t = 2$.

Test yourself (continued)

What to review

3 Find $\dfrac{dy}{dx}$ in terms of θ when:

Section 5.3

  (a) $x = 3 \sin \theta$, $y = 4 - 2 \cos \theta$,

  (b) $x = 2 \tan \theta$, $y = 6\theta + 4 \sec \theta$.

4 A curve is defined by the parametric equations

Section 5.4

$$x = 4 - 3t, \quad y = 9t^2 + 6t - 1.$$

Find a Cartesian equation for the curve.

5 A curve is defined by the parametric equations

Section 5.4

$$x = 1 + 2 \sin \theta, \quad y = 2 - 3 \cos \theta.$$

Find a Cartesian equation for the curve.

Test yourself **ANSWERS**

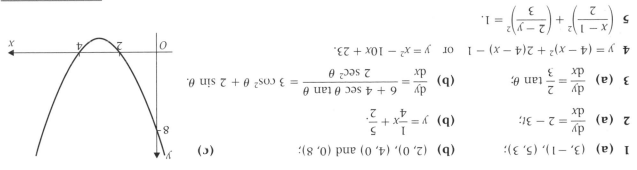

5 $\left(\dfrac{x-1}{2}\right)^2 + \left(\dfrac{y-2}{3}\right)^2 = 1.$

4 $y = (4-x)^2 + 2(4-x) - 1$ or $y = x^2 - 10x + 23.$

3 (a) $\dfrac{dy}{dx} = \dfrac{2}{3}\tan\theta;$ (b) $\dfrac{dy}{dx} = \dfrac{6 + 4\sec\theta\tan\theta}{2\sec^2\theta} = 3\cos^2\theta + 2\sin\theta.$

2 (a) $\dfrac{dy}{dx} = 2 - 3t;$ (b) $y = \dfrac{1}{4}x + \dfrac{5}{2}.$

1 (a) $(3,-1), (5,3);$ (b) $(2,0), (4,0)$ and $(0,8);$ (c)

C4: Further trigonometry with integration

Learning objectives

After studying this chapter, you should be able to:
- use the compound angle identities in trigonometry to prove other identities and to solve equations
- recall the double angle identities in trigonometry and use them to prove other identities and to solve equations
- use the double angle identities to find and evaluate integrals
- express $a \cos \theta + b \sin \theta$ as a single sine or cosine function
- solve equations of the form $a \cos \theta + b \sin \theta = c$.

6.1 Compound angle identities

In this section you will be shown how to obtain identities for trigonometric expressions of compound angles similar to $\sin(A + B)$.

A common error is to assume that $\sin (A + B) = \sin A + \sin B$. You can easily disprove this result by finding a counter example.

$\sin(30° + 60°) = \sin 90° = 1$

$\sin 30° + \sin 60° = 0.5 + 0.866... \neq 1$

so $\sin(A + B) \neq \sin A + \sin B$.

Consider a line of length 1 unit. Initially the line is in the position OB, where O is the origin. The line OB makes an angle β with the positive x-axis so the coordinates of B are ($\cos \beta$, $\sin \beta$). Rotate the line about O through an angle $(\alpha - \beta)$ so the line OA makes an angle α with the positive x-axis and the coordinates of A are $(\cos \alpha, \sin \alpha)$.

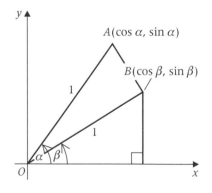

Using the formula for the distance between two points you get
$AB^2 = (\cos \alpha - \cos \beta)^2 + (\sin \alpha - \sin \beta)^2$.

Using the cosine rule for triangle OAB you get
$$AB^2 = 1^2 + 1^2 - 2(1)(1) \cos (\alpha - \beta).$$

$\Rightarrow \quad (\cos \alpha - \cos \beta)^2 + (\sin \alpha - \sin \beta)^2 = 2 - 2 \cos(\alpha - \beta)$

$\Rightarrow \quad \cos^2 \alpha + \cos^2 \beta - 2 \cos \alpha \cos \beta + \sin^2 \alpha + \sin^2 \beta - 2 \sin \alpha \sin \beta = 2 - 2 \cos (\alpha - \beta)$

$\Rightarrow \quad 2 - 2 \cos \alpha \cos \beta - 2 \sin \alpha \sin \beta = 2 - 2 \cos (\alpha - \beta)$

$\Rightarrow \quad \cos(\alpha - \beta) = \cos \alpha \cos \beta + \sin \alpha \sin \beta.$ [1]

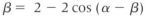

Use $\cos^2 x + \sin^2 x = 1$.

Letting $\alpha = A$ and $\beta = -B$ in [1] leads to

$$\cos(A + B) = \cos A \cos B - \sin A \sin B.$$

Letting $\alpha = \dfrac{\pi}{2} - A$ and $\beta = B$ in [1] gives

$$\cos\left[\frac{\pi}{2} - (A + B)\right] = \cos\left(\frac{\pi}{2} - A\right)\cos B + \sin\left(\frac{\pi}{2} - A\right)\sin B,$$

which leads to

$$\sin(A + B) = \sin A \cos B + \cos A \sin B.$$

Replacing B by $-B$ leads to

$$\sin(A - B) = \sin A \cos B - \cos A \sin B.$$

Writing $\tan(A + B) = \dfrac{\sin(A + B)}{\cos(A + B)}$ and using the previous

identities you can show that, provided $(A + B) \neq \left(k + \dfrac{1}{2}\right)\pi$,

$$\tan(A + B) = \frac{\tan A + \tan B}{1 - \tan A \tan B}.$$

Replacing B by $-B$ leads to

$$\tan(A - B) = \frac{\tan A - \tan B}{1 + \tan A \tan B}.$$

You now have the six compound angle identities which can be summarised as

$$\sin(A \pm B) = \sin A \cos B \pm \cos A \sin B$$
$$\cos(A \pm B) = \cos A \cos B \mp \sin A \sin B$$
$$\tan(A \pm B) = \frac{\tan A \pm \tan B}{1 \mp \tan A \tan B}, \quad \left[A \pm B \neq \left(k + \frac{1}{2}\right)\pi\right]$$

> Use $\cos(-\theta) = \cos\theta$
> and $\sin(-\theta) = -\sin\theta$.

> Use $\cos\left(\dfrac{\pi}{2} - \theta\right) = \sin\theta$
> and $\sin\left(\dfrac{\pi}{2} - \theta\right) = \cos\theta$.

> These identities are given in this form in the formulae booklet which you will have in your examination. You must ensure that you understand the signs in the identities.

6

Worked example 6.1

Given that $\tan 60° = \sqrt{3}$, show that $\tan 15° = 2 - \sqrt{3}$.

Solution

$$\tan 15° = \tan(60° - 45°) = \frac{\tan 60° - \tan 45°}{1 + \tan 60° \tan 45°}$$

$$= \frac{\sqrt{3} - 1}{1 + \sqrt{3}(1)} = \frac{(\sqrt{3} - 1)^2}{(\sqrt{3} - 1)(1 + \sqrt{3})} = \frac{3 - 2\sqrt{3} + 1}{3 - 1} = 2 - \sqrt{3}$$

Worked example 6.2

Prove the identity $\dfrac{\sin(A - B)}{\cos A \cos B} = \tan A - \tan B$.

Solution

$$\frac{\sin(A - B)}{\cos A \cos B} = \frac{\sin A \cos B - \cos A \sin B}{\cos A \cos B}$$

$$= \frac{\sin A \cos B}{\cos A \cos B} - \frac{\cos A \sin B}{\cos A \cos B}$$

$$= \frac{\sin A}{\cos A} - \frac{\sin B}{\cos B} = \tan A - \tan B$$

Worked example 6.3

Find all values of x in the interval $0° \leqslant x \leqslant 360°$ for which $\sin x = \cos (x + 30°)$.

Solution

$$\sin x = \cos (x + 30°)$$

$\Rightarrow \quad \sin x = \cos x \cos 30° - \sin x \sin 30°$

$\Rightarrow \quad \sin x + \sin x \sin 30° = \cos x \cos 30°$

$\Rightarrow \quad (1 + \sin 30°) \sin x = \cos x \cos 30°$

Divide both sides of the equation by $(1 + \sin 30°) \cos x$.

$\Rightarrow \quad \dfrac{\sin x}{\cos x} = \dfrac{\cos 30°}{1 + \sin 30°} \quad \Rightarrow \quad \tan x = \dfrac{0.866\ldots}{1.5} = 0.577\,35\ldots$

Now $\tan^{-1} (0.577\,35\ldots) = 30°$, so solutions of $\tan x = 0.577\,35\ldots$ in the interval $0° \leqslant x \leqslant 360°$ are $30°$ and $180° + 30°$.

The solutions of $\sin x = \cos (x + 30°)$ in the interval $0° \leqslant x \leqslant 360°$ are $30°$ and $210°$.

Worked example 6.4

By writing $\dfrac{1}{2} \cos x - \dfrac{\sqrt{3}}{2} \sin x$ as a single cosine, describe the transformation that maps the graph of $y = \cos x$ onto the graph of $y = \dfrac{1}{2} \cos x - \dfrac{\sqrt{3}}{2} \sin x$.

Solution

$$\frac{1}{2} \cos x - \frac{\sqrt{3}}{2} \sin x \equiv \cos x \cos 60° - \sin x \sin 60°$$

$$\equiv \cos (x + 60°)$$

> $\dfrac{1}{2} = \cos 60°, \dfrac{\sqrt{3}}{2} = \sin 60°$.
>
> **You will not need to remember the second of these two results in the exam.**

so you need the transformation that maps the graph of $y = \cos x$ onto the graph of $y = \cos(x + 60°)$.

The required transformation is the translation $\begin{bmatrix} -60° \\ 0 \end{bmatrix}$.

EXERCISE 6A

1 Given that $\tan 30° = \dfrac{1}{\sqrt{3}}$, show that $\tan 75° = 2 + \sqrt{3}$.

2 Write $\sin 45° \cos 30° + \cos 45° \sin 30°$ in the form:

(a) $\sin k°$, (b) $\cos k°$.

$\boxed{\sin \theta = \cos (90° - \theta)}$

3 Write $\dfrac{\tan 2^c - \tan 1.5^c}{1 + \tan 2^c \tan 1.5^c}$ in the form $\tan k$.

4 Write the following as $\sin k°$, where k is a constant to be determined:

(a) $\sin 40° \cos 20° + \cos 40° \sin 20°$,

(b) $\sin 40° \cos 20° - \cos 40° \sin 20°$,

(c) $\cos 40° \cos 30° + \sin 40° \sin 30°$,

(d) $\cos 40° \cos 30° - \sin 40° \sin 30°$.

5 Prove the identity $\dfrac{\sin(A + B)}{\cos A \cos B} = \tan A + \tan B$.

6 Prove the identity $\sin (x + y) - \sin(x - y) = 2 \cos x \sin y$.

7 Prove the identity $\tan (x + 45°) = \dfrac{1 + \tan x}{1 - \tan x}$.

8 Given that $\tan (x - y) = k$ and $\tan x = 1$, express $\tan y$ in terms of k.

9 Solve the equation $\cos x \sin \dfrac{\pi}{3} + \sin x \cos \dfrac{\pi}{3} = \dfrac{1}{4}$, for $0 \leqslant x \leqslant 2\pi$.

10 The angles x and y are acute and x is greater than y. Given that $\cos x \cos y = \dfrac{1}{4}$ and $\sin x \sin y = \dfrac{1}{4}$, find the values of $\cos (x + y)$ and $\cos (x - y)$ and hence, or otherwise, find the values of x and y.

11 Prove that $\cos \left(x + \dfrac{4\pi}{3}\right) + \cos \left(x + \dfrac{2\pi}{3}\right) + \cos x = 0$.

12 Find all values of x in the interval $0° \leqslant x \leqslant 360°$ for which $\cos (x + 60°) - \cos (x - 60°) = 1.5$.

13 By writing $\cos x \cos 60° + \sin x \sin 60°$ as a single cosine, describe the transformation that maps the graph of $y = \cos x$ onto the graph of $y = \cos x \cos 60° + \sin x \sin 60°$.

14 Describe the transformation that maps the graph of $y = \sin x$ onto the graph of $y = \cos \dfrac{\pi}{3} \sin x - \sin \dfrac{\pi}{3} \cos x$.

6

6.2 Double angle identities

In C2 you solved equations of the form $\cos ax = k$. In this section you will learn how to solve other equations including those of the form $\cos 2ax = \sin ax$ and $\sin 2ax = \cos ax$.

Using the identities in section 6.1 and replacing B by A you can show that

$$\sin 2A = 2 \sin A \cos A,$$
$$\cos 2A = \cos^2 A - \sin^2 A, \text{ or, using } \cos^2 A + \sin^2 A = 1,$$
$$\cos 2A = 2 \cos^2 A - 1,$$
$$\cos 2A = 1 - 2 \sin^2 A$$
$$\tan 2A = \frac{2 \tan A}{1 - \tan^2 A}$$

> These identities will **not** be given in the examination formulae booklet. They are important an should be learned.

Worked example 6.5

Show that $\sin 3x = 3 \sin x - 4 \sin^3 x$.

Solution

$$\begin{aligned}
\sin 3x = \sin (2x + x) &= \sin 2x \cos x + \cos 2x \sin x \\
&= (2 \sin x \cos x) \cos x + (1 - 2 \sin^2 x) \sin x \\
&= 2 \sin x \cos^2 x + \sin x - 2 \sin^3 x \\
&= 2 \sin x (1 - \sin^2 x) + \sin x - 2 \sin^3 x \\
\Rightarrow \sin 3x &= 3 \sin x - 4 \sin^3 x
\end{aligned}$$

> Need expression in terms of $\sin x$ only so use the form $1 - 2 \sin^2 x$ for $\cos 2x$.

> Use $\cos^2 x + \sin^2 x = 1$.

Worked example 6.6

Prove the identity $\dfrac{1}{\cos A + \sin A} + \dfrac{1}{\cos A - \sin A} = \dfrac{\tan 2A}{\sin A}$.

Solution

$$\begin{aligned}
\frac{1}{\cos A + \sin A} + \frac{1}{\cos A - \sin A} &= \frac{\cos A - \sin A + \cos A + \sin A}{(\cos A + \sin A)(\cos A - \sin A)} \\
&= \frac{2 \cos A}{\cos^2 A - \sin^2 A} = \frac{2 \cos A}{\cos 2A} \\
&= \frac{2 \cos A \sin 2A}{\cos 2A \sin 2A}
\end{aligned}$$

$$\begin{aligned}
\frac{1}{\cos A + \sin A} + \frac{1}{\cos A - \sin A} &= \frac{2 \cos A \tan 2A}{\sin 2A} \\
&= \frac{2 \cos A \tan 2A}{2 \sin A \cos A} \\
&= \frac{\tan 2A}{\sin A}.
\end{aligned}$$

> Need to get $\tan 2A = \dfrac{\sin 2A}{\cos 2A}$ so multiply numerator and denominator by $\sin 2A$.

> $\sin 2A = 2 \sin A \cos A$

> Cancel $2 \cos A$.

Worked example 6.7

Solve the equation $\sin 2x = \cos x$, for $0° \leqslant x < 360°$.

The given equation involves angles 2x and x. Make the angles the same if possible.

Solution

$\sin 2x = \cos x$

Do **not** cancel the cos x terms.

$\Rightarrow \quad 2 \sin x \cos x = \cos x$

$\Rightarrow \quad \cos x \, (2 \sin x - 1) = 0$

$\Rightarrow \quad \cos x = 0 \text{ or } \sin x = \dfrac{1}{2}$

$\cos x = 0 \Rightarrow x = 90°, 270°; \sin x = \dfrac{1}{2} \Rightarrow x = 30°, 150°$

The relevant solutions of $\sin 2x = \cos x$, are $30°, 90°, 150°, 270°$.

Worked example 6.8

Solve the equation $\cos 2\theta = 2 \cos \theta$ for $-\pi \leqslant \theta < \pi$.

The given equation involves angles 2θ and θ. Make the angles the same by using the double angle formulae.

Solution

$\cos 2\theta = 2 \cos \theta$

cos 2θ has three possible forms. Choose the form which only involves the remaining term, cos θ.

$\Rightarrow \quad 2 \cos^2 \theta - 1 = 2 \cos \theta$

$\Rightarrow \quad 2 \cos^2 \theta - 2 \cos \theta - 1 = 0$

So, $\cos \theta = \dfrac{2 \pm \sqrt{(4 + 8)}}{4} = \dfrac{2 \pm \sqrt{12}}{4} = 1.366\,02\ldots \text{ or } -0.366\,025\ldots$

Rearrange into a quadratic equation in cos θ and solve using the formula.

The greatest value of a cosine is 1 so you reject $\cos\theta = 1.366\,02\ldots$ and all solutions will come from $\cos\theta = -0.366\,025\ldots$.

$\text{Cos}^{-1}(0.366\,025\ldots) = 1.9455\ldots$ rads
so, the only solutions of $\cos 2\theta = 2 \cos\theta$ in the interval $-\pi \leqslant \theta < \pi$ are $\theta = -1.9455$ and $+1.9455\ldots$.

To three significant figures, $\theta = \pm 1.95$ radians.

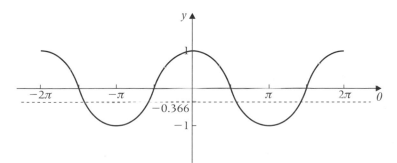

6

EXERCISE 6B

1 Prove the identity $\cos 3x = 4 \cos^3 x - 3 \cos x$.

2 Prove the identity $\dfrac{\sin 2x}{1 + \cos 2x} = \tan x$.

3 Solve the equation $2 \sin 2x = \sin x$, for $0° \leqslant x < 360°$.

4 Solve the equation $3 \sin 2\theta = \cos \theta$, for $0 \leqslant \theta < 2\pi$.

5 Solve the equation $\cos \theta + \cos 2\theta = 0$, for $0 \leqslant \theta \leqslant 2\pi$.

6 Prove the identity $\dfrac{1 - \cos 2A}{1 + \cos 2A} = \tan^2 A$.

7 Prove the identity $(\sin A + \cos A)^2 - 1 = \sin 2A$.

8 Prove the identity $\sqrt{\dfrac{1 + \sin 2A}{1 - \sin 2A}} \equiv \tan\left(A + \dfrac{\pi}{4}\right)$.

9 Solve the equation $\cos 2\theta + 5 \sin \theta = 3$, for $-\pi \leqslant \theta \leqslant \pi$.

10 Solve the equation $\tan 2\theta = 3 \tan \theta$, for $0 \leqslant \theta \leqslant 2\pi$.

11 Solve the equation $2 \cos^2 x = 2 \sin x \cos x + 1$, for $0° \leqslant x < 360°$.

12 Solve the equation $2 \sin 2x = \tan x$, for $0° \leqslant x < 360°$.

13 Express $\dfrac{\sin 2x}{1 - 2 \cos 2x}$ in terms of $\tan x$.

14 Prove the identity $\dfrac{1 + \sin 2A - \cos 2A}{1 + \sin 2A + \cos 2A} \equiv \tan A$.

15 Solve the equation $\cos 2\theta - 5 \cos \theta = 2$, for $-\pi \leqslant \theta \leqslant \pi$.

16 Given that $\cos 2A = \tan^2 x$, show that $\cos 2x = \tan^2 A$.

17 **(a)** Show that $\cos 2x - 3 \cos x - 1 = (2 \cos x + 1)(\cos x - 2)$.
   (b) Given that $0° \leqslant x < 180°$, solve the equation
   $\cos 2x - 3 \cos x - 1 = 0$.   [part A]

18 **(a)** Use the expansion of $\tan(A + B)$ to show that
   $$\tan 3x = \dfrac{3\tan x - \tan^3 x}{1 - 3 \tan^2 x}.$$

   (b) Use the result of **(a)** to find all solutions in the interval
   $0 < x < \pi$ of the equation
   $3\tan x - \tan^3 x = 1 - 3 \tan^2 x$.   [A]

19 Solve the equation $2 \sin \theta \cos 2\theta + \sin 2\theta = 0$, for
   $0 \leqslant \theta \leqslant 2\pi$.   [A]

6.3 Using double angle identities to solve integrals

In the previous section you were shown the double-angle identities:

$\sin 2A = 2 \sin A \cos A$

$\cos 2A = \cos^2 A - \sin^2 A$

$\cos 2A = 2 \cos^2 A - 1$

$\cos 2A = 1 - 2 \sin^2 A$

$\tan 2A = \dfrac{2 \tan A}{1 - \tan^2 A}$

You can use these identities to find more integrals.

> To integrate either $\sin^2 x$ or $\cos^2 x$ write each in terms of $\cos 2x$.

Worked example 6.9

Evaluate $\displaystyle\int_0^{\frac{5\pi}{12}} \sin^2 \theta \, d\theta$.

Solution

Using the identity $\cos 2A = 1 - 2\sin^2 A$ leads to

$$\int_0^{\frac{5\pi}{12}} \sin^2 \theta \, d\theta = \frac{1}{2} \int_0^{\frac{5\pi}{12}} (1 - \cos 2\theta) \, d\theta = \frac{1}{2} \left[\theta - \frac{1}{2} \sin 2\theta \right]_0^{\frac{5\pi}{12}}$$

$$= \frac{1}{2} \left(\frac{5\pi}{12} - \frac{1}{2} \sin \frac{5\pi}{6} \right) - \frac{1}{2}(0 - 0)$$

$$= \frac{5\pi}{24} - \frac{1}{4} \left(\frac{1}{2} \right) = \frac{5\pi - 3}{24}$$

Worked example 6.10

Find $\displaystyle\int \sin 3x \cos 3x \, dx$.

Solution

Using the identity $\sin 2A = 2 \sin A \cos A$, with $A = 3x$, gives $2 \sin 3x \cos 3x = \sin 6x$.

Integrating both sides with respect to x leads to

$$\int \sin 3x \cos 3x \, dx = \frac{1}{2} \int \sin 6x \, dx = \frac{1}{2} \left(-\frac{1}{6} \cos 6x \right) + c$$

$$\int \sin 3x \cos 3x \, dx = -\frac{1}{12} \cos 6x + c$$

6

Worked example 6.11 ———————————

Find $\int 6 \cos^2 3x \, dx$.

Solution

Using the identity $\cos 2A = 2 \cos^2 A - 1$ with $A = 3x$, leads to $2 \cos^2 3x = \cos 6x + 1$.

$$\int 6 \cos^2 3x \, dx = 3 \int (\cos 6x + 1) \, dx$$

$$= 3 \left(\frac{1}{6} \sin 6x + x + k \right)$$

$$= \frac{1}{2} \sin 6x + 3x + c$$

EXERCISE 6C ————————————————

1 Evaluate $\displaystyle\int_0^{\frac{\pi}{4}} 2 \sin x \cos x \, dx$.

2 Evaluate $\displaystyle\int_0^{\frac{\pi}{4}} 2 \cos^2 \theta \, d\theta$.

3 Evaluate $\displaystyle\int_0^{\frac{\pi}{9}} 2 \cos 3x \sin 3x \, dx$.

4 Evaluate $\displaystyle\int_0^{\frac{\pi}{6}} 4 \sin^2 \frac{x}{2} \, dx$.

5 Find $\displaystyle\int \frac{2 \tan x}{\tan 2x} \, dx$.

6 Show that $\displaystyle\int_{\frac{\pi}{4}}^{\frac{\pi}{2}} (\cos \theta - \sin \theta)(\cos \theta + \sin \theta) \, d\theta = -0.5$.

7 Evaluate $\displaystyle\int_0^{\frac{\pi}{4}} (\cos \theta + \sin \theta)^2 \, d\theta$.

8 Find the value of $\displaystyle\int_0^1 \cos^2 (2x + 1) \, dx$. Give your answer to three significant figures.

9 Find the value of $\displaystyle\int_0^{0.4} \frac{2 \tan x}{1 - \tan^2 x} \, dx$. Give your answer to three significant figures.

10 (a) By using the identity $\cos 2A = 1 - 2 \sin^2 A$ show that $\cos 4x = 1 - 8 \sin^2 x \cos^2 x$.

   (b) Hence show that $\displaystyle\int_0^{\frac{\pi}{4}} \sin^2 x \cos^2 x \, dx = \frac{\pi}{32}$.

6.4 Expressing $a \cos \theta + b \sin \theta$ as a single sine or cosine function

Using a graphics calculator you can see that the graph of $y = 3 \sin x + 4 \cos x$ has a similar shape to the basic curve $y = \sin x$ (and indeed to the curve $y = \cos x$). Since the stationary points do not occur at either $x = 90°$ or at $x = 270°$ and the maximum and minimum values are 5 and -5, respectively, you can deduce that a horizontal translation and a vertical stretch of scale factor 5 have been applied to the basic curve to give the curve $y = 3 \sin x + 4 \cos x$.

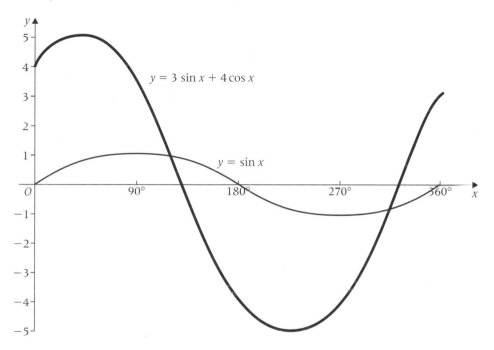

You studied such transformations in C2 chapter 5 from which you can deduce that the graph of $y = 3 \sin x + 4 \cos x$ is the same as the graph of a curve with equation of the form $y = 5 \sin (x + \alpha)$.

In section 6.1 you were shown the identities,

$$\sin (A \pm B) = \sin A \cos B \pm \cos A \sin B \quad \text{and}$$

$$\cos (A \pm B) = \cos A \cos B \mp \sin A \sin B.$$

The next few worked examples show you how to use these identities to write expressions of the form $a \cos \theta + b \sin \theta$ as a single sine or cosine function.

> These identities are also given in the examination formulae booklet.

Worked example 6.12

Express $\sqrt{3} \sin \theta - \cos \theta$ in the form $R \sin (\theta - \alpha)$, stating the value of R and giving the value of α to three significant figures in the interval $0 < \alpha < \dfrac{\pi}{2}$.

Solution

$R \sin (\theta - \alpha) \equiv \sqrt{3} \sin \theta - \cos \theta$

$R \sin \theta \cos \alpha - R \cos \theta \sin \alpha \equiv \sqrt{3} \sin \theta - \cos \theta$

	Use $\sin (A - B)$ $= \sin A \cos B - \cos A \sin B$.

$\Rightarrow \quad R \cos \alpha = \sqrt{3} \quad$ and $\quad R \sin \alpha = 1$

> Equating coefficients of $\sin \theta$ and then of $-\cos \theta$.

$\Rightarrow \quad (R \cos \alpha)^2 + (R \sin \alpha)^2 = (\sqrt{3})^2 + 1^2$

$\Rightarrow \quad R^2 (\cos^2 \alpha + \sin^2 \alpha) = (\sqrt{3})^2 + 1^2$

> These two lines may be omitted in an exam solution.

$\Rightarrow \quad R = \sqrt{(\sqrt{3})^2 + 1^2} = 2$

> Since $\cos^2 \alpha + \sin^2 \alpha = 1$.

$\Rightarrow \quad \cos \alpha = \dfrac{\sqrt{3}}{2} \quad$ and $\quad \sin \alpha = \dfrac{1}{2}$

> Or use $\dfrac{R \sin \alpha}{R \cos \alpha} = \tan \alpha = \dfrac{1}{\sqrt{3}}$.

$\Rightarrow \quad \alpha = 0.524$ (to 3 sf)

So $\sqrt{3} \sin \theta - \cos \theta \equiv 2 \sin (\theta - 0.524)$.

> Since $0 < \alpha < \dfrac{\pi}{2}$.

Worked example 6.13

(a) Express $\cos x + \sin x$ in the form $R \cos (x - \alpha)$, where R is a positive constant and $0° < \alpha < 90°$.

(b) Hence find the greatest and least values of $\cos x + \sin x$.

Solution

(a) $R \cos (x - \alpha) \equiv \cos x + \sin x$

$R \cos x \cos \alpha + R \sin x \sin \alpha \equiv \cos x + \sin x$

$\Rightarrow \quad R \cos \alpha = 1 \quad$ and $\quad R \sin \alpha = 1$

$\Rightarrow \quad R = \sqrt{1^2 + 1^2} = \sqrt{2}$

$\Rightarrow \quad \cos \alpha = \dfrac{1}{\sqrt{2}} = 0.7071\ldots \quad$ and $\quad \sin \alpha = 0.7071\ldots$

$\Rightarrow \quad \alpha = 45°$.

So $\cos x + \sin x \equiv \sqrt{2} \cos (x - 45°)$.

(b) The greatest value of $\cos (x - 45°)$ is 1

$\Rightarrow \quad$ the greatest value of $\sqrt{2} \cos (x - 45°)$ is $\sqrt{2}$

$\Rightarrow \quad$ the greatest value of $\cos x + \sin x$ is $\sqrt{2}$.

> Since $-1 \leqslant \cos \theta \leqslant 1$; see C2 chapter 4.

Similarly, since the least value of $\cos (x - 45°)$ is -1, the least value of $\cos x + \sin x$ is $-\sqrt{2}$.

In C3 you were shown how to differentiate $\sin x$ and $\cos x$. You could then use calculus, for example, to find the stationary points of the curve with equation $y = 3 \sin x + \sqrt{3} \cos x$. The next worked example shows you how to find the stationary points without the use of calculus.

Worked example 6.14

The curve $y = 3 \sin x + \sqrt{3} \cos x$ has domain $0 \leqslant x \leqslant 2\pi$.

Find the coordinates of the stationary points of the curve, giving your answers to three significant figures.

Solution

$3 \sin x + \sqrt{3} \cos x \equiv R \sin(x + \alpha) \equiv R \sin x \cos \alpha + R \cos x \sin \alpha$

$\Rightarrow \quad R = \sqrt{3^2 + (\sqrt{3})^2} = \sqrt{12} = 2\sqrt{3}$

> Since the coefficients of $\sin x$ and $\cos x$ are both positive if you use either $R \sin(x + \alpha)$ or $R \cos(x - \alpha)$, then α will be positive and acute.

$\Rightarrow \quad \cos \alpha = \dfrac{3}{2\sqrt{3}} = \dfrac{\sqrt{3}}{2} = 0.866\ldots \quad \text{and} \quad \sin \alpha = \dfrac{\sqrt{3}}{2\sqrt{3}} = \dfrac{1}{2}$

> Since the domain is in terms of π you work in radians.

$\Rightarrow \quad \alpha = 0.523\,598\ldots = 0.5236 \text{ (to 4 sf)}.$

So $3 \sin x + \sqrt{3} \cos x \equiv 2\sqrt{3} \sin(x + 0.5236)$.

> Since $-1 \leqslant \sin \theta \leqslant 1$,
> $-2\sqrt{3} \leqslant 2\sqrt{3} \sin \theta \leqslant 2\sqrt{3}$.

The maximum value of $2\sqrt{3} \sin(x + 0.5236)$ is $2\sqrt{3}$ and occurs when

$$\sin(x + 0.5236) = 1 \quad \Rightarrow \quad x + 0.5236 = \frac{\pi}{2},\ 2\pi + \frac{\pi}{2},\ \ldots\ .$$

> $\sin^{-1}(1) = 90° = \dfrac{\pi}{2}$ radians.

6

The only solution of this equation in the domain is

$x = \dfrac{\pi}{2} - 0.5236 = 1.047\ldots = 1.05 \text{ (to 3 sf)}$

$\Rightarrow \quad (1.05, 3.46)$ is the maximum point.

The minimum value of $2\sqrt{3} \sin(x + 0.5236)$ is $-2\sqrt{3}$ and occurs when

> $2\sqrt{3} = 3.4641\ldots$

$$\sin(x + 0.5236) = -1 \quad \Rightarrow \quad x + 0.5236 = -\frac{\pi}{2},\ 2\pi - \frac{\pi}{2},\ \ldots\ .$$

> $\sin^{-1}(-1) = -90° = -\dfrac{\pi}{2}$ radians

The only solution of this equation in the domain is

$x = \dfrac{3\pi}{2} - 0.5236 = 4.188\ldots = 4.19 \text{ (to 3 sf)}$

$\Rightarrow \quad (4.19, -3.46)$ is the minimum point.

The curve has two stationary points, whose coordinates, to three significant figures, are $(1.05, 3.46)$ and $(4.19, -3.46)$.

In general,

(a) $a \cos \theta + b \sin \theta \equiv R \cos(\theta - \alpha)$, where $R = \sqrt{a^2 + b^2}$ and $\cos \alpha = \dfrac{a}{R}$ and $\sin \alpha = \dfrac{b}{R}$.

(b) $a \cos \theta + b \sin \theta \equiv R \sin(\theta + \beta)$, where $R = \sqrt{a^2 + b^2}$ and $\sin \beta = \dfrac{a}{R}$ and $\cos \beta = \dfrac{b}{R}$.

In questions, $R > 0$ and α and β acute will be used.

(Note α and β can be negative in the case when a and b have different signs.)

The next worked example shows you how the work covered so far in this section can be used to integrate some more functions of the type $\dfrac{1}{a \cos x + b \sin x}$. You will also need some of the results from C3 section 8.3.

Worked example 6.15

Find $\displaystyle\int \dfrac{10}{3 \sin 2x - 4 \cos 2x} \, dx$.

Solution

Firstly you write $3 \sin 2x - 4 \cos 2x$ in the form $R \sin(2x - \alpha)$:

$3 \sin 2x - 4 \cos 2x \equiv R \sin(2x - \alpha) \equiv R \sin 2x \cos \alpha - R \cos 2x \sin \alpha$

$R = \sqrt{3^2 + (-4)^2} = 5$ and $3 = R \cos \alpha$ and $4 = R \sin \alpha$

$\Rightarrow \quad \cos \alpha = \dfrac{3}{5}$ and $\sin \alpha = \dfrac{4}{5} \Rightarrow \alpha = 0.9273^c$ (to 4 sf)

$\Rightarrow \quad 3 \sin 2x - 4 \cos 2x \equiv 5 \sin(2x - 0.9273)$

$\Rightarrow \displaystyle\int \dfrac{10}{3 \sin 2x - 4 \cos 2x} \, dx = \int \dfrac{10}{5 \sin(2x - 0.9273)} \, dx$

$\Rightarrow \displaystyle\int \dfrac{10}{3 \sin 2x - 4 \cos 2x} \, dx = \int 2 \operatorname{cosec}(2x - 0.9273) \, dx$

$\quad = -\ln |\operatorname{cosec}(2x - 0.9273) + \cot(2x - 0.9273)| + c$

> Differentiation and integration of trigonometric functions require angles to be in radians so α must be given in radians.

> The examination formulae booklet has
> $\displaystyle\int \operatorname{cosec} x \, dx$
> $= -\ln |\operatorname{cosec} x + \cot x| + c.$
> You could also use the chain rule (C3 section 8.3) or the substitution
> $u = \operatorname{cosec}(2x - 0.9273).$

EXERCISE 6D

1 Express each of the following in the form $R \cos(x - \alpha)$, where R is a positive constant and $0° < \alpha < 90°$.

   (a) $3 \cos x + \sqrt{3} \sin x$ **(b)** $3 \cos x + 4 \sin x$

   (c) $\sqrt{2} \cos x + \sin x$ **(d)** $\cos x + \sqrt{3} \sin x$

2 For each of the expressions in question **1** find **(i)** the greatest value, **(ii)** the least value.

3 Express f(x) in the form $R \sin(x + \alpha)$, where R is a positive constant and $0 < \alpha < \dfrac{\pi}{2}$:

   (a) $f(x) = 3 \sin x + \sqrt{3} \cos x$ **(b)** $f(x) = 5 \sin x + 12 \cos x$

   (c) $f(x) = \cos x + \sqrt{3} \sin x$ **(d)** $f(x) = \sin x + 2\sqrt{2} \cos x$

4 For each of the functions in question **3** find the range of f.

5 Express each of the following in the form $R \sin (x - \alpha)$, where R is a positive constant and $0° < \alpha < 90°$:

   (a) $3 \sin x - \sqrt{3} \cos x$ **(b)** $3 \sin x - 4 \cos x$

   (c) $\sqrt{2} \sin x - \cos x$ **(d)** $\sin x - \sqrt{3} \cos x$

6 For each of the expressions in question **5**, find the smallest positive value of x, in degrees, for which the expression takes its greatest value.

7 For each of the expressions in question **5**, find the smallest positive value of x, in degrees, for which the expression takes its least value.

8 (a) Express $\sqrt{3} \cos 2\theta - \sin 2\theta$ in the form $R \cos (2\theta + \alpha)$, where R is a positive constant and $0 < \alpha < \dfrac{\pi}{2}$.

   (b) The curve $y = \sqrt{3} \cos 2x - \sin 2x$ has domain $0 \leqslant x \leqslant \pi$. Find the coordinates of the stationary points of the curve.

9 The length of a diagonal of a rectangle is 10 cm. The diagonal is inclined at an angle $\theta°$ to a side of the rectangle. Show that the perimeter, P cm, of the rectangle is given by $P = 20\sqrt{2} \sin(\theta + 45)°$.

10 Find $\displaystyle\int_{-\frac{\pi}{4}}^{0} \frac{2}{\cos x - \sin x} \, dx$.

11 Find the value of $\displaystyle\int_{0}^{\frac{\pi}{6}} \frac{4}{\sqrt{3} \sin x + \cos x} \, dx$ to three significant figures.

6.5 Solving equations of the form $a \cos \theta + b \sin \theta = c$

In C2 you solved equations of the form $a \cos \theta = c$ and $b \sin \theta = c$. In C3 you used various identities to solve trigonometric equations. In this section you will learn how to solve equations of the form $a \cos \theta + b \sin \theta = c$ by using the earlier work in this chapter.

> To solve equations of the type $a \cos \theta + b \sin \theta = c$, as a first step, write $a \cos \theta + b \sin \theta$ either in the form $R \cos (\theta \pm \alpha)$ or in the form $R \sin (\theta \pm \alpha)$.

Worked example 6.16

Solve the equation $7 \cos x - 24 \sin x = 10$, for $0° < x < 360°$.

Solution

$7 \cos x - 24 \sin x \equiv R \cos (x + \alpha)$

$7 \cos x - 24 \sin x \equiv R \cos x \cos \alpha - R \sin x \sin \alpha.$

$\Rightarrow \quad R \cos \alpha = 7 \quad$ and $\quad R \sin \alpha = 24$

$\Rightarrow \quad R = \sqrt{7^2 + 24^2} = 25$

$\Rightarrow \quad \cos \alpha = \dfrac{7}{25} \quad$ and $\quad \sin \alpha = \dfrac{24}{25}$

$\Rightarrow \quad \alpha = 73.74°$

So $\quad 7 \cos x - 24 \sin x = 10,$

$\Rightarrow \quad 25 \cos (x + 73.74°) = 10$

$\Rightarrow \quad \cos (x + 73.74°) = 0.4$

$\Rightarrow \quad x + 73.74° = \pm 66.42°, 360° \pm 66.42°, \ldots$ | $\cos^{-1}(0.4) = 66.42\ldots°.$

$\Rightarrow \quad x = -140.16°, -7.32°, 219.84°, 352.68°\ldots$

So, within the range $0° < x < 360°$,

$x = 220°, 353°$, to the nearest degree.

Worked example 6.17

(a) Write $10 \sin x \cos x + 12 \cos 2x$ in the form $R \sin (2x + \alpha)$, where R is a positive constant and $0 < \alpha < \dfrac{\pi}{2}$.

(b) Hence solve the equation $10 \sin x \cos x + 12 \cos 2x + 7 = 0$, for $0 \leqslant x \leqslant \pi$.

Solution

(a) $10 \sin x \cos x + 12 \cos 2x = 5(2 \sin x \cos x) + 12 \cos 2x$

$\qquad\qquad\qquad\qquad\qquad = 5 \sin 2x + 12 \cos 2x$

$5 \sin 2x + 12 \cos 2x \equiv R \sin(2x + \alpha)$

$\qquad\qquad\qquad\qquad \equiv R \sin 2x \cos \alpha + R \cos 2x \sin \alpha$

$\Rightarrow \quad R = \sqrt{5^2 + 12^2} = 13$

$\Rightarrow \quad \cos \alpha = \dfrac{5}{13} \quad$ and $\quad \sin \alpha = \dfrac{12}{13}$

$\Rightarrow \quad \alpha = 1.176\ldots^c.$

So $10 \sin x \cos x + 12 \cos 2x \equiv 13 \sin (2x + 1.176\ldots^c).$

> You must first change the term $10 \sin x \cos x$ so that it is in terms of the angle $2x$.
> Use $\sin 2x = 2 \sin x \cos x$.
> (See section 6.2.)

(b) $10 \sin x \cos x + 12 \cos 2x = -7$

$\Rightarrow \quad 13 \sin (2x + 1.176...^c) = -7$

$\Rightarrow \quad \sin (2x + 1.176...^c) = -0.538\,46...$

$\Rightarrow \quad 2x + 1.176 = \pi + 0.5686...^c, \quad 2\pi - 0.5686...,$

$\Rightarrow \quad 2x = 2.5342..., \quad 4.5385..., \; ... \,.$

$\sin^{-1}(-0.538\,46...) = -0.5686...$ radians.

So, in the range $0 \leqslant x \leqslant \pi$, the solutions of the equation are $x = 1.27^c$ and 2.27^c to three significant figures.

EXERCISE 6E

1 (a) Write $6 \sin x + 8 \cos x$ in the form $R \sin (x + \alpha)$, where R is a positive constant and $0 < \alpha < \dfrac{\pi}{2}$.

(b) Hence solve the equation $6 \sin x + 8 \cos x = 7$, for $0 \leqslant x \leqslant 2\pi$.

2 Solve the equation $5 \cos x - 12 \sin x = 6.5$, for $0° \leqslant x \leqslant 360°$.

3 Solve the equation $4 \cos x + 3 \sin x = 1$ for $0° \leqslant x < 360°$.

4 Solve the equation $3 \sin \theta + 4 \cos \theta = 4$, for $0 \leqslant \theta < 2\pi$.

5 Solve the equation $12 \sin \theta - 9 \cos \theta = 10$, for $0 \leqslant \theta \leqslant 2\pi$.

6 Solve the equation $2 \cos 2\theta + \sqrt{5} \sin 2\theta = 3$, for $-\pi \leqslant \theta \leqslant \pi$.

7 Solve the equation $6 \sin \theta \cos \theta + 4 \cos 2\theta = 1$, for $0 \leqslant \theta \leqslant \pi$.

8 Solve the equation $\cos x = \sin x - 1$, for $0° \leqslant x < 360°$.

9 Solve the equation $8 \sin x = 6 \cos x - 3$, for $-180° \leqslant x < 180°$.

10 Solve the equation $2 \cos^2 x + \sin 2x = 1$, for $0° \leqslant x < 360°$.

MIXED EXERCISE

1 (a) Express $4 \sin \theta - 3 \cos \theta$ in the form $R \sin (\theta - \alpha)$, where R is a positive constant and $0° < \alpha < 90°$. Give the value of α to the nearest $0.1°$.

(b) Hence solve the equation $4 \sin \theta - 3 \cos \theta = 2$ for $0° < \theta < 360°$.

2 Solve the equation $3 \cos 2\theta - \cos \theta + 1 = 0$, giving all solutions in degrees to the nearest degree in the interval $0° \leqslant \theta \leqslant 360°$. [A]

3 (a) Express $6 \sin^2 \theta$ in the form $a + b \cos 2\theta$.

(b) Find the exact value of $\displaystyle\int_0^{\frac{\pi}{12}} 6 \sin^2 \theta \, d\theta$.

(c) Solve the equation $3 - 3 \cos 2\theta = 2 \operatorname{cosec} \theta$ giving all solutions in radians in the interval $0 < \theta < 2\pi$. [A]

6

4 The diagram shows a rectangle *OABC*.

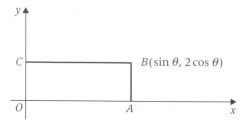

The point *B* has coordinates $(\sin\theta, 2\cos\theta)$, where $0 \leqslant \theta \leqslant \dfrac{\pi}{2}$.

The perimeter of the rectangle is *L*.

(a) **(i)** Write down the length *L* in terms of θ.

   (ii) Hence obtain an expression for *L* in the form

   $R\sin(\theta + \alpha)$, where $R > 0$ and $0 \leqslant \alpha \leqslant \dfrac{\pi}{2}$.

   Give your answer for α to three decimal places.

(b) Given that θ varies between 0 and $\dfrac{\pi}{2}$:

   (i) write down the maximum value of *L*,

   (ii) find the value of θ, to two decimal places, for which *L* is a maximum. [A]

5 **(a)** Given that $\tan x \neq 1$, show that

$$\frac{\cos 2x}{\cos x - \sin x} \equiv \cos x + \sin x.$$

(b) By expressing $\cos x + \sin x$ in the form $R\sin(x + \alpha)$,

solve for $0° \leqslant x \leqslant 360°$, $\dfrac{\cos 2x}{\cos x - \sin x} = \dfrac{1}{2}$. [A]

6 **(a)** Find the value of $\tan^{-1}2.4$ giving your answer in radians to three decimal places.

(b) Express $10\sin\theta + 24\cos\theta$ in the form $R\sin(\theta + \alpha)$,

where $R > 0$ and $0 < \alpha < \dfrac{\pi}{2}$.

(c) Hence:

   (i) write down the maximum value of $10\sin\theta + 24\cos\theta$;

   (ii) find a value of θ at which this maximum value occurs. [A]

7 **(a)** Show that $\dfrac{\cot^2 x}{1 + \cot^2 x} \equiv \cos^2 x$.

(b) Hence:

   (i) evaluate $\displaystyle\int_0^{\frac{\pi}{12}} \frac{\cot^2 x}{1 + \cot^2 x}\, dx$,

   (ii) solve $\dfrac{\cot^2 x}{1 + \cot^2 x} = 2\sin 2x$ for $0 < x < 2\pi$, giving your answers in radians to three significant figures.

Key point summary

1 $\sin(A \pm B) = \sin A \cos B \pm \cos A \sin B$ *p257*
$\cos(A \pm B) = \cos A \cos B \mp \sin A \sin B$

$\tan(A \pm B) = \dfrac{\tan A \pm \tan B}{1 \mp \tan A \tan B},\ \left[A \pm B \neq \left(k + \dfrac{1}{2}\right)\pi\right]$

2 $\sin 2A = 2 \sin A \cos A,$ *p260*
$\cos 2A = \cos^2 A - \sin^2 A,$ or, using $\cos^2 A + \sin^2 A = 1,$
$\cos 2A = 2 \cos^2 A - 1,$
$\cos 2A = 1 - 2 \sin^2 A$

$\tan 2A = \dfrac{2 \tan A}{1 - \tan^2 A}$

3 To integrate either $\sin^2 x$ or $\cos^2 x$ write each in terms   *p263*
of $\cos 2x$.

4 (a) $a \cos \theta + b \sin \theta \equiv R \cos(\theta - \alpha),$ where   *p267*
$R = \sqrt{a^2 + b^2}$ and $\cos \alpha = \dfrac{a}{R}$ and $\sin \alpha = \dfrac{b}{R}$

(b) $a \cos \theta + b \sin \theta \equiv R \sin(\theta + \beta),$ where   *p267*
$R = \sqrt{a^2 + b^2}$ and $\sin \beta = \dfrac{a}{R}$ and $\cos \beta = \dfrac{b}{R}$.

In questions, $R > 0$ and α and β acute will be used.
(Note α and β can be negative in the case when a
and b have different signs.)

5 To solve equations of the type $a \cos \theta + b \sin \theta = c$, as   *p269*
a first step, write $a \cos \theta + b \sin \theta$ either in the form
$R \cos(\theta \pm \alpha)$ or in the form $R \sin(\theta \pm \alpha)$.

These identities are in the
formulae booklet.

6

Test yourself	What to review
1 Solve the equation $\sin 2x \cos 20° + \cos 2x \sin 20° = 0.5$ for $0° \leqslant x \leqslant 360°$.	*Section 6.1*
2 (a) Given that $2 \sin\left(2x - \dfrac{\pi}{4}\right) = \cos\left(2x - \dfrac{\pi}{4}\right)$, show that $\tan 2x = 3$.	*Section 6.1*
(b) Given that x is an acute angle, find the exact value of $\tan x$ in surd form.	*Section 6.2*
3 Solve the equation $4 \sin 2x = \cos x$ for $0° < x < 360°$.	*Section 6.2*
4 Show that $\displaystyle\int_0^{\frac{\pi}{4}} (2 \cos^2 2x - 2 \sin^2 x)\,\mathrm{d}x = \dfrac{1}{2}$.	*Section 6.3*

Test yourself (continued)	What to review

5 (a) Express $2 \sin x + 3 \cos x$ in the form $R \sin(x + \alpha)$, where R is a positive constant and $0° < \alpha < 90°$.

Section 6.4

(b) Hence find the maximum and minimum values of $2 \sin x + 3 \cos x$ and find the smallest positive values of x for which these occur.

---

6 Solve the equation $2 \sin x - \cos x = 1$, for $0° < x < 360°$.

Section 6.5

---

7 Evaluate $\displaystyle\int_0^{\frac{\pi}{8}} \frac{2}{\sin 2\theta + \cos 2\theta}\, d\theta$ giving your answer to three significant figures.

Section 6.4

Test yourself **ANSWERS**

1 $5°, 65°, 185°, 245°$.

2 $\tan x = \dfrac{\sqrt{10} - 1}{3}$.

3 $7.2°, 90°, 172.8°, 270°$.

5 (a) $\sqrt{13} \sin(x + 56.3°)$;
(b) max. $(33.7°, \sqrt{13})$, min. $(213.7°, -\sqrt{13})$.

6 $53.1°, 180°$.

7 0.623 (to 3 sf).

CHAPTER 7

C4: Exponential growth and decay

Learning objectives

After studying this chapter, you should be able to:
- solve equations of the form $a^x = b$ using natural logarithms
- understand what is meant by exponential growth and decay
- realise that $\dfrac{dx}{dt} = kx$ has solution $x = Ae^{kt}$
- solve problems involving growth and decay.

7.1 Equations of the form $a^x = b$

It is sometimes possible to find an exact solution to an equation of the form $a^x = b$.

For instance, the equation $2^x = 512$ has solution $x = 9$.

However, more often equations of this type do not have exact solutions and they need to be solved using logarithms. Your calculator can find logarithms to base 10 and natural logarithms, to base e, and as the next worked example shows, either of these bases can be used.

Worked example 7.1

Solve each of the following equations, giving your answers to three significant figures:

(a) $2^x = 500$,　　　　　　　　(b) $7^y = 13.75$.

Solution

(a) $2^x = 500$

Taking natural logarithms of both sides gives

$$\ln (2^x) = \ln 500$$

$$\Rightarrow x \ln 2 = \ln 500$$

$$\Rightarrow x = \frac{\ln 500}{\ln 2} = 8.965\,784 \ldots$$

Therefore $x = 8.97$ (to three significant figures).

> Recall that $\ln (a^x) = x \ln a$.

> Since $2^x = 512$ has the solution $x = 9$, you would expect the answer to $2^x = 500$ to be a little less than 9.

7

> The solution to the equation $a^x = b$ is $x = \dfrac{\ln b}{\ln a}$, where logarithms are taken to base e.

(b) $7^y = 13.75$

You could take logarithms to base 10 of both sides

$$\log_{10}(7^y) = \log_{10} 13.75$$

$$\Rightarrow y \log_{10} 7 = \log_{10} 13.75$$

$$\Rightarrow y = \frac{\log_{10} 13.75}{\log_{10} 7} = 1.346\,947\,5\ldots$$

> A common mistake is to enter $\log_{10}\left(\dfrac{13.75}{7}\right)$ into a calculator instead of the correct expression.

Therefore $y = 1.35$ (to three significant figures).

> The solution to the equation $a^x = b$ is $x = \dfrac{\log_{10} b}{\log_{10} a}$, where logarithms are taken to base 10.

> You used logarithms to base 10 in unit C2.

7.2 More complicated equations

Sometimes the exponent is more complicated than simply x or y and so care needs to be taken with the use of brackets.

Worked example 7.2

Solve the following equations, giving answers to four significant figures:

(a) $3^{2x-5} = 79$, **(b)** $5^{3-4x} = 7^{x+2}$.

Solution

(a) $3^{2x-5} = 79$ can be solved by taking logarithms to base 10.

$$\ln(3^{2x-5}) = \ln 79$$

$$\Rightarrow (2x-5)\ln 3 = \ln 79$$

> Notice the need for brackets here.

$$\text{Hence, } (2x-5) = \frac{\ln 79}{\ln 3}$$

$$= 3.977\,242\,8\ldots$$

$$\Rightarrow 2x = 8.977\,242\,8\ldots$$

$$\Rightarrow x = 4.488\,621\,4\ldots$$

Therefore $x = 4.489$ (to four significant figures).

(b) $5^{3-4x} = 7^{x+2}$

This time there are exponents on both sides of the equation, but the method of solution, namely taking logarithms, is exactly the same. Taking natural logarithms of both sides,

$$\ln(5^{3-4x}) = \ln(7^{x+2})$$

> Again, notice the need for brackets.

$$\Rightarrow (3 - 4x)\ln 5 = (x + 2)\ln 7$$

$$\Rightarrow (3 - 4x) = (x + 2)\frac{\ln 7}{\ln 5} = (x + 2) \times 1.209\,06 \ldots$$

$$\Rightarrow 3 - (2 \times 1.209\,06 \ldots) = 4x + (x \times 1.209\,06 \ldots)$$

$$\Rightarrow 0.581\,876 \ldots = 5.209\,06 \ldots x$$

$$\Rightarrow x = 0.1117 \text{ (to four significant figures).}$$

EXERCISE 7A

1 Find the exact solutions to each of the following equations:

(a) $2^x = 64$,　　　　　　　　**(b)** $3^x = 2187$,

(c) $2^x = 0.125$,　　　　　　　**(d)** $5^x = 3125$.

2 Find the solutions to each of the following, giving your answers to three significant figures. *You may find the answers to question 1 helpful in confirming your results.*

(a) $2^x = 62$,　　　　　　　　**(b)** $3^x = 2200$,

(c) $2^x = 0.13$,　　　　　　　**(d)** $5^x = 3110$.

3 Solve each of the following, giving your answers to three significant figures.

(a) $7^x = 0.46$,　　　　　　　**(b)** $4^x = 200$,

(c) $3^x = 19$,　　　　　　　　**(d)** $12^x = 58.9$.

4 Given that $3^{2x+5} = 17$, show that $x = \dfrac{1}{2}\left(\dfrac{\ln 17}{\ln 3} - 5\right)$.

5 Solve the following equations, leaving your answers in terms of natural logarithms:

(a) $2^{x+4} = 6$,　　　　　　　**(b)** $3^{2x-1} = 17$,

(c) $2^{1-4x} = 5$,　　　　　　　**(d)** $5^{3x+4} = 31$.

6 Solve each of the following, giving your answers to three significant figures:

(a) $3^{2x-5} = 7$,　　　　　　　**(b)** $5^{7x-1} = 12$,

(c) $2^{5x-2} = 19$,　　　　　　**(d)** $7^{3-2x} = 19$.

7

7 Solve each of the following, giving your answers to four significant figures:

(a) $9^{x-7} = 17.5$, (b) $3^{4x-5} = 0.123$,

(c) $13^{x-5} = 1.345$, (d) $17^{5-4x} = 19.436$,

(e) $8^{6x-7} = 67.23$, (f) $6^{7x-4} = 23.89$.

8 Solve each of the following, giving your answers to three significant figures:

(a) $3^x = 7^{x+1}$, (b) $5^{x-1} = 4^{1-3x}$,

(c) $2^{x-2} = 9^{3-5x}$, (d) $7^{1-2x} = 4^{x+3}$,

(e) $7^{x+1} = 5^{1+3x}$, (f) $3^{2x-3} = 11^{2-3x}$.

9 Given that $2^{3x-5} = 3^{3-7x}$, show that $x = \dfrac{3k+5}{7k+3}$, where $k = \dfrac{\ln 3}{\ln 2}$.

10 State the condition that must be satisfied by b for the equation $2^x = b$ to have a solution.

7.3 Exponential growth

The graph of $N = 5000 \times 2^t$ is shown below for $t \dots 0$.

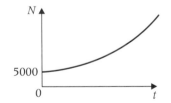

The graph increases very rapidly and demonstrates what is known as exponential growth. The graph might represent the growth of a certain strain of bacteria in a container which initially contains 5000 bacteria and t might represent the time in hours measured from the instant an experiment begins.

Suppose you wished to find the value of t when $N = 8000$.

You could write $8000 = 5000 \times 2^t$.

Then, $2^t = \dfrac{8}{5} = 1.6$,

and by taking natural logarithms

$$t = \frac{\ln 1.6}{\ln 2} \approx 0.678\,071\,9 \dots$$

So the time taken is approximately 0.678 hours or 41 minutes (to the nearest minute).

> The formula $x = a \times b^{kt}$, where a, b and k are positive constants, indicates that x is growing exponentially.

7.4 Exponential decay

Suppose the mass, M, measured in grams, of a block of ice when it is melting is given by the formula $M = 250 \times 3^{-2t}$, where t is the time in hours measured from a particular instant. The graph is shown below.

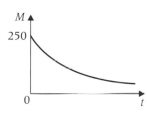

This is an example of exponential decay, where eventually M would reduce to the value zero.

How long would it take for the block of ice to reduce to half its original size?

Writing $125 = 250 \times 3^{-2t}$ gives $3^{-2t} = \dfrac{1}{2}$.

Taking natural logarithms, $-2t \ln 3 = \ln 0.5$.

Therefore, $2t = -\dfrac{\ln 0.5}{\ln 3} \approx 0.6309 \ldots$.

So that t is approximately 0.315 and hence the time taken is about 0.315 hours or 19 minutes, to the nearest minute.

> The formula $x = a \times b^{-kt}$, where a, b and k are positive constants, indicates that x is decaying exponentially.

Worked example 7.3

A radioactive substance decays so that its mass, m grams, after time t years from a given instant is given by $m = m_0 e^{-\frac{t}{50}}$.

(a) Find the percentage change in mass after 2 years.

(b) Find the time taken for the mass to reduce to a quarter of its original size.

> The symbol m_0 is often used by scientists to represent the mass when $t = 0$.

Solution

(a) When $t = 2$, $m = m_0 e^{-\frac{1}{25}} = m_0 \times 0.960\,789 \ldots$.
   The mass is now 96% of its original value so the percentage change is a decrease of 4%.

7

(b) When $m = \dfrac{1}{4} m_0$, the equation becomes $\dfrac{1}{4} m_0 = m_0 e^{-\frac{t}{50}}$.

Hence, $\dfrac{1}{4} = e^{-\frac{t}{50}}$ and taking natural logarithms gives

$$-\frac{t}{50} = \ln \frac{1}{4} \approx -1.386\ 29 \ldots .$$

So that $t = 50 \times 1.386\ 29 \ldots \approx 69.3147 \ldots .$

Which means it will take approximately 69 years to reduce to a quarter of its original size.

EXERCISE 7B

1 Given that $N = 300 \times 2^t$, sketch the graph of N against t for $t \geqslant 0$. Find the value of t when:

   (a) $N = 500$,   **(b)** $N = 1000$,   **(c)** $N = 1800$.

2 Given that $x = 500 \times 3^{2t}$, state the value of x when $t = 0$. Find the value of t when:

   (a) $x = 1500$,   **(b)** $x = 4000$,   **(c)** $x = 10^{10}$.

3 Given that $W = 70 \times 5^{-t}$, sketch the graph of W against t for $t \geqslant 0$. Find the value of t when:

   (a) $W = 70$,   **(b)** $W = 14$,   **(c)** $W = 0.001$.

4 The temperature $T\,°C$ of liquid in a container m minutes after a particular instant is given by $T = 20 + 60 \times 3^{-\frac{m}{5}}$.

   (a) State the initial temperature.

   (b) Calculate the temperature after 5 minutes.

   (c) Find the time taken for the temperature to fall to 25°C.

5 The amount of money, £P, in a bank account at time t months after the account is opened is given by $P = 500 \times 1.002^t$.

   (a) Find the initial amount of money in the account.

   (b) Find the amount of money, to the nearest penny, in the account after one year.

   (c) Determine the length of time, in complete months, when there first will be £600 in the account.

   (d) Comment on whether the formula is likely to be valid when t is not a positive integer.

6 The population, P million, for England and Wales from 1841 to 1901 can be modelled by the equation $P = 15.9 \times 1.012^t$, where t is the number of years after 1 January 1841.

   (a) Use the model to estimate:

       (i) the population on 1 January 1861,

       (ii) the year during which the population first reached 26 million.

   (b) Find P when $t = 160$. Discuss what this predicts and comment on whether this is likely to be accurate.

7.5 Rate of change

When $x = a \times b^{kt}$ it is not easy to find an expression for $\dfrac{dx}{dt}$.

However, any expression of the form $a \times b^{kt}$ can be changed into the form $a \times e^{ct}$, where $c = k \ln b$, and this expression can be easily differentiated with respect to t.

Worked example 7.4

Given that $x = 10e^{5t}$, show that $\dfrac{dx}{dt} = kx$, stating the value of k.

Solution

$x = 10e^{5t} \implies \dfrac{dx}{dt} = 5 \times 10e^{5t} = 5x.$

Therefore, $\dfrac{dx}{dt} = kx$, where $k = 5$.

This example illustrates an important aspect of exponential growth, namely that the rate of increase of x is actually proportional to x.

In general, $x = Ae^{kt} \implies \dfrac{dx}{dt} = kx.$

7

Worked example 7.5

(a) Given that $N = Ae^{kt}$, where A and k are constants, show that $\dfrac{dN}{dt} = kN$.

(b) The number of bacteria, N, in a colony is such that the rate of increase of N is proportional to N. The time, t, is measured in hours from the instant that $N = 2$ million. When $t = 3$, $N = 5$ million. Find the value of t when $N = 8$ million. [A]

Solution

(a) Differentiating $N = Ae^{kt}$ with respect to t gives

$$\frac{dN}{dt} = k \times Ae^{kt}.$$

Therefore, $\dfrac{dN}{dt} = kN$.

(b) The statement 'the rate of increase of N is proportional to N',

is interpreted mathematically as $\dfrac{dN}{dt} = kN$.

This means that you need to use the result from **(a)** and therefore $N = Ae^{kt}$.

When $t = 0$, $N = 2$ million $\Rightarrow A = 2 \times 10^6$.

When $t = 3$, $N = 5$ million $\Rightarrow 5 \times 10^6 = 2 \times 10^6 \times e^{3k}$

$\Rightarrow e^{3k} = 2.5 \Rightarrow 3k = \ln 2.5 \Rightarrow k = 0.3054\ldots$

So when $N = 8$ million, you have $8 \times 10^6 = 2 \times 10^6 \times e^{kt}$

$\Rightarrow e^{kt} = 4 \Rightarrow kt = \ln 4$

$\Rightarrow t = \dfrac{\ln 4}{k}$

$\Rightarrow t \approx 4.54$ (to three significant figures).

Worked example 7.6

(a) Given that $T = Ae^{-kx}$, show that $\dfrac{dT}{dx} = -kT$, where k and A are positive constants.

(b) The rate of decrease of temperature, $T\,°C$, of the liquid in a container is proportional to the temperature of the liquid at any instant. Initially the temperature is $70°C$, and after 10 minutes the temperature has fallen to $50°C$. Find the temperature after a further 10 minutes.

Solution

(a) $T = Ae^{-kx} \Rightarrow \dfrac{dT}{dx} = -k \times Ae^{-kx} \Rightarrow \dfrac{dT}{dx} = -k \times T.$

(b) The statement 'the rate of decrease of temperature of the liquid in a container is proportional to the temperature of the liquid' can be interpreted as $\dfrac{dT}{dx} = -kT$.

Using **(a)** with T for temperature and x minutes for the time of cooling, the temperature T must satisfy $T = Ae^{-kx}$.

When $x = 0$, $T = 70 \Rightarrow 70 = Ae^0 = A$.

When $x = 10$, $T = 50 \Rightarrow 50 = Ae^{-10k} = 70e^{-10k}$.

$\Rightarrow e^{-10k} = \dfrac{5}{7} \Rightarrow e^{10k} = \dfrac{7}{5} = 1.4 \Rightarrow 10k = \ln 1.4 \approx 0.33647\ldots$

Hence $k \approx 0.033\,647\ldots$.

After a further 10 minutes, $x = 20$, so $T = 70e^{-20k} = 35.714\ldots$.

The temperature has fallen to about $35.7°C$.

EXERCISE 7C

1 Express each of the following in the form e^{kx}:

   (a) 3^x,   (b) 2^x,   (c) 5^{2x},   (d) 4^{-x},

   (e) 7^{-3x},   (f) $6^{\frac{x}{2}}$,   (g) $5^{-\frac{x}{3}}$.

2 Given that $y = 4e^{3x}$, show that $\dfrac{dy}{dx} = ky$, stating the value of the constant k. Find the value of $\dfrac{dy}{dx}$ when $x = 0$.

3 Given that $y = 5e^{-2x}$, show that $\dfrac{dy}{dx} = py$, stating the value of the constant p. Find the exact value of y when $x = \ln 3$ and hence find the corresponding value of $\dfrac{dy}{dx}$.

4 You are given that $y = Ae^{kx}$, where A and k are constants, and $y = 5$ when $x = 0$.

   (a) Find the value of A.

   (b) In addition, when $x = 2$, $y = 15$. Find the exact value of k and hence find the value of y when $x = 3$.

5 (a) Given that $P = Ae^{kt}$, where A and k are constants, show that $\dfrac{dP}{dt} = kP$.

   (b) The population, P, of insects in a colony is such that the rate of increase of P is proportional to P. The time, t, is measured in months from the instant that $P = 3000$. When $t = 5$, $P = 8000$. Find the value of t when $P = 15\,000$.

6 (a) Given that $m = m_0 e^{-kt}$, show that $\dfrac{dm}{dt} = -km$, where k and m_0 are positive constants.

   (b) The rate of decrease of mass of a ball of snow is proportional to the mass of the snow remaining at any instant. Initially the mass is 500 g and after 4 minutes the mass has fallen to 300 g. Find the mass after a further 6 minutes.

7 (a) Given that $N = N_0 e^{kt}$, show that $\dfrac{dN}{dt} = kN$, where N_0 and k are constants.

   (b) The number, N, of bacteria in a tank of water increases at a rate proportional to the number present at any instant. After timing for 3 hours, the number is 5000 and after 5 hours the number is 10 000. Find the initial number of bacteria and the time taken to reach 20 000.

7.6 General solution to the rate of change equation

Problems involved with growth and decay often involve a statement such as 'the rate of change of the quantity q is proportional to the quantity present at a given time t.'

This can be interpreted as $\dfrac{dq}{dt} = k \times q$.

In the previous section, you verified that when $q = Ae^{kt}$, it follows by differentiation that $\dfrac{dq}{dt} = k \times q$.

In this section you will prove the converse of this result.

Worked example 7.7

Given that $\dfrac{dy}{dt} = 5y$ and $y = 2$ when $t = 0$, prove that $y = 2e^{5t}$.

Solution

Since $\dfrac{dy}{dt} = 5y$, and from earlier work you know that $\dfrac{dt}{dy} = \dfrac{1}{\frac{dy}{dt}}$,

therefore you can write $\dfrac{dt}{dy} = \dfrac{1}{5y}$.

It now follows that $t = \displaystyle\int \dfrac{1}{5y}\,dy = \dfrac{1}{5}\int \dfrac{1}{y}\,dy = \dfrac{1}{5}\ln y + C$, where C is a constant.

Putting $t = 0$ and $y = 2$, gives $0 = \dfrac{1}{5}\ln 2 + C$

or $C = -\dfrac{1}{5}\ln 2$.

Hence, $t = \dfrac{1}{5}\ln y - \dfrac{1}{5}\ln 2 = \dfrac{1}{5}(\ln y - \ln 2) = \dfrac{1}{5}\ln\left(\dfrac{y}{2}\right)$

$\Rightarrow 5t = \ln\left(\dfrac{y}{2}\right)$

Taking logarithms of both sides leads to $e^{5t} = \dfrac{y}{2} \Rightarrow y = 2e^{5t}$.

Worked example 7.8

Given that $\dfrac{dQ}{dx} = -4Q$ and $Q = 3$ when $x = 0$, find an expression for Q in terms of x.

Solution

$\dfrac{dQ}{dx} = -4Q \Rightarrow \dfrac{dx}{dQ} = \dfrac{-1}{4Q}$

Integrating with respect to Q gives

$x = \displaystyle\int \dfrac{-1}{4Q}\,dQ = -\dfrac{1}{4}\int \dfrac{1}{Q}\,dQ = -\dfrac{1}{4}\ln Q + C$

When $x = 0$, $Q = 3$. Therefore, $0 = -\dfrac{1}{4}\ln 3 + C$ or $C = \dfrac{1}{4}\ln 3$

Hence, $x = -\dfrac{1}{4}(\ln Q - \ln 3) = -\dfrac{1}{4}\ln\left(\dfrac{Q}{3}\right)$

Multiplying both sides by -4

$$-4x = \ln\left(\dfrac{Q}{3}\right)$$

Hence, $e^{-4x} = \dfrac{Q}{3} \Rightarrow Q = 3e^{-4x}$

Although you need to be able to derive solutions as in the previous two worked examples, it is important that you are aware of the following key result.

> The general solution of $\dfrac{dx}{dt} = kx$ is $x = Ae^{kt}$, where A is an arbitrary constant.

Worked example 7.9

Write down the general solutions of the following rate of change equations:

(a) $\dfrac{dN}{dt} = 4N,$ **(b)** $\dfrac{dP}{dt} = -3P,$ **(c)** $5\dfrac{dQ}{dt} = Q.$

Solution

(a) Using the result in the previous Key Point with $k = 4$ gives the general solution as

$$N = Ae^{4t},$$

where A is an arbitrary constant.

(b) The value of k is negative in this case, $k = -3$. The general solution is $P = Be^{-3t}$, where B is an arbitrary constant.

(c) You need to rearrange the equation so it is of the form given in the Key Point above.

$$5\dfrac{dQ}{dt} = Q \Rightarrow \dfrac{dQ}{dt} = \dfrac{1}{5}Q$$

So $k = \dfrac{1}{5}$, and the general solution is $Q = Ce^{\frac{t}{5}}$, where C is an arbitrary constant.

Clearly any letter other than A could have been used for the arbitrary constant.

Worked example 7.10

The yearly rate of increase of a population, P, of birds in a colony is proportional to P. On 1 July 2000 there were 300 birds and a year later there were 380 in the colony.

(a) Find an expression for P in terms of the time in years, t, measured from 1 July 2000.

(b) Predict the number of birds in the colony on 1 July 2008.

Solution

(a) $\dfrac{\mathrm{d}P}{\mathrm{d}t} = kP$, which has general solution $P = Ae^{kt}$.

When $t = 0$, $P = 300$. So $300 = Ae^0 = A$.

Therefore, $P = 300e^{kt}$.

Also, $P = 380$, when $t = 1 \Rightarrow 380 = 300e^k$.

Hence, $e^k = \dfrac{380}{300} = \dfrac{19}{15}$.

Since $e^{kt} = (e^k)^t$, it follows that

$$P = 300 \times \left(\frac{19}{15}\right)^t$$

> Notice that this model takes no account of the reduction of numbers in the colony as birds die or leave the colony, etc.

(b) When $t = 8$, $P = 300 \times \left(\dfrac{19}{15}\right)^8 \approx 1988$.

EXERCISE 7D

1 Given that $\dfrac{\mathrm{d}y}{\mathrm{d}t} = 7y$ and $y = 4$ when $t = 0$, prove that $y = 4e^{7t}$.

2 Given that $\dfrac{\mathrm{d}x}{\mathrm{d}t} = -3x$ and $x = 9$ when $t = 0$, prove that

$x = 9e^{-3t}$.

3 Given that $\dfrac{\mathrm{d}N}{\mathrm{d}x} = -2N$ and $N = 8$ when $x = 0$, find an

expression for N in terms of x.

4 Given that $\dfrac{\mathrm{d}P}{\mathrm{d}t} = 8P$ and $P = 4$ when $t = 0$, find an expression
for P in terms of t.

5 Write down the general solutions of the following rate of change equations:

(a) $\dfrac{\mathrm{d}N}{\mathrm{d}t} = 5N$,   **(b)** $\dfrac{\mathrm{d}P}{\mathrm{d}t} = -7P$,   **(c)** $6\dfrac{\mathrm{d}Q}{\mathrm{d}t} = 5Q$.

6 The yearly rate of increase of the population, P, in a certain country is proportional to P. On 1 January 2000 there were 7 million and the population rose to 8 million six years later.

(a) Find an expression for P in terms of the time t measured in years from 1 January 2000.

(b) Predict the population of the country on 1 January 2015.

7 A radioactive substance decays at a rate proportional to its mass m. Initially its mass was 2×10^{-3} g and after 50 days its mass reduced to 1.5×10^{-3} g.

  (a) Find an expression for m in terms of the time t, in days, from when the mass was 2×10^{-3} g.

  (b) How long would it take for the mass to reach half its original size?

8 The amount of money, £Q, in a special deposit account grows at a rate proportional to the amount in that account at any moment in time. Initially £20 000 is deposited and after 8 years it grows to £26 000.

  (a) Find an expression for Q in terms of t, the time in years of the investment.

  (b) Determine how long it takes to have £30 000 in the account.

9 The population, P, of insects in a colony is given by $P = Ae^{kt}$, where A and k are constants and the time t is measured in months.

  (a) Given that $P = 500$ when $t = 0$ and that $P = 750$ when $t = 10$, find the value of k.

  (b) Find the value of t when $P = 1500$, giving your answer to three significant figures.   [A]

10 The amount of money, £P, in a special savings account at time t years after 1 January 2000 is given by $P = 100 \times 1.05^t$.

  (a) State the amount of money in the account on 1 January 2000.

  (b) Calculate, to the nearest penny, the amount of money in the account on 1 January 2004.

  (c) Find the value of t when $P = 150$, giving your answer to three significant figures.   [A]

11 The decay of a radioactive substance can be modelled by the equation $m = m_0e^{-kt}$, where m grams is the mass at time t years, m_0 grams is the initial mass and k is a constant.

  (a) The time taken for a sample of the radioactive substance strontium 90 to decay to half its initial mass is 28 years. Show that the value of k is approximately 0.024 755.

  (b) A sample of strontium 90 has a mass of 1 gram. Assuming this mass has resulted from radioactive decay, use the model to find the mass this sample would have had 100 years ago. Give your answer to three significant figures.   [A]

12 (a) The quantity N varies with t so that $\dfrac{dN}{dt} = 4N$, and $N = 200$ when $t = 0$.

    (i) Write down an expression for $\dfrac{dt}{dN}$ in terms of N.

    (ii) Given that $4t = \displaystyle\int \dfrac{1}{N}\, dN$ and using the fact that $N = 200$ when $t = 0$, find an expression for t in terms of N.

    (iii) Hence, show that $N = 200e^{4t}$.

(b) The number of bacteria, N, in a colony is such that the **hourly** rate of increase of N is equal to four times the number of bacteria present. Initially there are 200 bacteria present. Use the results from **(a)** to find the time taken for the number of bacteria to grow to 700, giving your answer to the nearest minute.   **[A]**

13 Observations were made of the number of bacteria in a certain specimen. The number N present after t minutes is modelled by the formula $N = Ac^t$ where A and c are constants. Initially there are 1000 bacteria in the specimen.

(a) Write down the value of A.

(b) Given that there are 12 000 bacteria after 60 minutes, show that the value of c is 1.0423 to four decimal places.

(c) **(i)** Express t in terms of N.

    (ii) Calculate, to the nearest minute, the time taken for the number of bacteria to increase from one thousand to one million.   **[A]**

14 (a) A car has a value of £15 000 when new and a value of £11 000 exactly 2 years later. The value of the car as it depreciates is modelled by $V = Pe^{-kt}$, where £V is the value of the car t years after it is sold as new, and P and k are constants.

    (i) State the value of P.

    (ii) Find the value of k, giving your answer to three decimal places.

(b) Another car depreciates according to the model $W = 18\,000e^{-0.175t}$, where £W is its value t years after it is sold as new. Assuming that both cars were sold as new on 1 January 2005, calculate the year during which they will have depreciated to the same value.   **[A]**

Key point summary

1 The solution to the equation $a^x = b$ is $x = \dfrac{\ln b}{\ln a}$, when   *p276*
logarithms are taken to base e.

2 The solution to the equation $a^x = b$ is $x = \dfrac{\log_{10} b}{\log_{10} a}$,   *p276*
when logarithms are taken to base 10.

3 The formula $x = a \times b^{kt}$, where a, b and k are positive   *p279*
constants, indicates that x is growing exponentially.

4 The formula $x = a \times b^{-kt}$, where a, b and k are positive *p279*
constants, indicates that x is decaying exponentially.

5 In general, $x = Ae^{kt} \Rightarrow \dfrac{dx}{dt} = kx$.   *p281*

6 The general solution of $\dfrac{dx}{dt} = kx$ is $x - Ae^{kt}$, where A *p285*
is an arbitrary constant.

Test yourself	**What to review**

1 Solve the equation $6^x = 19$, giving your answer to three
significant figures.

Section 7.1

2 Given that $5^{4x-1} - 19$, show that $x = \dfrac{1}{4}\left(\dfrac{\ln 19}{\ln 5} + 1\right)$.

Section 7.2

3 The population, P million, for a particular country is given
by $P = 12.3 \times 1.02^t$, where t is the number of years after
1 January 2003. Use the formula to predict:

(a) the population on 1 January 2010,

(b) the year when the population will first reach 20 million.

Section 7.3

4 Given that $N = 300e^{-\frac{t}{12}}$:

(a) show that $\dfrac{dN}{dt} = kN$ and state the value of k,

(b) use the formula to find the value of t when $N = 200$.

Section 7.5

5 An experiment reveals that the number, N, of bacteria in a
test tube increases at a rate proportional to the number present
at any instant. Initially there are 2000 present and after 10 days
the number has risen to 3000.

(a) Find an expression for N in terms of the time, t, that has
elapsed in days since the experiment began.

(b) Calculate the time taken for the number of bacteria to
increase to 5000.

Section 7.6

7

1 1.64.

3 (a) 14.1 million; **(b)** 2027.

4 (a) $k = -\dfrac{1}{12}$; **(b)** 4.87.

5 (a) $N = 2000e^{kt}$, where $k = \dfrac{\ln 1.5}{10} \approx 0.0405$; **(b)** 22.6 days.

CHAPTER 8

C4: Differential equations

Learning objectives

After studying this chapter, you should be able to:
- formulate first order differential equations
- solve analytically first order differential equations with separable variables.

8.1 Introduction

In this chapter you will learn how to formulate a differential equation using the knowledge you acquired in earlier modules, particularly in C4 chapter 7, where you considered growth and decay rates.

> A **differential equation** is an equation which involves at least one derivative of a variable with respect to another variable.
> For example,
>
> $$\frac{dy}{dx} = 2x - 3, \quad x\frac{dx}{dt} = e^t \sin x, \quad \frac{d^2m}{dt^2} + 3\frac{dm}{dt} + 4m = e^t.$$

In this chapter you will only consider first order differential equations.

> **First order differential equations** are differential equations in which the highest derivative is the first.
> For example,
>
> $$\frac{dy}{dx} = 2x^4 + 3, \quad x^2\frac{dx}{dt} = \ln t.$$

$\dfrac{d^2m}{dt^2} + 3\dfrac{dm}{dt} + 4m = e^t$ is a second order differential equation because its highest derivative is $\dfrac{d^2m}{dt^2}$.

8.2 Forming differential equations

Differential equations are frequently used to describe physical laws. The most widely used first order differential equations describe exponential growth or decay. You were introduced to this topic in the previous chapter.

If P is proportional to Q,
$P \propto Q \Rightarrow P = kQ$, where k is a constant of proportionality.

In this section you will need to convert the following statements, given in words, to the corresponding mathematical differential equations:

- The rate of **increase** of x is proportional to x

$$\Rightarrow \frac{dx}{dt} = kx;$$

- The rate of **decrease** of x is proportional to x

$$\Rightarrow \frac{dx}{dt} = -kx;$$

where k is a constant of proportionality.

Worked example 8.1

Form a differential equation for the information in each of the following:

(a) The rate of decrease of mass m of a ball of snow is proportional to the mass of snow remaining at any instant.

(b) The rate of spread of a rumour is proportional to the product of the fraction of the population who have heard the rumour and the fraction of the population who have not heard the rumour.

(c) The gradient of a curve at each point (x, y) on the curve is equal to the square of the sum of the coordinates of that point.

Solution

(a) Rate of change of m is $\dfrac{dm}{dt}$,

so rate of **decrease** of m is $-\dfrac{dm}{dt} \Rightarrow -\dfrac{dm}{dt} \propto m$

$$\Rightarrow \frac{dm}{dt} = -km, \text{ where } k \text{ is a positive constant.}$$

(b) Rate of spread of a rumour at time t is the rate of increase of y.
The fraction of the population who have not heard the rumour is $1 - y$.

Hence,

$$\frac{dy}{dt} \propto y(1 - y)$$

$$\Rightarrow \frac{dy}{dt} = ky(1 - y), \text{ where } k \text{ is a positive constant.}$$

For example if $\frac{2}{5}$ of the population have heard the rumour then $\frac{3}{5}$ of the population have not heard it.

$k > 0$ since 'spread' $\Rightarrow$ 'increase'

(c) The gradient of a curve at the point (x, y) is given by $\dfrac{dy}{dx}$. The sum of the coordinates of the point is $(x + y)$.

$$\Rightarrow \frac{dy}{dx} = (x + y)^2.$$

The next worked example shows you how to find a first order differential equation which is equivalent to an equation relating two variables and containing one arbitrary constant.

> $y = x^2 + A$, where A is an arbitary constant, and $\dfrac{dy}{dx} = 2x$ are equivalent.

Worked example 8.2

By eliminating the arbitrary constant A find a first order differential equation that is equivalent to $x^2 + y^2 = Ay$.

Solution

Differentiating both sides of $x^2 + y^2 = Ay$ with respect to x gives

$$2x + 2y\frac{dy}{dx} = A\frac{dy}{dx}.$$

> Use implicit differentiation. See section 4.2.

Multiplying throughout by $y \Rightarrow 2xy + 2y^2\dfrac{dy}{dx} = Ay\dfrac{dy}{dx}$

and using the given equation $\Rightarrow 2xy + 2y^2\dfrac{dy}{dx} = (x^2 + y^2)\dfrac{dy}{dx}$

and so the required first order differential equation is

$(x^2 - y^2)\dfrac{dy}{dx} = 2xy.$

The next worked example shows you how a given substitution can be used to write a first order differential equation into a more suitable form.

8

Worked example 8.3

The gradient of a curve at each point (x, y) on the curve is equal to four times the x-coordinate of the point minus twice the y-coordinate of the point.

(a) Express this information in the form of a differential equation.

(b) Given that $y = z + 2x$, show that your answer to part **(a)** can be written as $\dfrac{dz}{dx} = -2(1 + z)$.

Solution

(a) The gradient of the curve at (x, y) is $\dfrac{dy}{dx}$.

Four times the x-coordinate of the point minus twice the y-coordinate of the point is $4x - 2y$.

So the required differential equation is $\dfrac{dy}{dx} = 4x - 2y$.

(b) Differentiate $y = z + 2x$ with respect to x

$$\frac{d}{dx}(y) = \frac{d}{dx}(z + 2x) = \frac{d}{dx}(z) + \frac{d}{dx}(2x)$$

$$\Rightarrow \frac{dy}{dx} = \frac{dz}{dx} + 2.$$

So answer to **(a)** becomes $\frac{dz}{dx} + 2 = 4x - 2(z + 2x)$

$$\Rightarrow \frac{dz}{dx} = 4x - 2z - 4x - 2$$

$$\Rightarrow \frac{dz}{dx} = -2(1 + z) \text{ as required.}$$

EXERCISE 8A

1 Form a differential equation for the information in each of the following:

(a) A radioactive substance decays at a rate proportional to the mass m at time t.

(b) The yearly rate of increase of a population, P, of birds in a colony is proportional to P.

(c) The value, £V, of a car at age t months depreciates at a rate which is proportional to V.

(d) The rate of cooling of a body is proportional to the difference between the temperature, T, of the body and the constant temperature T_0 of its surroundings at time t.

(e) The gradient of a curve at each point (x, y) on the curve is equal to twice the product of the coordinates of that point.

2 By eliminating the arbitrary constant A find a first order differential equation that is equivalent to each of the following:

(a) $y = Ax$	**(b)** $y = Ax + 2$	**(c)** $y^2 = Ax$
(d) $y = Ae^{2x}$	**(e)** $y = A \sin x$	**(f)** $Ay = x - 2$
(g) $y = Ax + \ln x$	**(h)** $x^2 + y^2 = A$	**(i)** $x^3 + y^3 = Ay$

3 The gradient of a curve at each point (x, y) on the curve is equal to the sum of twice the x-coordinate of the point and four times the y-coordinate of the point.

(a) Express this information in the form of a differential equation.

(b) Given that $2y = 2z - x$, show that your answer to part **(a)** can be written as $\frac{dz}{dx} = \frac{1}{2}(8z + 1)$.

8.3 Finding the general solution of a first order differential equation by separation of variables

First order differential equations of the form $\dfrac{dy}{dx} = f(x, y)$ can be

solved by separation of variables if $f(x, y)$ can be written as $g(x)h(y)$.

For example, $\dfrac{dy}{dx} = \dfrac{x^2 y}{y^2 + 1}$ can be written as $\dfrac{dy}{dx} = (x^2)\left(\dfrac{y}{y^2 + 1}\right)$, so

$\dfrac{dy}{dx} = \dfrac{x^2 y}{y^2 + 1}$ can be solved by separation of variables, but

$\dfrac{dy}{dx} = y^2 + \sin x$ cannot be solved by separation of variables because

$y^2 + \sin x$ cannot be split into the form $g(x)h(y)$.

Consider the general differential equation $\dfrac{dy}{dx} = g(x)h(y)$.

Dividing both sides by $h(y)$ gives $\dfrac{1}{h(y)}\dfrac{dy}{dx} = g(x)$.

Integrating both sides with respect to $x \Rightarrow \displaystyle\int \dfrac{1}{h(y)}\dfrac{dy}{dx}\, dx = \int g(x)\, dx$

$\Rightarrow \displaystyle\int \dfrac{1}{h(y)}\, dy = \int g(x)\, dx$

> The variables y and x are now separated.

Integrating both sides and combining the constants of integration leads to

> Some differential equations, just like some integrals, cannot be solved analytically. They have to be solved numerically. Solving differential equations numerically involves a level of mathematics beyond this module.

The **general solution** of the first order differential

equation $\dfrac{dy}{dx} = g(x)h(y)$ is given by

$$\int \dfrac{1}{h(y)}\, dy = \int g(x)\, dx + A, \text{ where } A$$

is an arbitrary constant.

> For example, $\displaystyle\int 2y\, dy = \int 2x\, dx$
>
> $y^2 + c_1 = x^2 + c_2 \Rightarrow y^2 = x^2 + A$, where $A = c_2 - c_1$.

> Note that the general solution of a first order differential equation must contain one arbitrary constant. The arbitrary constant can be written in different forms. If both integrals give logarithmic terms it is often more convenient to write the arbitrary constant in the form ln B and then to use the laws of logarithms (see C2).

Worked example 8.4

Find the general solution of the differential equation

$\dfrac{1}{2x}\dfrac{dy}{dx} = 2y + 3$ giving your answer in the form $y = f(x)$.

Solution

$\dfrac{1}{2x}\dfrac{dy}{dx} = (2y + 3)$

Rewrite the differential equation as $\dfrac{1}{(2y + 3)}\dfrac{dy}{dx} = 2x$.

Integrate both sides with respect to x $\Rightarrow \displaystyle\int \dfrac{1}{2y + 3}\dfrac{dy}{dx}\,dx = \int 2x\,dx$

$\Rightarrow \displaystyle\int \dfrac{1}{2y + 3}\,dy = \int 2x\,dx$

Integrating both sides and combining the constants of integration
gives $\dfrac{1}{2}\ln|2y + 3| = x^2 + A$, where A is an arbitrary constant

$\Rightarrow \ln|2y + 3| = 2x^2 + 2A$

$\Rightarrow 2y + 3 = e^{2x^2 + 2A}$

$\Rightarrow 2y + 3 = e^{2x^2}e^{2A}$

But e^{2A} is a constant, call it B, so
the general solution of $\dfrac{1}{2x}\dfrac{dy}{dx} = 2y + 3$ is $y = \dfrac{1}{2}(Be^{2x^2} - 3)$,
where B is an arbitrary constant.

> Insert brackets if none appear on a right-hand side with more than one term.

> The variables y and x are now separated.

> Taking the exponential of both sides and using $e^{\ln z} = z$.

> Note that the solution may be written in the form $y = Ce^{2x^2} - \dfrac{3}{2}$, where $C\left(=\dfrac{B}{2}\right)$ is an arbitrary constant, so take care when checking answers to the exercises.

Worked example 8.5

Find the general solution of the differential equation $\dfrac{dN}{dt} = -\dfrac{N}{12}$
giving your answer in the form $N = f(t)$.

> Compare Ex7C Q7.

Solution

$\dfrac{dN}{dt} = -\dfrac{N}{12}$

Rewrite the differential equation as $\dfrac{1}{N}\dfrac{dN}{dt} = -\dfrac{1}{12}$.

Integrate both sides with respect to t $\Rightarrow \displaystyle\int \dfrac{1}{N}\dfrac{dN}{dt}\,dt = \int -\dfrac{1}{12}\,dt$

$\Rightarrow \displaystyle\int \dfrac{1}{N}\,dN = \int -\dfrac{1}{12}\,dt$

Integrating both sides and combining the constants of integration
gives $\ln|N| = -\dfrac{1}{12}t + A$, where A is an arbitrary constant.

$\Rightarrow N = e^{-\frac{1}{12}t + A}$

$\Rightarrow N = e^{-\frac{1}{12}t}e^{A}$

But e^{A} is a constant, call it B, so the general solution of $\dfrac{dN}{dt} = -\dfrac{N}{12}$
is $N = Be^{-\frac{1}{12}t}$, where B is an arbitrary constant.

> Easier to keep the $-\dfrac{1}{12}$ on the RHS.

> The variables N and t are now separated.

> If the required form had not been given in the question you could have left the general solution in this form.

> Taking the exponential of both sides and using $e^{\ln z} = z$.

> Compare Chapter 7 Test Yourself Q4. $B = 300$. This is true if $N = 300$ when $t = 0$. You will be shown how to find the value of arbitrary constants in the next section.

EXERCISE 8B

Find the general solution of these differential equations. Give your answers in explicit form where possible.

For example, in the form $y = f(x)$.

1 $\dfrac{dy}{dx} = 4x$

2 $\dfrac{dy}{dx} = y$

3 $\dfrac{dy}{dx} = y + 2$

4 $\dfrac{dy}{dx} = x(y + 1)$

5 $\dfrac{dy}{dx} = e^y$

6 $\dfrac{dy}{dx} = \sec y$

7 $\dfrac{dx}{dt} = x + 2$

8 $\dfrac{dy}{dx} = \dfrac{4x}{y}$

9 $\dfrac{dy}{dx} = \dfrac{x + 4}{y}$

10 $\dfrac{dy}{dx} - \dfrac{y}{x}$

11 $\dfrac{dy}{dx} = \dfrac{y}{x + 4}$

12 $2\dfrac{dy}{dx} = y + 2$

13 $2y\dfrac{dy}{dx} = \dfrac{y^2 + 1}{4x}$

14 $\dfrac{dx}{dt} = \dfrac{e^t}{x}$

15 $\dfrac{dx}{dt} = \dfrac{x + 2}{t + 1}$

16 $\dfrac{dx}{dt} = \cos^2 x \sin t$

17 $\dfrac{dx}{dt} = \dfrac{e^x}{e^t}$

18 $\dfrac{dx}{dt} = \dfrac{t(x^2 + 3)}{2x}$

19 $\dfrac{dx}{dt} = e^{x + t}$

20 $2\sqrt{xy}\,\dfrac{dy}{dx} = 2 + x$

8.4 Finding the solution of a simple differential equation which satisfies a given condition

This section shows you how to find the arbitrary constant in the general solution of a first order differential equation which is solved using separation of variables. The solution which satisfies the differential equation and the given condition is sometimes called the **particular solution**.

> To find a solution of a differential equation which satisfies a given condition, firstly find the general solution of the differential equation and then substitute the given condition to find the arbitrary constant.

Worked example 8.6

Find, in the form $y = f(x)$, the solution of the differential equation $x\dfrac{dy}{dx} = y + 3$ given that $y = 3$ when $x = 2$.

8

Solution

$x\dfrac{dy}{dx} = (y + 3)$

> Insert brackets on the right-hand side.

Rewrite the differential equation as $\dfrac{1}{(y+3)}\dfrac{dy}{dx} = \dfrac{1}{x}$.

Integrate both sides with respect to $x \Rightarrow \displaystyle\int \dfrac{1}{y+3}\dfrac{dy}{dx}\,dx = \int\dfrac{1}{x}\,dx$

$\Rightarrow \displaystyle\int\dfrac{1}{y+3}\,dy = \int\dfrac{1}{x}\,dx$

> The variables y and x are now separated.

Integrating both sides and combining the constants of integration gives

> If you were finding the general solution you would replace A by $\ln B$ which then leads to $\ln(y+3) = \ln(Bx)$ or $y = Bx - 3$.

$\ln|y + 3| = \ln x + A$, where A is an arbitrary constant.

When $x = 2$, $y = 3$

> When not finding the general solution substitute the given values immediately after integration has taken place.

$\Rightarrow \ln(3 + 3) = \ln 2 + A$

$\Rightarrow A = \ln 6 - \ln 2 \quad \Rightarrow \quad A = \ln\dfrac{6}{2} = \ln 3$

> Using the second law of logarithms (C2 section 11.3).

$\Rightarrow \ln|y + 3| = \ln x + \ln 3$

> Using the first law of logarithms (C2 section 11.3).

$\Rightarrow \ln|y + 3| = \ln 3x$

Taking exponentials of both sides gives $y + 3 = 3x$ and so the solution of the differential equation $x\dfrac{dy}{dx} = y + 3$ such that $y = 3$ when $x = 2$ is $y = 3(x - 1)$.

> You can check the answer satisfies the given condition and the differential equation. Clearly when $x = 2$, $y = 3(2 - 1) = 3$ and differentiating $y = 3(x - 1)$ gives $\dfrac{dy}{dx} = 3 \Rightarrow x\dfrac{dy}{dx} = 3x = y + 3$.

Worked example 8.7

Find, in the form $x = f(t)$, the solution of the differential equation $\dfrac{t}{t+1}\dfrac{dx}{dt} = \tan x$, given that $x = \dfrac{\pi}{6}$ when $t = 1$.

Solution

$\dfrac{t}{t+1}\dfrac{dx}{dt} = \tan x$

Rewrite the differential equation as $\dfrac{1}{\tan x}\dfrac{dx}{dt} = \dfrac{t+1}{t}$

$\Rightarrow \cot x\dfrac{dx}{dt} = \dfrac{t+1}{t}$

Integrate both sides with respect to $t \Rightarrow \displaystyle\cot x\dfrac{dx}{dt}\,dt = \int\dfrac{t+1}{t}\,dt$

$\Rightarrow \displaystyle\int\cot x\,dx = \int\dfrac{t}{t} + \dfrac{1}{t}\,dt = \int 1 + \dfrac{1}{t}\,dt$

Integrating both sides and combining the constants of integration gives

> From the formulae booklet:
> $\displaystyle\int \cot x\,dx = \ln|\sin x| + c.$

$\ln|\sin x| = t + \ln t + A$, where A is an arbitrary constant.

When $t = 1$, $x = \dfrac{\pi}{6}$

$\Rightarrow \ln\left|\sin\dfrac{\pi}{6}\right| = 1 + \ln 1 + A \quad \Rightarrow \quad A = \ln\dfrac{1}{2} - 1$

$\Rightarrow \ln|\sin x| = t + \ln t + \ln\dfrac{1}{2} - 1$

$\Rightarrow \ln|\sin x| - \ln\dfrac{1}{2}t = t - 1$

$\Rightarrow \ln\left|\dfrac{\sin x}{\frac{1}{2}t}\right| = t - 1 \quad \Rightarrow \quad \ln\left|\dfrac{2\sin x}{t}\right| = t - 1$

$\Rightarrow \dfrac{2\sin x}{t} = e^{t-1} \quad \Rightarrow \quad \sin x = \dfrac{t}{2}e^{t-1}$

$\Rightarrow x = \sin^{-1}\left(\dfrac{t}{2}e^{t-1}\right)$

> When not finding the general solution substitute the given values immediately after integration has taken place.
>
> $$\ln t + \ln\dfrac{1}{2} = \ln\left(t \times \dfrac{1}{2}\right)$$

EXERCISE 8C

In each question find the solution of the differential equation that satisfies the given condition. Where possible give your answers in a simplified explicit form.

1 $\dfrac{dy}{dx} = 8x + 2$, $y = 8$ when $x = 1$.

2 $\dfrac{dy}{dx} = 4y$, $y = 1$ when $x = -1$.

3 $\dfrac{dN}{dt} = 1 - N$, $N = \dfrac{1}{2}$ when $t = 0$.

4 $\dfrac{dP}{dt} = 2P$, $P = 200$ when $t = 0$.

5 $\dfrac{dy}{dx} = 2xe^{-y}$, $y = 0$ when $x = 1$.

6 $\dfrac{dy}{dx} = (y + 1)(2x + 3)$, $y = 0$ when $x = 1$.

7 $\dfrac{dx}{dt} = \dfrac{x}{t + 1}$, $x = 2$ when $t = 3$.

8 $\dfrac{dy}{dx} = \cos^2 y \cos x$, $y = \dfrac{\pi}{4}$ when $x = \dfrac{\pi}{2}$.

9 $v\dfrac{dv}{dx} = 2(16 + v^2)$, $v = 0$ when $x = 0$.

10 $\dfrac{dy}{dx} = e^{x-y}$, $y = 0$ when $x = 0$.

11 $\dfrac{dy}{dx} = e^{x+2y}$, $y = -\dfrac{1}{2}\ln 2$ when $x = 0$.

12 $\dfrac{dy}{dx} = \dfrac{y}{(x + 1)(x + 2)}$, $y = 2$ when $x = 1$.

13 $\dfrac{dy}{dx} = \dfrac{xy}{x + 1}$, $y = 2$ when $x = 0$.

8

14 $\dfrac{dy}{dx} - 2 = y,$ $\qquad\qquad$ $y = 3$ when $x = 0.$

15 $\dfrac{dy}{dx} = \cot y \cot x,$ $\qquad$ $y = 0$ when $x = \dfrac{\pi}{2}.$

16 $x^2 \dfrac{dy}{dx} + y = 1,$ $\qquad\quad$ $y = 1 - e$ when $x = 1.$

17 $x \dfrac{dy}{dx} = y - y^2,$ $\qquad\quad$ $y = \dfrac{1}{2}$ when $x = 2.$

18 $x \dfrac{dy}{dx} + 4 = y^2,$ $\qquad\quad$ $y = 3$ when $x = 1.$

Worked examination question

A radioactive substance decays at a rate proportional to its mass. The mass of the substance remaining after t days is x grams. At time $t = 0$ the mass of the substance is A grams.

> The third sentence could have been 'Initially the mass of the substance is A grams.' 'Initially' $\Rightarrow$ 'At $t = 0$'.

(a) Write down a differential equation relating x and t.

(b) By solving the differential equation show that

$t = \dfrac{1}{k} \ln\left(\dfrac{A}{x}\right)$, where k is a positive constant.

(c) Given that $x = \dfrac{1}{2}A$ when $t = 8$, calculate the time taken for the mass of the substance to equal $\dfrac{1}{5}A$, giving your answer to the nearest day.

(d) Find the mass of the substance at the time when the rate of decay is 0.5 grams per day, giving your answer to three significant figures. $\qquad$ [A]

Solution

(a) At time t the mass of the substance is x and the rate of change of x is $\dfrac{dx}{dt}$.

Since x is decreasing you have $-\dfrac{dx}{dt} \propto x$

$\Rightarrow \dfrac{dx}{dt} = -kx$, where k is a positive constant.

> See worked example 8.1(a).

(b) $\dfrac{dx}{dt} = -kx \quad \Rightarrow \quad \dfrac{1}{x}\dfrac{dx}{dt} = -k$

$\Rightarrow \displaystyle\int \dfrac{1}{x}\dfrac{dx}{dt}\, dt = \int -k\, dt$

> Separating the variables. Since A is in the question you use a different letter for the arbitrary constant.

$\Rightarrow \displaystyle\int \dfrac{1}{x}\, dx = \int -k\, dt \quad \Rightarrow \quad \ln x = -kt + C$

When $t = 0,\, x = A \quad \Rightarrow \quad \ln A = -0 + C$

> Using the given condition to find the arbitrary constant C.

$\Rightarrow \quad \ln x = -kt + \ln A \quad \Rightarrow \quad kt = \ln A - \ln x$

$\Rightarrow \quad kt = \ln\left(\dfrac{A}{x}\right) \quad \Rightarrow \quad t = \dfrac{1}{k}\ln\left(\dfrac{A}{x}\right)$, where k is a positive constant as required.

(c) $t = \dfrac{1}{k} \ln\left(\dfrac{A}{x}\right)$

When $t = 8$, $x = \dfrac{1}{2}A$ $\Rightarrow$ $8 = \dfrac{1}{k} \ln\left(\dfrac{A}{\frac{1}{2}A}\right) = \dfrac{1}{k} \ln 2$

$\Rightarrow$ $\dfrac{1}{k} = \dfrac{8}{\ln 2}$

When $x = \dfrac{1}{5}A$, $t = \dfrac{8}{\ln 2} \ln\left(\dfrac{A}{\frac{1}{5}A}\right) = \dfrac{8}{\ln 2} \ln 5 = 18.57\ldots$

So, to the nearest day, $t = 19$.

(d) The rate of **decay** is $-\dfrac{dx}{dt} = kx$

$\Rightarrow$ $0.5 = kx$ $\Rightarrow$ $x = \dfrac{1}{k} \times 0.5 = \dfrac{4}{\ln 2} = 5.77$ (to 3 sf)

> In the solution to **(b)**, the term involving k is in the form $\dfrac{1}{k}$.
> You don't need the value of k yet.

The next exercise contains questions which combine the topics covered in this chapter. It also illustrates how topics covered in this chapter are linked with those covered in other chapters of the C3 book, in particular chapters on methods of integration.

MIXED EXERCISE

1 The gradient of a curve at each point (x, y) on the curve is equal to a quarter of the sum of the x-coordinate of the point and twice the y-coordinate of the point.

(a) Express this information in the form of a differential equation.

(b) Given that $2y = z - x$, show that your answer to **(a)** can be written as $\dfrac{dz}{dx} = \dfrac{z + 2}{2}$.

(c) Given that $z = 0$ when $x = 0$ find z in terms of x.

(d) Hence find the equation of the curve in the form $y = f(x)$.

2 (a) Use integration by parts to find $\displaystyle\int x\, e^x\, dx$.

(b) Hence find the solution of the differential equation

$\dfrac{dy}{dx} = yx\, e^x$, given that $y = e$ when $x = 1$.　　　　[A]

3 (a) Find $\displaystyle\int x\, e^{-3x}\, dx$.

(b) Solve the differential equation $e^{3x} \dfrac{dy}{dx} = x \cos^2 y$ given that $y = 0$ when $x = 0$.　　　　[A]

8

4 (a) Solve the differential equation $\dfrac{dy}{dx} = \dfrac{1}{y^2}$, giving the general solution for y in terms of x.

(b) Find the particular solution of this differential equation for which $y = -1$ when $x = 1$. [A]

5 (a) Use integration by parts to find $\displaystyle\int x^{\frac{1}{2}} \ln x \, dx$.

(b) Find the solution of the differential equation

$$\frac{dy}{dx} = (xy)^{\frac{1}{2}} \ln x, \text{ for which } y = 1 \text{ when } x = 1.$$ [A]

6 (a) Solve the differential equation $\dfrac{dx}{dt} = \dfrac{10 - x}{5}$, given that $x = 1$ when $t = 0$, giving your answer in the form:

    (i) t in terms of x,   **(ii)** x in terms of t.

(b) Find the value of t when $x = 2$, giving your answer to three decimal places.

7 Before the start of a memory test a student has to memorise a set of 60 facts. According to psychologists the rate at which a person can memorise a set of facts is proportional to the number of facts remaining to be memorised. So if a student memorises x facts in t minutes, taking both x and t as continuous variables, this situation can be modelled by the differential equation $\dfrac{dx}{dt} = k(60 - x)$, where k is a positive constant and $x \leqslant 60$ for all $t \geqslant 0$.

(a) By solving the differential equation and assuming that initially no facts are memorised, show that $x = 60(1 - e^{-kt})$.

(b) Given that the student memorises fifteen facts in the first twenty minutes, show that $20k = \ln \dfrac{4}{3}$.

(c) To obtain the top grade the student needs to have memorised 57 facts in 3 hours. Show that, on the basis of this model, the student will not obtain a top grade in this memory test. [A]

8 (a) Use the substitution $u = 3 + \cos x$, or otherwise, to determine $\displaystyle\int \frac{\sin x}{(3 + \cos x)^2} \, dx$.

(b) Hence find the solution of the differential equation

$$(3 + \cos x)^2 \frac{dy}{dx} = y \sin x, \text{ given that } y = e \text{ when } x = 0. \quad [A]$$

9 A pond covers an area of 300 m². A specimen of pondweed grows on the surface of the pond. At time t days after the weed is first discovered, it covers an area of A m².
The area of the pond covered by the weed increases at a rate which is proportional to the square root of the area of the pond already covered by the weed.
Initially, the area covered by the weed is 0.25 m² and its rate of growth is 1 m² per day.

   (a) Show that $\dfrac{\mathrm{d}A}{\mathrm{d}t} = 2A^{\frac{1}{2}}$.

   (b) Find a relationship between t and A.

   (c) Deduce, to the nearest day, the time taken for the pond's surface to be completely covered by this weed.

10 (a) Find $\int x\,\mathrm{e}^{3x}\,\mathrm{d}x$.

   (b) Solve the differential equation $2y\dfrac{\mathrm{d}y}{\mathrm{d}x} = 3x\mathrm{e}^{3x}$ given

      that $y = 1$ when $x = 0$.   [A part]

11 A college supplies packs of stationery at cut-price rates to its 1200 students, each student being allowed one pack. After t days x students have bought their packs. The number of students buying their packs per day is assumed to be one-fifth of the number of students who have yet to buy their packs. Taking x and t to be continuous variables, this situation can be modelled by the differential equation

$\dfrac{\mathrm{d}x}{\mathrm{d}t} = k(c - x)$, where k and c are constants.

   (a) State the values of k and c.

   (b) Solve the differential equation given that $x = 0$ when $t = 0$.

   (c) Find, on the basis of this model, the number of students who have bought their packs at the end of the tenth day.   [A]

12 The amount, x, of a certain radioactive substance present at time t in a reaction is given by the differential equation

$\dfrac{\mathrm{d}x}{\mathrm{d}t} = a - bx$, where a and b are positive constants.

   (a) Solve the differential equation for x in terms of t given that when $t = 0$ the amount of the substance present is $\dfrac{a}{4b}$.

   (b) Find the time taken for x to attain the value $\dfrac{a}{2b}$.

   (c) Find the limiting value of x as $t \to \infty$.   [A]

13 In a model to estimate the increase in population of a small island it is assumed that the population, P, at time t years after 1 January 1991, is a continuous variable which increases at a rate proportional to P.

(a) (i) Write down a differential equation relating P and t.

(ii) Given that the population of the island was 50 000 on 1 January 1991, solve the differential equation to show that $P = 50\,000\,e^{kt}$, where k is a constant.

(iii) The population of the island was 50 500 on 1 January 1992. Calculate the value of k, giving your answer to three significant figures.

(iv) Calculate the population of the island on 1 January 2001, giving your answer to three significant figures.

(b) On 1 January 1981 the population of the island was 45 300. Investigate whether this information is consistent with the model used in (a). [A]

Key point summary

1 A **differential equation** is an equation which involves at least one derivative of a variable with respect to another variable. For example,   *p291*

$$\frac{dy}{dx} = 2x - 3, \quad x\frac{dx}{dt} = e^t \sin x, \quad \frac{d^2m}{dt^2} + 3\frac{dm}{dt} + 4m = e^t.$$

2 **First order differential equations** are differential equations in which the highest derivative is the first. For example,   *p291*

$$\frac{dy}{dx} = 2x^4 + 3, \quad x^2\frac{dx}{dt} = \ln t.$$

3 ● The rate of **increase** of x is proportional to x *p292*

$$\Rightarrow \frac{dx}{dt} = kx;$$

● The rate of **decrease** of x is proportional to x

$$\Rightarrow \frac{dx}{dt} = -kx;$$

where k is a constant of proportionality.

4 The **general solution** of the first order differential equation $\frac{dy}{dx} = g(x)h(y)$ is given by   *p295*

$$\int \frac{1}{h(y)}\,dy = \int g(x)\,dx + A, \text{ where } A$$

is an arbitrary constant.

5 To find a solution of a differential equation which satisfies a given condition, firstly find the general solution of the differential equation and then substitute the given condition to find the arbitrary constant.

p297

Test yourself	What to review
1 The gradient of a curve at the point $P(x, y)$ is four times the gradient of the line joining the point $(1, 2)$ to P. Express this information in the form of a differential equation.	*Section 8.2*
2 Find the general solution of the differential equation $$2y\frac{dy}{dx} = xy^2 + 2x.$$	*Section 8.3*
3 By finding the solution of the differential equation $\frac{dy}{dx} = e^{2x+3y}$ such that $y = \frac{1}{3}\ln 2$ when $x = 0$, show that $x = \frac{1}{2}\ln\frac{2}{3}$ when $y = 0$.	*Section 8.4*

Test yourself   ANSWERS

2 $y^2 = Ae^2 - 2$, where A is an arbitrary constant.

1 $\frac{dy}{dx} = \frac{4(y-2)}{x-1}$.

8

CHAPTER 9

C4: Vector equations of lines

Learning objectives

After studying this chapter, you should be able to:
- use vector notation
- find the magnitude of a vector
- understand the term unit vector
- find a vector equation of a line
- determine whether two lines intersect and find points of intersection when lines do intersect
- calculate the scalar product and use it to find the angle between two vectors or two lines
- calculate the perpendicular distance from a point to a line.

9.1 Vector notation

A **vector** is a quantity that has both magnitude and a direction. It is usually drawn with an arrowhead to indicate its direction. The vector shown is rather like a journey from A to B and can be represented by $\overrightarrow{AB}$.

This notation distinguishes the vector from the length of the line segment which is written as AB. Alternatively, the magnitude of the vector can be written as $|\overrightarrow{AB}|$. Quantities such as lengths and magnitudes, where there is no associated direction, are known as **scalar** quantities.

When O is the origin, the vector $\overrightarrow{OP}$ is said to be the **position vector** of the point P. The position vector of P is often represented by $\mathbf{p}$ (a lower case letter printed in bold).

When writing by hand, it is difficult to represent bold print and so you need to distinguish a vector from a scalar by writing the letter representing a vector with a wiggly line underneath it such as $\underset{\sim}{p}$ or $\underset{\sim}{q}$.

Without the bold print (or the wiggly underlining) the symbol p means the magnitude of the vector $\mathbf{p}$. The magnitude of $\mathbf{p}$ can also be written as $|\mathbf{p}|$.

A real effort must be made to make it clear in your working when you are using vectors or simply trying to represent the length of a line segment.

When describing translations in two dimensions, you used vectors to represent how one curve could be translated to give another curve. You used a **column vector** such as $\begin{bmatrix} 3 \\ 5 \end{bmatrix}$ to indicate a translation of 3 units in the x-direction and 5 units in the y-direction.

It is possible to have vectors in three dimensions and so the vector $\begin{bmatrix} 3 \\ 1 \\ 8 \end{bmatrix}$ could represent a translation of 3 units in the x-direction, 1 unit in the y-direction and 8 units in the z-direction.

> A vector is represented by bold type such as **v**, or by using an arrow above the letters such as $\overrightarrow{AB}$ to indicate the direction.

> The magnitude of the vector $\overrightarrow{AB}$ is written as $|\overrightarrow{AB}|$ or as AB. The magnitude of the vector **v** is written as $|\mathbf{v}|$ or as v.

> In two dimensions, when a vector **v** is written as $\mathbf{v} = \begin{bmatrix} a \\ b \end{bmatrix}$, the quantities a and b are the components of the vector in the x- and y-directions, respectively.

> In three dimensions, when a vector **v** is written as $\mathbf{v} = \begin{bmatrix} a \\ b \\ c \end{bmatrix}$, the quantities a and b and c are the components of the vector in the x-, y- and z-directions, respectively.

9

9.2 Adding and subtracting vectors

When the vectors **a** and **b** have the same number of components it is possible to add the two vectors by adding the respective components so that, for example, when $\mathbf{a} = \begin{bmatrix} 2 \\ 1 \end{bmatrix}$ and $\mathbf{b} = \begin{bmatrix} 5 \\ -4 \end{bmatrix}$,

then $\mathbf{a} + \mathbf{b} = \begin{bmatrix} 2 \\ 1 \end{bmatrix} + \begin{bmatrix} 5 \\ -4 \end{bmatrix} = \begin{bmatrix} 2 + 5 \\ 1 - 4 \end{bmatrix} = \begin{bmatrix} 7 \\ -3 \end{bmatrix}.$

Geometrically, you have the situation shown opposite, where the vectors **a**, **b** and **a** + **b** form a triangle.

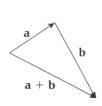

When you multiply a vector **a** by a scalar quantity t, you multiply each of the components by t.

Hence when $\mathbf{a} = \begin{bmatrix} 2 \\ 1 \end{bmatrix}$, the vector $3\mathbf{a}$ is given by $3\mathbf{a} = \begin{bmatrix} 6 \\ 3 \end{bmatrix}$, which is

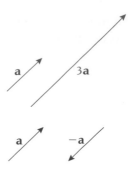

a vector in the same direction as **a** but three times as long.

As the diagram opposite shows, the vector $-\mathbf{a}$ has the same length as vector **a** but is in the opposite direction to **a**.

The vector $\overrightarrow{AB}$ can be found in terms of the position vectors of A and B.

Suppose the position vectors of the points A and B are **a** and **b**, respectively. These are indicated in the diagram below, together with the origin, O.

The vector 'journey' from A to B can be seen to consist of the sum of the two vectors **b** and $-\mathbf{a}$ by following in the direction of the arrows along the broken lines so that

$$\overrightarrow{AB} = \mathbf{b} + (-\mathbf{a}) = \mathbf{b} - \mathbf{a}.$$

> If the point A has position vector **a** and the point B has position vector **b**, then $\overrightarrow{AB} = \mathbf{b} - \mathbf{a}$.

Worked example 9.1

Given that $\mathbf{p} = \begin{bmatrix} 6 \\ -2 \end{bmatrix}$ and $\mathbf{q} = \begin{bmatrix} 5 \\ 3 \end{bmatrix}$, find

(a) $\mathbf{p} - \mathbf{q}$,   (b) $2\mathbf{p} + 3\mathbf{q}$,   (c) $4\mathbf{q} - 3\mathbf{p}$.

Solution

(a) Subtracting the values of each component gives

$$\mathbf{p} - \mathbf{q} = \begin{bmatrix} 1 \\ -5 \end{bmatrix}.$$

(b) $2\mathbf{p} = \begin{bmatrix} 12 \\ -4 \end{bmatrix}$ and $3\mathbf{q} = \begin{bmatrix} 15 \\ 9 \end{bmatrix}$

Hence, $2\mathbf{p} + 3\mathbf{q} = \begin{bmatrix} 27 \\ 5 \end{bmatrix}$.

(c) $4\mathbf{q} = \begin{bmatrix} 20 \\ 12 \end{bmatrix}$ and $3\mathbf{p} = \begin{bmatrix} 18 \\ -6 \end{bmatrix}$. Therefore $4\mathbf{q} - 3\mathbf{p} = \begin{bmatrix} 2 \\ 18 \end{bmatrix}$.

Worked example 9.2

In three dimensional space, the point C has coordinates $(2, -1, 7)$ and the point D has coordinates $(4, 5, 3)$.

(a) Write down the position vectors of C and D.

(b) Find the vector $\overrightarrow{CD}$.

Solution

(a) The position vector of C is the vector that represents the journey from the origin $(0, 0, 0)$ to the point $C(2, -1, 7)$. It is usually represented by **c**.

Hence the position vector of C is given by $\mathbf{c} = \begin{bmatrix} 2 \\ -1 \\ 7 \end{bmatrix}$.

Similarly, the position vector of D is given by $\mathbf{d} = \begin{bmatrix} 4 \\ 5 \\ 3 \end{bmatrix}$.

(b) Using the result in the previous Key Point

$$\overrightarrow{CD} = \mathbf{d} - \mathbf{c},$$

$$\mathbf{d} - \mathbf{c} = \begin{bmatrix} 4 \\ 5 \\ 3 \end{bmatrix} - \begin{bmatrix} 2 \\ -1 \\ 7 \end{bmatrix} = \begin{bmatrix} 2 \\ 6 \\ -4 \end{bmatrix} \text{ so } \overrightarrow{CD} = \begin{bmatrix} 2 \\ 6 \\ -4 \end{bmatrix}.$$

Alternative solution

An alternative approach is to say that in finding the vector $\overrightarrow{CD}$, the x-component changes from 2 to 4, (an increase of 2), the y-component changes from -1 to 5 (an increase of 6) and the z-component changes from 7 to 3 (a decrease of 4).

Therefore, $\overrightarrow{CD} = \begin{bmatrix} 2 \\ 6 \\ -4 \end{bmatrix}$.

9

9.3 The magnitude of a vector

In two dimensions, suppose the vector $\mathbf{p} = \begin{bmatrix} a \\ b \end{bmatrix}$ has its x-component equal to a and its y-component equal to b, where a and b are positive.

By Pythagoras' Theorem, the length of the vector is $\sqrt{a^2 + b^2}$.

We write $|\mathbf{p}| = \sqrt{a^2 + b^2}$.

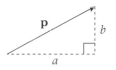

In a similar way, even when a and b are not positive, because the square of a negative quantity is always positive, the length of the vector $\begin{bmatrix} a \\ b \end{bmatrix}$ will still be $\sqrt{a^2 + b^2}$.

> In two dimensions, the magnitude of the vector $\begin{bmatrix} a \\ b \end{bmatrix}$ is $\sqrt{a^2 + b^2}$.

Imagine a cuboid with sides of length a, b and c as shown below.

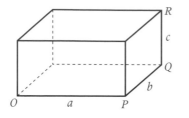

The base is rectangular so that angle OPQ is a right angle. Hence the length of OQ is $\sqrt{a^2 + b^2}$.

The triangle OQR is right-angled at Q.

Hence, by Pythagoras, $OR^2 = OQ^2 + QR^2 = (a^2 + b^2) + c^2$.

Therefore, the length of OR is $\sqrt{a^2 + b^2 + c^2}$.

This result can be applied to vectors in three dimensions.

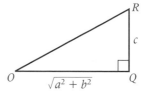

> In three dimensions, the magnitude of the vector $\begin{bmatrix} a \\ b \\ c \end{bmatrix}$ is $\sqrt{a^2 + b^2 + c^2}$.

Worked example 9.3

Given that $\mathbf{p} = \begin{bmatrix} -1 \\ 2 \\ 2 \end{bmatrix}$ and $\mathbf{q} = \begin{bmatrix} 12 \\ -3 \\ 4 \end{bmatrix}$, find the magnitude of the vectors   **(a) p,**   **(b) q,**   **(c) p + q.**

Solution

(a) The magnitude of **p** is $\sqrt{(-1)^2 + 2^2 + 2^2} = \sqrt{9} = 3$.

Using the notation introduced in section 9.1, we can write
$$|\mathbf{p}| = 3 \text{ or } p = 3.$$

(b) Similarly $|\mathbf{q}| = \sqrt{12^2 + (-3)^2 + 4^2} = \sqrt{169} = 13$.

(c) Since $\mathbf{p} + \mathbf{q} = \begin{bmatrix} 11 \\ -1 \\ 6 \end{bmatrix}$, the magnitude of the vector is

$$|\mathbf{p} + \mathbf{q}| = \sqrt{11^2 + (-1)^2 + 6^2} = \sqrt{158}.$$

9.4 The distance between two points in three dimensions

Suppose the coordinates of the point P are (x_1, y_1, z_1) and that the point Q has coordinates (x_2, y_2, z_2), then the vector $\overrightarrow{PQ}$ is given by

$$\overrightarrow{PQ} = \begin{bmatrix} x_2 - x_1 \\ y_2 - y_1 \\ z_2 - z_1 \end{bmatrix}$$ and the distance between the two points P and Q

is equal to the magnitude of the vector $\overrightarrow{PQ}$.

But $|\overrightarrow{PQ}|^2 = (x_2 - x_1)^2 + (y_2 - y_1)^2 + (z_2 - z_1)^2$.

Hence,

> The distance between the points (x_1, y_1, z_1) and (x_2, y_2, z_2) is
> $\sqrt{(x_2 - x_1)^2 + (y_2 - y_1)^2 + (z_2 - z_1)^2}$.

Worked example 9.4

Find the distance between the points P and Q with coordinates $(2, -3, 7)$ and $(4, 1, 5)$, respectively.

Solution

The distance PQ is given by $\sqrt{(4 - 2)^2 + (1 - (-3))^2 + (5 - 7)^2}$
$$= \sqrt{2^2 + 4^2 + (-2)^2} = \sqrt{4 + 16 + 4} = \sqrt{24}.$$

EXERCISE 9A

1 Given that $\mathbf{p} = \begin{bmatrix} 3 \\ -4 \end{bmatrix}$ and $\mathbf{q} = \begin{bmatrix} 5 \\ 12 \end{bmatrix}$, find:

  (a) $\mathbf{p} + \mathbf{q}$,   (b) $3\mathbf{p} + \mathbf{q}$,   (c) $3\mathbf{q} - 5\mathbf{p}$.

2 Given that $\mathbf{r} = \begin{bmatrix} 5 \\ -4 \end{bmatrix}$ and $\mathbf{s} = \begin{bmatrix} -1 \\ -3 \end{bmatrix}$, find:

  (a) $\mathbf{r} - \mathbf{s}$,   (b) $\mathbf{r} + 5\mathbf{s}$,   (c) $4\mathbf{s} - 3\mathbf{r}$.

3 Given that $\mathbf{a} = \begin{bmatrix} 4 \\ -3 \\ 0 \end{bmatrix}$ and $\mathbf{b} = \begin{bmatrix} -1 \\ -2 \\ 2 \end{bmatrix}$, find:

  (a) $\mathbf{a} + 4\mathbf{b}$,   (b) $2\mathbf{a} - 3\mathbf{b}$,   (c) $5\mathbf{b} - 4\mathbf{a}$.

4 Given that $\mathbf{c} = \begin{bmatrix} 4 \\ -3 \\ 12 \end{bmatrix}$ and $\mathbf{d} = \begin{bmatrix} -3 \\ -2 \\ 6 \end{bmatrix}$, find:

  (a) $\mathbf{c} - 2\mathbf{d}$,   (b) $3\mathbf{c} + 4\mathbf{d}$,   (c) $2\mathbf{c} - 3\mathbf{d}$.

5 In two dimensional space, the point E has coordinates $(-1, 4)$ and the point F has coordinates $(5, -3)$.

(a) Write down the position vectors of E and F.

(b) Find the vector $\overrightarrow{EF}$.

(c) Hence find the distance between the points E and F.

6 In three dimensional space, the point R has coordinates $(1, -2, 4)$ and the point S has coordinates $(-1, 3, 1)$.

(a) Write down the position vectors of R and S.

(b) Find the vector $\overrightarrow{RS}$.

(c) Hence find the distance between the points R and S.

7 Given that $\mathbf{p} = \begin{bmatrix} 3 \\ -4 \end{bmatrix}$ and $\mathbf{q} = \begin{bmatrix} 5 \\ 12 \end{bmatrix}$, find the magnitude of:

(a) $\mathbf{p}$,   (b) $\mathbf{q}$,   (c) $\mathbf{p} + \mathbf{q}$.

8 Given that $\mathbf{r} = \begin{bmatrix} 5 \\ -4 \end{bmatrix}$ and $\mathbf{s} = \begin{bmatrix} -1 \\ -3 \end{bmatrix}$, find the magnitude of:

(a) $\mathbf{r}$,   (b) $\mathbf{s}$,   (c) $4\mathbf{s} - 3\mathbf{r}$.

9 Given that $\mathbf{a} = \begin{bmatrix} 4 \\ -3 \\ 0 \end{bmatrix}$ and $\mathbf{b} = \begin{bmatrix} -1 \\ -2 \\ 2 \end{bmatrix}$, find the magnitude of:

(a) $\mathbf{a}$,   (b) $\mathbf{b}$,   (c) $\mathbf{a} + \mathbf{b}$.

10 Given that $\mathbf{c} = \begin{bmatrix} 4 \\ -3 \\ 12 \end{bmatrix}$ and $\mathbf{d} = \begin{bmatrix} -3 \\ -2 \\ 6 \end{bmatrix}$, find the magnitude of:

(a) $\mathbf{c}$,   (b) $\mathbf{d}$,   (c) $\mathbf{c} - \mathbf{d}$.

11 Find the distance between the points P and Q in the following cases:

(a) $P(3, 1, 5)$ and $Q(7, 1, 2)$,

(b) $P(-2, 0, 1)$ and $Q(-4, -1, 4)$,

(c) $P(1, 3, -1)$ and $Q(5, -2, 2)$,

(d) $P(-2, -4, 3)$ and $Q(-3, -5, -4)$.

9.5 Unit vectors

A unit vector is a vector that has magnitude 1.

In two dimensions, the following vectors are examples of unit vectors: $\begin{bmatrix} 1 \\ 0 \end{bmatrix}, \begin{bmatrix} 0 \\ -1 \end{bmatrix}, \begin{bmatrix} 0.6 \\ -0.8 \end{bmatrix}$.

> The third vector is a unit vector because
> $\sqrt{0.6^2 + (-0.8)^2} = \sqrt{0.36 + 0.64} = \sqrt{1} = 1$.

In three dimensional space, examples of unit vectors
are $\begin{bmatrix} 0 \\ 0 \\ 1 \end{bmatrix}$, $\begin{bmatrix} \frac{1}{3} \\ \frac{2}{3} \\ \frac{2}{3} \end{bmatrix}$ and $\begin{bmatrix} -1 \\ 0 \\ 0 \end{bmatrix}$.

> The second vector is a unit vector because
> $$\sqrt{\left(\tfrac{1}{3}\right)^2 + \left(\tfrac{2}{3}\right)^2 + \left(\tfrac{2}{3}\right)^2} = \sqrt{\tfrac{1}{9} + \tfrac{4}{9} + \tfrac{4}{9}} = \sqrt{1} = 1.$$

You can always find a unit vector in the same direction as a given vector.

Worked example 9.5

Find a unit vector in the same direction as the vector $\begin{bmatrix} -1 \\ 3 \\ -2 \end{bmatrix}$.

Solution

The given vector has magnitude
$$\sqrt{(-1)^2 + 3^2 + (-2)^2} = \sqrt{1 + 9 + 4} = \sqrt{14}.$$

Hence, by dividing each component of the vector by $\sqrt{14}$ you will have a unit vector in the same direction as the original vector.

The required unit vector is $\dfrac{1}{\sqrt{14}}\begin{bmatrix} -1 \\ 3 \\ -2 \end{bmatrix}$ or, if you prefer, $\begin{bmatrix} \frac{-1}{\sqrt{14}} \\ \frac{3}{\sqrt{14}} \\ \frac{-2}{\sqrt{14}} \end{bmatrix}$.

> You can see that the first form of the answer is much easier to write.

9.6 Unit base vectors

The vector $\begin{bmatrix} 3 \\ 5 \\ 7 \end{bmatrix}$ could be written as $\begin{bmatrix} 3 \\ 0 \\ 0 \end{bmatrix} + \begin{bmatrix} 0 \\ 5 \\ 0 \end{bmatrix} + \begin{bmatrix} 0 \\ 0 \\ 7 \end{bmatrix}$ which in

turn is equal to $3\begin{bmatrix} 1 \\ 0 \\ 0 \end{bmatrix} + 5\begin{bmatrix} 0 \\ 1 \\ 0 \end{bmatrix} + 7\begin{bmatrix} 0 \\ 0 \\ 1 \end{bmatrix}$.

In three dimensions, the vectors $\begin{bmatrix} 1 \\ 0 \\ 0 \end{bmatrix}$, $\begin{bmatrix} 0 \\ 1 \\ 0 \end{bmatrix}$ and $\begin{bmatrix} 0 \\ 0 \\ 1 \end{bmatrix}$ are known

as the unit base vectors and are usually denoted by the special vectors **i**, **j** and **k**, respectively.

Consequently, the vector $\begin{bmatrix} 3 \\ 5 \\ 7 \end{bmatrix}$ can be conveniently written as

$3\mathbf{i} + 5\mathbf{j} + 7\mathbf{k}$.

9

Worked example 9.6

The vectors **a** and **b** are given by

$$\mathbf{a} = 12\mathbf{i} - 4\mathbf{j} + 3\mathbf{k} \quad \text{and} \quad \mathbf{b} = 2\mathbf{i} + \mathbf{j} - 2\mathbf{k}.$$

(a) Find: **(i)** $2\mathbf{a} + 3\mathbf{b}$, **(ii)** $4\mathbf{a} - 5\mathbf{b}$.

(b) Find the magnitude of: **(i)** **a**, **(ii)** **b**, **(iii)** $\mathbf{a} - 4\mathbf{b}$.

(c) Find a unit vector in the direction of **b**.

Solution

(a) **(i)** $\begin{aligned}2\mathbf{a} + 3\mathbf{b} &= 2(12\mathbf{i} - 4\mathbf{j} + 3\mathbf{k}) + 3(2\mathbf{i} + \mathbf{j} - 2\mathbf{k}) \\ &= 24\mathbf{i} - 8\mathbf{j} + 6\mathbf{k} + 6\mathbf{i} + 3\mathbf{j} - 6\mathbf{k} \\ &= 30\mathbf{i} - 5\mathbf{j} + 0\mathbf{k} = 30\mathbf{i} - 5\mathbf{j}\end{aligned}$

    (ii) $\begin{aligned}4\mathbf{a} - 5\mathbf{b} &= 4(12\mathbf{i} - 4\mathbf{j} + 3\mathbf{k}) - 5(2\mathbf{i} + \mathbf{j} - 2\mathbf{k}) \\ &= 48\mathbf{i} - 16\mathbf{j} + 12\mathbf{k} - 10\mathbf{i} - 5\mathbf{j} + 10\mathbf{k} \\ &= 38\mathbf{i} - 21\mathbf{j} + 22\mathbf{k}\end{aligned}$

(b) **(i)** The components of **a** are 12, -4 and 3. Hence the magnitude of **a** is

$$|\mathbf{a}| = \sqrt{12^2 + (-4)^2 + 3^2} = \sqrt{144 + 16 + 9} = \sqrt{169} = 13$$

    (ii) Similarly, $|\mathbf{b}| = \sqrt{2^2 + 1^2 + (-2)^2} = \sqrt{4 + 1 + 4} = \sqrt{9} = 3$

    (iii) $\begin{aligned}\mathbf{a} - 4\mathbf{b} &= (12\mathbf{i} - 4\mathbf{j} + 3\mathbf{k}) - 4(2\mathbf{i} + \mathbf{j} - 2\mathbf{k}) \\ &= 12\mathbf{i} - 4\mathbf{j} + 3\mathbf{k} - 8\mathbf{i} - 4\mathbf{j} + 8\mathbf{k} \\ &= 4\mathbf{i} - 8\mathbf{j} + 11\mathbf{k}\end{aligned}$

    Hence, $|\mathbf{a} - 4\mathbf{b}| = \sqrt{4^2 + (-8)^2 + 11^2} = \sqrt{16 + 64 + 121} = \sqrt{201}$

(c) Since the magnitude of **b** is 3, a unit vector in the direction of **b** is $\frac{1}{3}(2\mathbf{i} + \mathbf{j} - 2\mathbf{k})$ or $\frac{2}{3}\mathbf{i} + \frac{1}{3}\mathbf{j} - \frac{2}{3}\mathbf{k}$.

9.7 The scalar product

Two vectors can be combined in a kind of multiplication so as to produce a scalar quantity using what is called the scalar product. The scalar product of two vectors **a** and **b** is written with a dot between the two vectors and **a.b** is usually read as 'a dot b'. It is therefore sometimes referred to as the dot product rather than the scalar product.

> You may meet another product of two vectors in your further studies which is called the vector product.

The scalar product of **a** and **b**, written as **a.b**, has value equal to $|\mathbf{a}| \times |\mathbf{b}| \times \cos\theta$, where θ is the angle between the two vectors. You can write this simply as $\mathbf{a}.\mathbf{b} = ab\cos\theta$, where a is the magnitude of **a** and b is the magnitude of **b**.

Since the unit base vectors **i**, **j** and **k** are mutually perpendicular and since $\cos 90° = 0$, it follows that $\mathbf{i}.\mathbf{j} = \mathbf{j}.\mathbf{i} = 0$ and also $\mathbf{k}.\mathbf{j} = \mathbf{j}.\mathbf{k} = 0$ and $\mathbf{i}.\mathbf{k} = \mathbf{k}.\mathbf{i} = 0$.

Also since the angle between a vector and itself is $0°$ and because $\cos 0° = 1$, it follows that $\mathbf{i}.\mathbf{i} = 1 \times 1 \times \cos 0° = 1$. Similarly $\mathbf{j}.\mathbf{j} = 1$ and $\mathbf{k}.\mathbf{k} = 1$.

This enables you to produce a general formula for the scalar product of two vectors in three dimensions.

Suppose $\mathbf{p} = a\mathbf{i} + b\mathbf{j} + c\mathbf{k}$ and $\mathbf{q} = d\mathbf{i} + e\mathbf{j} + f\mathbf{k}$, then

$$\mathbf{p} \cdot \mathbf{q} = (a\mathbf{i} + b\mathbf{j} + c\mathbf{k}) \cdot (d\mathbf{i} + e\mathbf{j} + f\mathbf{k})$$

$$= a\mathbf{i} \cdot (d\mathbf{i} + e\mathbf{j} + f\mathbf{k}) + b\mathbf{j} \cdot (d\mathbf{i} + e\mathbf{j} + f\mathbf{k}) + c\mathbf{k} \cdot (d\mathbf{i} + e\mathbf{j} + f\mathbf{k})$$

$$= ad \times 1 + ae \times 0 + af \times 0 + bd \times 0 + be \times 1$$
$$+ cd \times 0 + ce \times 0 + cf \times 1$$

$$= ad + be + cf.$$

Writing the vectors as columns: $\begin{bmatrix} a \\ b \\ c \end{bmatrix} \cdot \begin{bmatrix} d \\ e \\ f \end{bmatrix} = ad + be + cf.$

> The scalar product of $\mathbf{p}$ and $\mathbf{q}$, written as $\mathbf{p.q}$, has value $|\mathbf{p}| \times |\mathbf{q}| \times \cos \theta$ or $pq \cos \theta$, where θ is the angle between the two vectors.

> If $\mathbf{p} = \begin{bmatrix} a \\ b \\ c \end{bmatrix}$ and $\mathbf{q} = \begin{bmatrix} d \\ e \\ f \end{bmatrix}$ then $\mathbf{p.q}$, is evaluated as
>
> $$\begin{bmatrix} a \\ b \\ c \end{bmatrix} \cdot \begin{bmatrix} d \\ e \\ f \end{bmatrix} = ad + be + cf.$$

Worked example 9.7

(a) Find $\mathbf{p.q}$ when $\mathbf{p} = \begin{bmatrix} 5 \\ -3 \end{bmatrix}$ and $\mathbf{q} = \begin{bmatrix} -2 \\ -7 \end{bmatrix}$.

(b) Find the scalar product of $\mathbf{a}$ and $\mathbf{b}$ where $\mathbf{a} = 3\mathbf{i} - 2\mathbf{j} + 5\mathbf{k}$ and $\mathbf{b} = 4\mathbf{i} + 3\mathbf{j} - 7\mathbf{k}$.

Solution

(a) The scalar product in two dimensions is found by adding the products of corresponding components of the vectors, so

$$\mathbf{p.q} = 5 \times (-2) + (-3) \times (-7) = -10 + 21 = 11.$$

> Notice that the answer is a scalar quantity (i.e. a number and not a vector).

(b) $\mathbf{a.b} = 3 \times 4 + (-2) \times 3 + 5 \times (-7) = 12 - 6 - 35 = -29.$
This illustrates that the scalar product can actually be a negative quantity.

9.8 The angle between two vectors

You can use the two different forms of the scalar product to find the angle between two vectors.

You know that $\mathbf{p}.\mathbf{q} = |\mathbf{p}| \times |\mathbf{q}| \times \cos\theta$ and also that if $\mathbf{p} = \begin{bmatrix} a \\ b \\ c \end{bmatrix}$

and $\mathbf{q} = \begin{bmatrix} d \\ e \\ f \end{bmatrix}$ then $\mathbf{p}.\mathbf{q} = \begin{bmatrix} a \\ b \\ c \end{bmatrix}.\begin{bmatrix} d \\ e \\ f \end{bmatrix} = ad + be + cf.$

Worked example 9.8

Find the angle between the vectors $\mathbf{p}$ and $\mathbf{q}$, where $\mathbf{p} = \begin{bmatrix} 2 \\ -3 \\ 5 \end{bmatrix}$

and $\mathbf{q} = \begin{bmatrix} 4 \\ 1 \\ -7 \end{bmatrix}$, giving your answer to the nearest 0.1°.

Solution

$|\mathbf{p}| = \sqrt{2^2 + (-3)^2 + 5^2} = \sqrt{4 + 9 + 25} = \sqrt{38}$

and $|\mathbf{q}| = \sqrt{4^2 + 1^2 + (-7)^2} = \sqrt{16 + 1 + 49} = \sqrt{66}.$

> Notice that the minus sign in the scalar product indicates that the angle is obtuse.

Therefore $\mathbf{p}.\mathbf{q} = |\mathbf{p}| \times |\mathbf{q}| \times \cos\theta = \sqrt{38} \times \sqrt{66} \times \cos\theta$, where θ is the angle between the two vectors.

But also $= \begin{bmatrix} 2 \\ -3 \\ 5 \end{bmatrix}.\begin{bmatrix} 4 \\ 1 \\ -7 \end{bmatrix} = (2 \times 4) + (-3 \times 1) + (5 \times -7)$

$$= 8 - 3 - 35 = -30$$

Hence, $\sqrt{38} \times \sqrt{66} \times \cos\theta = -30$

$$\Rightarrow \cos\theta = -\frac{30}{\sqrt{38} \times \sqrt{66}} = -0.599\,04\ldots$$

Hence, $\theta = 126.801\ldots°.$

The angle between the two vectors is 126.8° (to the nearest 0.1°).

Worked example 9.9

Given that $\mathbf{a} = 3\mathbf{i} - 2\mathbf{j} + 5\mathbf{k}$ and $\mathbf{b} = 2\mathbf{i} - 2\mathbf{j} - 2\mathbf{k}$.

(a) Find **a.b**.

(b) What can you deduce about the vectors **a** and **b**?

> This is an important idea which you will use in later sections of this chapter.

Solution

(a) $\mathbf{a.b} = (3 \times 2) + (-2 \times -2) + (5 \times -2) = 6 + 4 - 10 = 0$

(b) Since $\mathbf{a.b} = |\mathbf{a}| \times |\mathbf{b}| \times \cos\theta$ and neither **a** nor **b** is the zero vector, it follows that $\cos\theta = 0$. Hence $\theta = 90°$. The vectors **a** and **b** are perpendicular.

If $\mathbf{a.b} = 0$, where **a** and **b** are non-zero vectors, then **a** and **b** are perpendicular.

EXERCISE 9B

1 Show that each of the following are unit vectors:

(a) $\begin{bmatrix} -\dfrac{3}{5} \\ \dfrac{4}{5} \end{bmatrix}$ 　　**(b)** $\begin{bmatrix} \dfrac{5}{13} \\ \dfrac{12}{13} \end{bmatrix}$ 　　**(c)** $\begin{bmatrix} \dfrac{1}{\sqrt{5}} \\ -\dfrac{2}{\sqrt{5}} \end{bmatrix}$

(d) $\begin{bmatrix} -\dfrac{1}{3} \\ \dfrac{2}{3} \\ -\dfrac{2}{3} \end{bmatrix}$ 　　**(e)** $\begin{bmatrix} \dfrac{3}{7} \\ \dfrac{6}{7} \\ \dfrac{2}{7} \end{bmatrix}$

2 Find a unit vector in the same direction as each of the following vectors:

(a) $\begin{bmatrix} -5 \\ 12 \end{bmatrix}$ 　**(b)** $\begin{bmatrix} 7 \\ -24 \end{bmatrix}$ 　**(c)** $\begin{bmatrix} 4 \\ 3 \\ 0 \end{bmatrix}$ 　**(d)** $\begin{bmatrix} 2 \\ 1 \\ 2 \end{bmatrix}$ 　**(e)** $\begin{bmatrix} 6 \\ 3 \\ -2 \end{bmatrix}$

3 Find a unit vector in the same direction as each of the following vectors:

(a) $\begin{bmatrix} -4 \\ 1 \end{bmatrix}$ 　**(b)** $\begin{bmatrix} -3 \\ -2 \end{bmatrix}$ 　**(c)** $\begin{bmatrix} 1 \\ 3 \\ -1 \end{bmatrix}$ 　**(d)** $\begin{bmatrix} 2 \\ -4 \\ 3 \end{bmatrix}$ 　**(e)** $\begin{bmatrix} 5 \\ -3 \\ 2 \end{bmatrix}$

9

4 The vectors **a** and **b** are given by

$\quad$ **a** = 3**i** − 2**j** + 6**k** $\quad$ and $\quad$ **b** = 2**i** − 2**j** − **k**.

$\quad$ **(a)** Find: **(i)** 3**a** + 2**b**, **(ii)** 4**a** − 3**b**.

$\quad$ **(b)** Find the magnitude of:

$\qquad$ **(i) a**, $\quad$ **(ii) b**, $\quad$ **(iii) a** + **b**.

$\quad$ **(c)** Find a unit vector in the direction of **a**.

5 The vectors **p** and **q** are given by

$\quad$ **p** = 2**i** − 2**j** + **k** $\quad$ and $\quad$ **q** = 3**i** − 2**j** − 5**k**.

$\quad$ **(a)** Find: **(i)** 3**p** + 4**q**, **(ii)** 2**p** − 5**q**.

$\quad$ **(b)** Find the magnitude of: **(i) p**, $\quad$ **(ii) q**, $\quad$ **(iii) p** − **q**.

$\quad$ **(c)** Find a unit vector in the direction of **p**.

6 In two dimensions the point D has coordinates $(3, -1)$ and the point E has coordinates $(-2, 2)$. Find a unit vector in the direction of $\overrightarrow{DE}$.

7 In three dimensional space the point G has coordinates $(2, 1, 4)$ and the point H has coordinates $(3, 0, 7)$. Find a unit vector in the direction of $\overrightarrow{GH}$.

8 Find **p.q** for the following pairs of vectors **p** and **q**:

$\quad$ **(a)** $\mathbf{p} = \begin{bmatrix} 2 \\ -3 \end{bmatrix}$ and $\mathbf{q} = \begin{bmatrix} 2 \\ 4 \end{bmatrix}$, $\quad$ **(b)** $\mathbf{p} = \begin{bmatrix} -4 \\ -2 \end{bmatrix}$ and $\mathbf{q} = \begin{bmatrix} 3 \\ 5 \end{bmatrix}$,

$\quad$ **(c)** $\mathbf{p} = \begin{bmatrix} 2 \\ 5 \end{bmatrix}$ and $\mathbf{q} = \begin{bmatrix} -1 \\ -2 \end{bmatrix}$, $\quad$ **(d)** $\mathbf{p} = \begin{bmatrix} -5 \\ 6 \end{bmatrix}$ and $\mathbf{q} = \begin{bmatrix} -4 \\ 3 \end{bmatrix}$,

$\quad$ **(e)** $\mathbf{p} = \begin{bmatrix} 1 \\ 2 \\ 3 \end{bmatrix}$ and $\mathbf{q} = \begin{bmatrix} 4 \\ 3 \\ 2 \end{bmatrix}$, $\quad$ **(f)** $\mathbf{p} = \begin{bmatrix} -4 \\ 5 \\ 2 \end{bmatrix}$ and $\mathbf{q} = \begin{bmatrix} -3 \\ -2 \\ 1 \end{bmatrix}$,

$\quad$ **(g)** $\mathbf{p} = \begin{bmatrix} -4 \\ 1 \\ -1 \end{bmatrix}$ and $\mathbf{q} = \begin{bmatrix} 0 \\ -2 \\ 3 \end{bmatrix}$, **(h)** $\mathbf{p} = \begin{bmatrix} 0 \\ -2 \\ 1 \end{bmatrix}$ and $\mathbf{q} = \begin{bmatrix} 5 \\ 2 \\ 4 \end{bmatrix}$.

9 Find the scalar product of **a** and **b** in each of these cases:

$\quad$ **(a)** **a** = 4**i** − 3**j** + 2**k** $\quad$ and $\quad$ **b** = 7**i** + 2**j** − 2**k**,

$\quad$ **(b)** **a** = 3**i** + 2**j** + **k** $\quad$ and $\quad$ **b** = 5**i** − 3**j** − 4**k**,

$\quad$ **(c)** **a** = 5**i** − **j** + 4**k** $\quad$ and $\quad$ **b** = −2**i** + 5**j** + 3**k**,

$\quad$ **(d)** **a** = 4**i** + 6**j** − 7**k** $\quad$ and $\quad$ **b** = 10**i** − 7**j** + 8**k**.

10 Determine whether or not the following pairs of vectors are perpendicular:

$\quad$ **(a)** $\mathbf{p} = \begin{bmatrix} 2 \\ -3 \end{bmatrix}$ and $\mathbf{q} = \begin{bmatrix} 6 \\ 4 \end{bmatrix}$, $\quad$ **(b)** $\mathbf{p} = \begin{bmatrix} -4 \\ -2 \end{bmatrix}$ and $\mathbf{q} = \begin{bmatrix} 3 \\ -6 \end{bmatrix}$,

$\quad$ **(c)** $\mathbf{p} = \begin{bmatrix} 4 \\ 3 \end{bmatrix}$ and $\mathbf{q} = \begin{bmatrix} -3 \\ -4 \end{bmatrix}$, $\quad$ **(d)** $\mathbf{p} = \begin{bmatrix} 1 \\ -2 \end{bmatrix}$ and $\mathbf{q} = \begin{bmatrix} -8 \\ -4 \end{bmatrix}$.

11 Determine whether or not the following pairs of vectors are perpendicular:

(a) $\mathbf{r} = 4\mathbf{i} - 3\mathbf{j} + 11\mathbf{k}$ and $\mathbf{s} = 7\mathbf{i} + 2\mathbf{j} - 2\mathbf{k}$,

(b) $\mathbf{a} = 3\mathbf{i} + 2\mathbf{j} + \mathbf{k}$ and $\mathbf{b} = 5\mathbf{i} - 3\mathbf{j} - 9\mathbf{k}$,

(c) $\mathbf{p} = 5\mathbf{i} - \mathbf{j} + 5\mathbf{k}$ and $\mathbf{q} = -2\mathbf{i} + 5\mathbf{j} + 3\mathbf{k}$,

(d) $\mathbf{u} = 4\mathbf{i} + 6\mathbf{j} - 2\mathbf{k}$ and $\mathbf{v} = 10\mathbf{i} - 7\mathbf{j} + \mathbf{k}$.

12 Find the angle between the following pairs of vectors, giving your answer to the nearest $0.1°$:

(a) $\mathbf{u} = \begin{bmatrix} 2 \\ -5 \end{bmatrix}$ and $\mathbf{v} = \begin{bmatrix} 3 \\ 1 \end{bmatrix}$,   (b) $\mathbf{p} = \begin{bmatrix} 4 \\ -1 \end{bmatrix}$ and $\mathbf{q} = \begin{bmatrix} -1 \\ -2 \end{bmatrix}$.

13 Find the cosine of the angle between the vectors:

(a) $\begin{bmatrix} 2 \\ -1 \\ 2 \end{bmatrix}$ and $\begin{bmatrix} 2 \\ 2 \\ 1 \end{bmatrix}$,   (b) $\begin{bmatrix} 3 \\ -2 \\ 6 \end{bmatrix}$ and $\begin{bmatrix} -1 \\ -2 \\ 2 \end{bmatrix}$,

(c) $\begin{bmatrix} 2 \\ 2 \\ 1 \end{bmatrix}$ and $\begin{bmatrix} -3 \\ 12 \\ 4 \end{bmatrix}$,

(d) $3\mathbf{i} - 4\mathbf{j} + 12\mathbf{k}$ and $2\mathbf{i} + 3\mathbf{j} - 6\mathbf{k}$,

(e) $\mathbf{i} - 3\mathbf{j} + 2\mathbf{k}$ and $\mathbf{i} - 2\mathbf{j} - 3\mathbf{k}$.

14 Find the angle between the following pairs of vectors, giving your answer to the nearest $0.1°$:

(a) $\mathbf{p} = \begin{bmatrix} 2 \\ -1 \\ 2 \end{bmatrix}$ and $\mathbf{q} = \begin{bmatrix} 1 \\ 2 \\ 2 \end{bmatrix}$,   (b) $\mathbf{u} = \begin{bmatrix} -4 \\ -3 \\ 2 \end{bmatrix}$ and $\mathbf{v} = \begin{bmatrix} 3 \\ -1 \\ 1 \end{bmatrix}$,

(c) $\mathbf{r} = 3\mathbf{i} - 2\mathbf{j} + \mathbf{k}$ and $\mathbf{s} = 2\mathbf{i} + \mathbf{j} - 2\mathbf{k}$,

(d) $\mathbf{c} = 5\mathbf{i} + 3\mathbf{j} + 2\mathbf{k}$ and $\mathbf{d} = 3\mathbf{i} - 2\mathbf{j} - 5\mathbf{k}$.

9.9 The vector equation of a straight line

Consider the straight line l in the direction of the vector $\mathbf{d}$, passing through the point A, as shown opposite. Denote the position vector of A by $\mathbf{a}$.

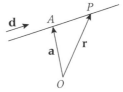

Let the general point P on the line, with position vector $\mathbf{r}$, have coordinates (x, y, z) so that $\mathbf{r} = \begin{bmatrix} x \\ y \\ z \end{bmatrix}$.

Since the vector sum $\overrightarrow{OA} + \overrightarrow{AP} = \overrightarrow{OP}$, you can write $\mathbf{a} + \overrightarrow{AP} = \mathbf{r}$.

But $\overrightarrow{AP}$ is some scalar multiple of vector $\mathbf{d}$, so $\overrightarrow{AP} = \lambda\mathbf{d}$.

Hence, the vector equation of the line is $\mathbf{r} = \mathbf{a} + \lambda\mathbf{d}$, where, in three dimensions, $\mathbf{r} = \begin{bmatrix} x \\ y \\ z \end{bmatrix}$, and λ is a scalar parameter.

In terms of three dimensional column vectors, when $\mathbf{a} = \begin{bmatrix} a_1 \\ a_2 \\ a_3 \end{bmatrix}$,

and $\mathbf{d} = \begin{bmatrix} d_1 \\ d_2 \\ d_3 \end{bmatrix}$, the equation of the line is $\begin{bmatrix} x \\ y \\ z \end{bmatrix} = \begin{bmatrix} a_1 \\ a_2 \\ a_3 \end{bmatrix} + \lambda\begin{bmatrix} d_1 \\ d_2 \\ d_3 \end{bmatrix}$.

Worked example 9.10

Find a vector equation of the line passing through the points $(1, 3, 7)$ and $(3, 5, 8)$.

Solution

Using the notation above, you can choose the point A to be

$(1, 3, 7)$ and so $\mathbf{a} = \begin{bmatrix} 1 \\ 3 \\ 7 \end{bmatrix}$.

Writing B as $(3, 5, 8)$, the direction of the line is parallel to the

vector $\overrightarrow{AB}$ and $\overrightarrow{AB} = \begin{bmatrix} 2 \\ 2 \\ 1 \end{bmatrix}$.

The equation of the line is therefore

$$\mathbf{r} = \begin{bmatrix} 1 \\ 3 \\ 7 \end{bmatrix} + \lambda\begin{bmatrix} 2 \\ 2 \\ 1 \end{bmatrix} \quad \text{or} \quad \begin{bmatrix} x \\ y \\ z \end{bmatrix} = \begin{bmatrix} 1 \\ 3 \\ 7 \end{bmatrix} + \lambda\begin{bmatrix} 2 \\ 2 \\ 1 \end{bmatrix}.$$

You could have written this more compactly as
$$\mathbf{r} = \mathbf{i} + 3\mathbf{j} + 7\mathbf{k} + \lambda(2\mathbf{i} + 2\mathbf{j} + \mathbf{k}).$$

9.10 The intersection of two lines

In two dimensions, two non-identical lines that are not parallel will intersect.

Worked example 9.11

Two lines in two dimensions have equations $\mathbf{r} = \begin{bmatrix} 3 \\ 2 \end{bmatrix} + s\begin{bmatrix} -1 \\ 4 \end{bmatrix}$ and

$\mathbf{r} = \begin{bmatrix} -2 \\ 0 \end{bmatrix} + t\begin{bmatrix} 3 \\ -1 \end{bmatrix}$, where s and t are scalar parameters.

(a) Find the point of intersection of the two lines.

(b) Find the acute angle between the two lines.

Solution

(a) By equating the two expressions for **r**, you can find the point of intersection:

$$\begin{bmatrix} 3 \\ 2 \end{bmatrix} + s\begin{bmatrix} -1 \\ 4 \end{bmatrix} = \begin{bmatrix} -2 \\ 0 \end{bmatrix} + t\begin{bmatrix} 3 \\ -1 \end{bmatrix}$$

Therefore $3 - s = -2 + 3t$ and $2 + 4s = 0 - t$
$\Rightarrow s + 3t = 5$ and $4s + t = -2$ which can be solved simultaneously to give $s = -1$ and $t = 2$.

Substituting $s = -1$ into $\mathbf{r} = \begin{bmatrix} 3 \\ 2 \end{bmatrix} + s\begin{bmatrix} -1 \\ 4 \end{bmatrix}$ gives $\mathbf{r} = \begin{bmatrix} 4 \\ -2 \end{bmatrix}$.

Or substituting $t = 2$ into $\mathbf{r} = \begin{bmatrix} -2 \\ 0 \end{bmatrix} + t\begin{bmatrix} 3 \\ -1 \end{bmatrix}$ also gives
$\mathbf{r} = \begin{bmatrix} 4 \\ -2 \end{bmatrix}$.

The point of intersection has coordinates $(4, -2)$.

(b) The angle between the two lines is the angle θ between the two direction vectors $\begin{bmatrix} -1 \\ 4 \end{bmatrix}$ and $\begin{bmatrix} 3 \\ -1 \end{bmatrix}$.

The scalar product is $\begin{bmatrix} -1 \\ 4 \end{bmatrix} \cdot \begin{bmatrix} 3 \\ -1 \end{bmatrix} = -3 - 4 = -7$.

But the scalar product is also given by $pq \cos\theta$, where p is the magnitude of $\begin{bmatrix} -1 \\ 4 \end{bmatrix}$, so $p = \sqrt{17}$ and q is the magnitude of $\begin{bmatrix} 3 \\ -1 \end{bmatrix}$, so $q = \sqrt{10}$.

Hence $\sqrt{17} \times \sqrt{10} \times \cos\theta = -7$, or

$$\cos\theta = \frac{-7}{\sqrt{17} \times \sqrt{10}} \approx -0.536\,875\ldots$$

Therefore $\theta = 122.5°$, to the nearest $0.1°$.

But when two straight lines intersect, there are two possible angles between them. Hence the acute angle between the lines is $57.5°$, to the nearest $0.1°$.

In order to find the angle between two lines you find the angle between the direction vectors of the lines.

The acute angle, θ, between $\mathbf{r} = \mathbf{a} + \lambda\mathbf{p}$ and $\mathbf{r} = \mathbf{b} + \mu\mathbf{q}$ is given by $\cos\theta = \dfrac{|\mathbf{p.q}|}{pq}$.

In three dimensions, lines that are not parallel do not always intersect. When they are neither parallel nor intersect each other, they are said to be **skew**.

Worked example 9.12

The line l_1 has equation $\mathbf{r} = \begin{bmatrix} 1 \\ 3 \\ 7 \end{bmatrix} + \lambda \begin{bmatrix} 2 \\ 2 \\ 1 \end{bmatrix}$ and the line l_2 has

equation $\mathbf{r} = \begin{bmatrix} 5 \\ 1 \\ 4 \end{bmatrix} + \mu \begin{bmatrix} 3 \\ 0 \\ -1 \end{bmatrix}$.

(a) Show that l_1 and l_2 intersect and find the coordinates of the point of intersection.

(b) Find the acute angle between l_1 and l_2.

Solution

(a) For any points of intersection

$$\begin{bmatrix} 1 \\ 3 \\ 7 \end{bmatrix} + \lambda \begin{bmatrix} 2 \\ 2 \\ 1 \end{bmatrix} = \begin{bmatrix} 5 \\ 1 \\ 4 \end{bmatrix} + \mu \begin{bmatrix} 3 \\ 0 \\ -1 \end{bmatrix}.$$

There are therefore three equations to be satisfied:

$$2\lambda - 3\mu = 4$$
$$2\lambda + 0\mu = -2$$
$$\lambda + \mu = -3$$

The second of these equations gives $\lambda = -1$.

Substituting into the last equation gives $\mu = -2$.

You need to check that all three equations are satisfied so you check in the first equation and $(2 \times -1) - (3 \times -2) = 4$ so all three equations are satisfied by $\lambda = -1$ and $\mu = -2$.

Therefore the two lines intersect.

Substituting in either line equation gives the position vector

of the point of intersection as $\mathbf{r} = \begin{bmatrix} 1 \\ 3 \\ 7 \end{bmatrix} + (-1) \begin{bmatrix} 2 \\ 2 \\ 1 \end{bmatrix} = \begin{bmatrix} -1 \\ 1 \\ 6 \end{bmatrix}$.

So the lines intersect at the point $(-1, 1, 6)$.

> Note that there are clearly values of λ and μ which satisfy any pair of these equations. The important thing is to verify that these values also satisfy the third equation.

(b) The angle θ between the two lines is the angle between the

two direction vectors $\begin{bmatrix} 2 \\ 2 \\ 1 \end{bmatrix}$ and $\begin{bmatrix} 3 \\ 0 \\ -1 \end{bmatrix}$.

The scalar product is $\begin{bmatrix} 2 \\ 2 \\ 1 \end{bmatrix} . \begin{bmatrix} 3 \\ 0 \\ -1 \end{bmatrix} = 6 - 1 = 5$.

> Since the scalar product is positive, there is no need to take its modulus in order to find the acute angle.

But the scalar product is also given by $pq \cos \theta$, where p is the magnitude of $\begin{bmatrix} 2 \\ 2 \\ 1 \end{bmatrix}$, namely $\sqrt{1 + 4 + 4} = 3$

and q is the magnitude of $\begin{bmatrix} 3 \\ 0 \\ -1 \end{bmatrix}$, namely $\sqrt{10}$.

Hence $3 \times \sqrt{10} \times \cos \theta = 5$, or $\cos \theta = \dfrac{5}{3 \times \sqrt{10}} \approx 0.527$.

The angle is $58.2°$ (to the nearest $0.1°$).

Worked example 9.13

The line l_1 has equation $\mathbf{r} = \begin{bmatrix} 1 \\ 5 \\ 4 \end{bmatrix} + t \begin{bmatrix} 2 \\ -1 \\ 1 \end{bmatrix}$.

The line l_2 has equation $\mathbf{r} = \begin{bmatrix} 0 \\ 3 \\ 5 \end{bmatrix} + s \begin{bmatrix} 1 \\ 0 \\ 1 \end{bmatrix}$.

(a) Verify that the vector $\mathbf{v} = \mathbf{i} + \mathbf{j} - \mathbf{k}$ is perpendicular to each of the lines l_1 and l_2.

(b) Show that the lines l_1 and l_2 are skew lines.

(c) Find the acute angle between the two lines l_1 and l_2.

Solution

(a) $\mathbf{v} \cdot \begin{bmatrix} 2 \\ -1 \\ 1 \end{bmatrix} = \begin{bmatrix} 1 \\ 1 \\ -1 \end{bmatrix} \cdot \begin{bmatrix} 2 \\ -1 \\ 1 \end{bmatrix} = 2 - 1 - 1 = 0$

so $\mathbf{v}$ is perpendicular to the line l_1.

Also $\mathbf{v} \cdot \begin{bmatrix} 1 \\ 0 \\ 1 \end{bmatrix} = \begin{bmatrix} 1 \\ 1 \\ -1 \end{bmatrix} \cdot \begin{bmatrix} 1 \\ 0 \\ 1 \end{bmatrix} = 1 + 0 - 1 = 0$

so $\mathbf{v}$ is perpendicular to the line l_2.

(b) Any point of intersection will satisfy $\begin{bmatrix} 1 \\ 5 \\ 4 \end{bmatrix} + t \begin{bmatrix} 2 \\ -1 \\ 1 \end{bmatrix} = \begin{bmatrix} 0 \\ 3 \\ 5 \end{bmatrix} + s \begin{bmatrix} 1 \\ 0 \\ 1 \end{bmatrix}$.

Writing as three simultaneous equations: $\quad 2t - s = -1$
$$t + 0s = 2$$
$$t - s = 1$$

The second equation gives $t = 2$, therefore substituting into the third equation gives $s = 1$. But that would mean that $2t - s = 3$ so the first equation would not be satisfied.

Alternatively, subtracting the second equation from the first gives $t - s = -3$ which is inconsistent with the third equation $t - s = 1$.

The two lines do **not** intersect.

Since $\begin{bmatrix} 2 \\ -1 \\ 1 \end{bmatrix}$ is not parallel to $\begin{bmatrix} 1 \\ 0 \\ 1 \end{bmatrix}$ the lines l_1 and l_2 are skew lines.

(c) Taking the scalar product of the two direction vectors gives

$$\begin{bmatrix} 2 \\ -1 \\ 1 \end{bmatrix} . \begin{bmatrix} 1 \\ 0 \\ 1 \end{bmatrix} = 2 + 1 = 3.$$

The magnitude of $\begin{bmatrix} 2 \\ -1 \\ 1 \end{bmatrix}$ is $\sqrt{6}$ and the magnitude of $\begin{bmatrix} 1 \\ 0 \\ 1 \end{bmatrix}$ is $\sqrt{2}$.

The angle θ between the two lines must satisfy

$\sqrt{6} \times \sqrt{2} \times \cos \theta = 3$ so that $\cos \theta = \dfrac{1}{2}$ and hence $\theta = 60°$.

Therefore the angle between the two lines is $60°$.

EXERCISE 9C

1 Two lines in two dimensions have equations $\mathbf{r} = \begin{bmatrix} 1 \\ -2 \end{bmatrix} + s \begin{bmatrix} -1 \\ 3 \end{bmatrix}$

and $\mathbf{r} = \begin{bmatrix} -2 \\ 3 \end{bmatrix} + t \begin{bmatrix} 1 \\ -1 \end{bmatrix}$, where s and t are scalar parameters.

(a) Find the point of intersection of the two lines.

(b) Find the acute angle between the two lines.

2 Two lines have equations $\mathbf{r} = \begin{bmatrix} 5 \\ -2 \end{bmatrix} + s \begin{bmatrix} -3 \\ 2 \end{bmatrix}$ and

$\mathbf{r} = \begin{bmatrix} -7 \\ -7 \end{bmatrix} + t \begin{bmatrix} 2 \\ 3 \end{bmatrix}$, where s and t are scalar parameters.

(a) Show that the two lines are perpendicular.

(b) Find the point of intersection of the two lines.

3 The line l_1 has equation $\begin{bmatrix} x \\ y \\ z \end{bmatrix} = \begin{bmatrix} 3 \\ 1 \\ 0 \end{bmatrix} + t \begin{bmatrix} 0 \\ 2 \\ 1 \end{bmatrix}$.

The line l_2 has equation $\begin{bmatrix} x \\ y \\ z \end{bmatrix} = \begin{bmatrix} 1 \\ 1 \\ 3 \end{bmatrix} + s \begin{bmatrix} 1 \\ 1 \\ -1 \end{bmatrix}$.

(a) Show that the lines l_1 and l_2 intersect and find the coordinates of their point of intersection.

(b) Find the cosine of the acute angle between l_1 and l_2.

4 Show that the two lines $\mathbf{r} = 2\mathbf{i} + 3\mathbf{j} - \mathbf{k} + s(\mathbf{i} + 2\mathbf{j} - 4\mathbf{k})$ and $\mathbf{r} = 3\mathbf{i} + 2\mathbf{k} + t(3\mathbf{i} - 5\mathbf{j} + 2\mathbf{k})$ are skew.

5 The line l_1 has equation $\mathbf{r} = \begin{bmatrix} 2 \\ 0 \\ -1 \end{bmatrix} + s\begin{bmatrix} 3 \\ 4 \\ 5 \end{bmatrix}$.

   (a) Find the acute angle between the line l_1 and the line l_2 with

      equation $\mathbf{r} = \begin{bmatrix} 5 \\ -2 \\ 7 \end{bmatrix} + t\begin{bmatrix} 3 \\ -1 \\ -2 \end{bmatrix}$

   (b) The line l_3 passes through the points $A(3, -2, 4)$ and $B(5, 1, 7)$. Show that the lines l_1 and l_3 intersect and find the coordinates of their point of intersection.

6 The line l_1 has equation $\begin{bmatrix} x \\ y \\ z \end{bmatrix} = \begin{bmatrix} 3 \\ -2 \\ 1 \end{bmatrix} + t\begin{bmatrix} 4 \\ 4 \\ 3 \end{bmatrix}$.

   The line l_2 has equation $\begin{bmatrix} x \\ y \\ z \end{bmatrix} = \begin{bmatrix} 8 \\ -1 \\ 2 \end{bmatrix} + s\begin{bmatrix} -1 \\ 3 \\ 2 \end{bmatrix}$.

   (a) Show that the lines l_1 and l_2 intersect and find the coordinates of their point of intersection.

   (b) Show that the vector $\begin{bmatrix} 1 \\ 11 \\ -16 \end{bmatrix}$ is perpendicular to both l_1 and l_2.
                                                                [A]

7 The line l_1 has equation $\mathbf{r} = \begin{bmatrix} 1 \\ 0 \\ -2 \end{bmatrix} + \lambda\begin{bmatrix} 1 \\ 4 \\ 3 \end{bmatrix}$.

   The line l_2 has equation $\mathbf{r} = \begin{bmatrix} 5 \\ 5 \\ 10 \end{bmatrix} + \mu\begin{bmatrix} 2 \\ -3 \\ 6 \end{bmatrix}$.

   (a) Show that the lines l_1 and l_2 intersect at a point P and find the position vector of P.

   (b) Find the acute angle between the lines l_1 and l_2, giving your answer to the nearest degree.

   (c) The line l_3 passes through the points $(0, 0, 0)$ and $(2, 8, 6)$. Show that l_1 and l_3 are **not** skew lines.   [A]

9.11 The perpendicular distance from a point to a line

The point P on the line $\mathbf{r} = \mathbf{a} + \lambda\mathbf{d}$ closest to the point Q is found by determining the value of $\lambda = p$ so that the vector PQ is perpendicular to the line. This is done by making use of the fact that $\overrightarrow{PQ}.\mathbf{d} = 0$.

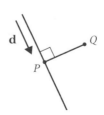

9

Worked example 9.14

In two dimensions, find the point on the line $\mathbf{r} = \begin{bmatrix} 5 \\ -1 \end{bmatrix} + \lambda \begin{bmatrix} 1 \\ -3 \end{bmatrix}$

that is closest to the point $Q(4, 7)$.

Solution

The point P is a general point where $\lambda = p$ on the line and so its x-coordinate is $5 + p$ and its y-coordinate is $-1 - 3p$.

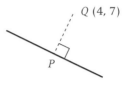

The vector $\overrightarrow{PQ} = \begin{bmatrix} 4 - (5 + p) \\ 7 - (-1 - 3p) \end{bmatrix} = \begin{bmatrix} -1 - p \\ 8 + 3p \end{bmatrix}$.

If Q is the closest point to the line, then PQ must be perpendicular to the line.

Therefore $\overrightarrow{PQ} \cdot \begin{bmatrix} 1 \\ -3 \end{bmatrix} = 0$.

Hence, $(-1 - p) - 3(8 + 3p) = 0$ or $-25 - 10p = 0$.

So $p = -2.5$.

The position vector of P is $\mathbf{r} = \begin{bmatrix} 5 \\ -1 \end{bmatrix} - 2.5 \begin{bmatrix} 1 \\ -3 \end{bmatrix} = \begin{bmatrix} 2.5 \\ 6.5 \end{bmatrix}$.

The point P, the point on the line closest to Q, has coordinates $(2.5, 6.5)$.

Worked example 9.15

The point A has coordinates $(3, 3, 6)$.

The line l has equation $\mathbf{r} = \begin{bmatrix} 5 \\ 3 \\ -2 \end{bmatrix} + \lambda \begin{bmatrix} 2 \\ 1 \\ -3 \end{bmatrix}$.

(a) The point P on the line l is where $\lambda = p$. Show that

$$\overrightarrow{AP} \cdot \begin{bmatrix} 2 \\ 1 \\ -3 \end{bmatrix} = 28 + 14p.$$

(b) Hence find the coordinates of the foot of the perpendicular from the point A to the line l.

(c) Determine the shortest distance from A to the line l.

Solution

(a) The point P has position vector $\mathbf{p} = \begin{bmatrix} 5 \\ 3 \\ -2 \end{bmatrix} + p \begin{bmatrix} 2 \\ 1 \\ -3 \end{bmatrix} = \begin{bmatrix} 5 + 2p \\ 3 + p \\ -2 - 3p \end{bmatrix}$.

The point A has coordinates $(3, 3, 6)$.

Hence $\overrightarrow{AP} = \begin{bmatrix} 5 + 2p \\ 3 + p \\ -2 - 3p \end{bmatrix} - \begin{bmatrix} 3 \\ 3 \\ 6 \end{bmatrix} = \begin{bmatrix} 2 + 2p \\ p \\ -8 - 3p \end{bmatrix}$.

$\overrightarrow{AP} . \begin{bmatrix} 2 \\ 1 \\ -3 \end{bmatrix} = \begin{bmatrix} 2 + 2p \\ p \\ -8 - 3p \end{bmatrix} . \begin{bmatrix} 2 \\ 1 \\ -3 \end{bmatrix} = 4 + 4p + p + 24 + 9p = 28 + 14p$

(b) In order for AP to be perpendicular to the line, this scalar product must be equal to zero.

Therefore $28 + 14p = 0$. Hence $p = -2$.

The foot of the perpendicular from the point A to the line l has

position vector $\begin{bmatrix} 5 \\ 3 \\ -2 \end{bmatrix} + (-2) \begin{bmatrix} 2 \\ 1 \\ -3 \end{bmatrix} = \begin{bmatrix} 1 \\ 1 \\ 4 \end{bmatrix}$.

Its coordinates are $(1, 1, 4)$.

(c) The distance between $A(3, 3, 6)$ and the point $(1, 1, 4)$ is the shortest distance between A and the line.

Shortest distance $= \sqrt{(3 - 1)^2 + (3 - 1)^2 + (6 - 4)^2} = \sqrt{12} = 2\sqrt{3}$.

EXERCISE 9D

1 In two dimensions, the point A has coordinates $(3, 8)$ and the

point B lies on the line $\mathbf{r} = \begin{bmatrix} 2 \\ -5 \end{bmatrix} + \lambda \begin{bmatrix} 5 \\ -3 \end{bmatrix}$, where $\lambda = b$.

(a) Show that $\overrightarrow{AB} . \begin{bmatrix} 5 \\ -3 \end{bmatrix} = 34(1 + b)$.

(b) Hence find the point on the line closest to A.

2 Find the closest point on the line $\mathbf{r} = \begin{bmatrix} 2 \\ -4 \end{bmatrix} + \lambda \begin{bmatrix} 1 \\ 2 \end{bmatrix}$ to the point

$B(2, 1)$. Hence find the shortest distance from B to the line.

3 In two dimensions, find the foot of the perpendicular from the

point $C(7, 9)$ to the line $\mathbf{r} = \begin{bmatrix} 0 \\ 6 \end{bmatrix} + \lambda \begin{bmatrix} 2 \\ 5 \end{bmatrix}$.

9

4 The point A has coordinates $(2, 1, 3)$ and the line l has equation

$$\mathbf{r} = \begin{bmatrix} 2 \\ 5 \\ 8 \end{bmatrix} + \lambda \begin{bmatrix} 1 \\ 0 \\ 2 \end{bmatrix}.$$

(a) The point P on the line l is where $\lambda = p$. Show that

$$\overrightarrow{AP} \cdot \begin{bmatrix} 1 \\ 0 \\ 2 \end{bmatrix} = 5p + 10.$$

(b) Hence find the coordinates of the foot of the perpendicular from the point A to the line l.

(c) Determine the shortest distance from A to the line l.

5 The point B has coordinates $(5, -2, 2)$ and the line l has equation $\mathbf{r} = \begin{bmatrix} -1 \\ 4 \\ 2 \end{bmatrix} + \lambda \begin{bmatrix} 2 \\ 1 \\ -1 \end{bmatrix}.$

(a) The point P on the line l is where $\lambda = p$. Show that

$$\overrightarrow{BP} \cdot \begin{bmatrix} 2 \\ 1 \\ -1 \end{bmatrix} = 6p - 6.$$

(b) Hence find the coordinates of the foot of the perpendicular from the point B to the line l.

(c) Determine the shortest distance from B to the line l.

6 (a) Find the vector equation of the line l_1 which passes through the point $A(3, -1, 2)$ and $B(2, 0, 2)$.

(b) The line l_2 has equation $\mathbf{r} = \begin{bmatrix} 4 \\ 1 \\ -1 \end{bmatrix} + \mu \begin{bmatrix} 1 \\ 0 \\ -1 \end{bmatrix}.$ Show that the

lines l_1 and l_2 intersect and find the coordinates of their point of intersection.

(c) Show that the point $C(9, 1, -6)$ lies on the line l_2.

(d) Find the coordinates of the point D on l_1 such that CD is perpendicular to l_1. [A]

7 The points $A(5, 1, 2)$ and $B(-1, 7, 8)$ have position vectors **a** and **b**, respectively.

   (a) Show that the two lines $\mathbf{r} = \mathbf{a} + \lambda(4\mathbf{i} - \mathbf{j} + \mathbf{k})$ and $\mathbf{r} = \mathbf{b} + \mu(2\mathbf{i} - 5\mathbf{j} - 7\mathbf{k})$ intersect and find the coordinates of the point of intersection C.

   (b) Find a vector equation of AB.

   (c) The point D has coordinates $(3, 7, -2)$. Show that CD is perpendicular to AB.

8 The line l_1 passes through the point $A(2, 0, 2)$ and has equation

$$\mathbf{r} = \begin{bmatrix} 2 \\ 0 \\ 2 \end{bmatrix} + t \begin{bmatrix} 2 \\ 6 \\ -3 \end{bmatrix}.$$

   (a) Show that the line l_2 which passes through the points $B(4, 4, -5)$ and $C(0, -8, 1)$ is parallel to the line l_1.

   (b) P is the point on line l_1 such that angle CPA is a right angle.

      (i) Find the value of the parameter t at P.

      (ii) Hence show that the shortest distance between the lines l_1 and l_2 is $2\sqrt{5}$.   [A]

Key point summary

1 A vector is represented by bold type such as **v**, or by   *p307* using an arrow above the letters such as $\overrightarrow{AB}$ to indicate the direction.

2 The magnitude of the vector $\overrightarrow{AB}$ is written as $|\overrightarrow{AB}|$ or   *p307* as AB. The magnitude of the vector **v** is written as $|\mathbf{v}|$ or as v.

3 In two dimensions, when a vector **v** is written as   *p307* $\mathbf{v} = \begin{bmatrix} a \\ b \end{bmatrix}$, the quantities a and b are the components of the vector in the x- and y-directions, respectively.

4 In three dimensions, when a vector **v** is written as   *p307* $\mathbf{v} = \begin{bmatrix} a \\ b \\ c \end{bmatrix}$, the quantities a and b and c are the components of the vector in the x-, y- and z-directions, respectively.

5 If the point A has position vector **a** and the point B *p308* has position vector **b**, then $\overrightarrow{AB} = \mathbf{b} - \mathbf{a}$.

6 In two dimensions, the magnitude of the vector $\begin{bmatrix} a \\ b \end{bmatrix}$ is   *p310* $\sqrt{a^2 + b^2}$.

9

7 In three dimensions, the magnitude of the vector $\begin{bmatrix} a \\ b \\ c \end{bmatrix}$ *p310*
is $\sqrt{a^2 + b^2 + c^2}$.

8 The distance between the points (x_1, y_1, z_1) and *p311*
(x_2, y_2, z_2) is $\sqrt{(x_2 - x_1)^2 + (y_2 - y_1)^2 + (z_2 - z_1)^2}$.

9 The scalar product of **p** and **q**, written as **p.q**, has *p315*
value $|\mathbf{p}| \times |\mathbf{q}| \times \cos \theta$ or $pq \cos \theta$, where θ is the
angle between the two vectors.

10 If $\mathbf{p} = \begin{bmatrix} a \\ b \\ c \end{bmatrix}$ and $\mathbf{q} = \begin{bmatrix} d \\ e \\ f \end{bmatrix}$ then **p.q** is evaluated as *p315*

$$\begin{bmatrix} a \\ b \\ c \end{bmatrix} \cdot \begin{bmatrix} d \\ e \\ f \end{bmatrix} = ad + be + cf.$$

11 If $\mathbf{a.b} = 0$, where **a** and **b** are non-zero vectors, then *p317*
a and **b** are perpendicular.

12 In order to find the angle between two lines you find *p321*
the angle between the direction vectors of the lines.

The acute angle, θ, between $\mathbf{r} = \mathbf{a} + \lambda\mathbf{p}$ and $\mathbf{r} = \mathbf{b} + \mu\mathbf{q}$

is given by $\cos \theta = \dfrac{|\mathbf{p.q}|}{pq}$.

Test yourself	**What to review**
1 Find the magnitude of the vector $\begin{bmatrix} 2 \\ -1 \end{bmatrix}$.	*Section 9.3*
2 Find the distance between the points $(3, 1, 4)$ and $(1, 0, 2)$.	*Section 9.4*
3 Calculate the scalar product $\begin{bmatrix} 2 \\ -1 \\ 3 \end{bmatrix} \cdot \begin{bmatrix} 5 \\ -4 \\ -2 \end{bmatrix}$	*Section 9.7*
4 Find the angle between the two vectors $\begin{bmatrix} 2 \\ -3 \end{bmatrix}$ and $\begin{bmatrix} -3 \\ 5 \end{bmatrix}$, giving your answer to the nearest $0.1°$.	*Section 9.8*

Test yourself (continued)	**What to review**

5 Show that the line l_1 with equation $\mathbf{r} = \begin{bmatrix} 1 \\ 4 \\ 5 \end{bmatrix} + \lambda \begin{bmatrix} 2 \\ 2 \\ 1 \end{bmatrix}$ and the

Section 9.10

line l_2 with equation $\mathbf{r} = \begin{bmatrix} 5 \\ 0 \\ 2 \end{bmatrix} + \mu \begin{bmatrix} 3 \\ -1 \\ -1 \end{bmatrix}$ intersect and find the

coordinates of their point of intersection.

6 The point A has coordinates $(1, 0, 2)$ and the line l has

Section 9.11

equation $\mathbf{r} = \begin{bmatrix} 1 \\ 2 \\ 3 \end{bmatrix} + \lambda \begin{bmatrix} 1 \\ 1 \\ 0 \end{bmatrix}$.

(a) The point P on the line l is where $\lambda = p$. Show that

$$\overrightarrow{AP} \cdot \begin{bmatrix} 1 \\ 1 \\ 0 \end{bmatrix} = 2p + 2.$$

(b) Hence find the coordinates of the foot of the perpendicular from the point A to the line l.

(c) Determine the shortest distance from A to the line l.

Test yourself **ANSWERS**

6 **(b)** $(0, 1, 3)$; **(c)** $\sqrt{3}$.

5 $(-1, 2, 4)$.

4 $177.3°$.

3 8.

2 3.

1 $\sqrt{5}$.

9

C4: Exam style practice paper

Time allowed 1 hour 30 minutes

Answer **all** questions

1 The polynomial p(x) is defined by p(x) = $4x^3 - 6x^2 - 4x + 3$.
   (a) Prove that $(2x - 1)$ is a factor of p(x). (2 marks)
   (b) Hence simplify the expression $\dfrac{\text{p}(x)}{1 - 4x^2}$. (3 marks)
   (c) Find the remainder when p(x) is divided by $(2x + 1)$. (2 marks)

2 (a) Express $\dfrac{28x^2}{(2x + 1)(3 - x)}$ in the form

   $$A + \frac{B}{2x + 1} + \frac{C}{3 - x}.$$ (3 marks)

   (b) Hence find $\displaystyle\int_0^1 \frac{28x^2}{(2x + 1)(3 - x)}\, dx$, giving your
   answer in the form $p \ln 3 + q \ln 2 + r$, where p, q and r
   are integers. (5 marks)

3 (a) Express $\sqrt{3} \cos \theta + \sin \theta$ in the form $R \cos(\theta - \alpha)$, where
   R is a positive constant and $0° < \theta < 90°$. (3 marks)

   (b) Hence find the solutions in the interval $0° \leqslant \theta \leqslant 360°$
   of the equation
   $$\sqrt{3} \cos \theta + \sin \theta + 1 = 0.$$ (4 marks)

4 (a) The quantity N varies with t so that $\dfrac{dN}{dt} = 3N$, and
   $N = 400$ when $t = 0$.
   (i) Solve the differential equation expressing t in terms
       of N. (4 marks)
   (ii) Hence, show that $n = 400\,e^{3t}$. (2 marks)

   (b) The number of bacteria, N, in a colony is such that the
   hourly rate of increase of N is equal to three times the
   number of bacteria present. Initially there are 400
   bacteria present. Use the results from **(a)** to find:
   (i) the number of bacteria present after 2 hours, giving
       your answer to the nearest thousand; (1 mark)
   (ii) the time taken for the number of bacteria to
        grow to 700, giving your answer to the nearest
        minute. (2 marks)

5 A curve is defined by the parametric equations $x = t^2 + \dfrac{2}{t}$, $y = t^2 - \dfrac{2}{t}$. The point P on the curve is where $t = 2$.

(a) Find the coordinates of the point P. (1 mark)

(b) Find $\dfrac{dy}{dx}$ at the point P. (4 marks)

(c) Find an equation for the normal to the curve at P. (4 marks)

(d) Verify that the Cartesian equation of the curve is $(x + y)(x - y)^2 = k$, stating the value of the constant k. (3 marks)

6 (a) Find the binomial expansion in ascending powers of x up to and including the term in x^3 of:

(i) $(1 - x)^{-2}$, (3 marks)

(ii) $\dfrac{1}{(3 + x)}$. (3 marks)

(b) Show that $\dfrac{1}{3 + x} + \dfrac{2}{(1 - x)^2} \equiv \dfrac{x^2 + 7}{(3 + x)(1 - x)^2}$. (2 marks)

(c) Hence find the binomial expansion in ascending powers of x up to and including the term in x^3 of $\dfrac{x^2 + 7}{(3 + x)(1 - x)^2}$. (3 marks)

7 A curve has equation $y^3 + 3y + 3\cos 2x + 3\sin x = 0$.

(a) Show that $\dfrac{dy}{dx} = \dfrac{2\sin 2x - \cos x}{y^2 + 1}$. (4 marks)

(b) Hence find the coordinates of all the stationary points of the curve in the interval $0 < x < 2\pi$, giving your answers to three significant figures. (5 marks)

8 The points A and B have coordinates $(3, -1, 4)$ and $(-3, 2, 2)$, respectively. The line l has vector equation $\mathbf{r} = \begin{bmatrix} 6 \\ 3 \\ 3 \end{bmatrix} + \lambda \begin{bmatrix} 3 \\ -1 \\ -2 \end{bmatrix}$.

(a) Find the distance between the points A and B. (2 marks)

(b) Find a vector equation of the line passing through the points A and B. (2 marks)

(c) Find the acute angle between the line l and the line AB. (3 marks)

(d) The point P on the line l is where $\lambda = p$. Show that $\overrightarrow{AP} \cdot \begin{bmatrix} 3 \\ -1 \\ -2 \end{bmatrix} = 7 + 14p$ and hence find the coordinates of the foot of the perpendicular from the point A to the line l. (5 marks)

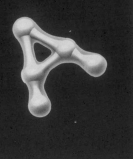

Answers

Core maths 3

EXERCISE 1A

1 (a) -2;   (b) 54.

2 (a) -8;   (b) 216.

3 (a) 3;   (b) 5;   (c) 4;   (d) 1.

4 (a) 4;   (b) 3;   (c) 4;   (d) 5.

5 (a) $-8, 9$;   (b) -6.

6 (a) $17, -23$;   (b) $-0.5, 3$.

7 (a) -1;   (b) -1;   (c) -1;   (d) -7;   (e) -25.

8 (a) $\frac{1}{27}$;   (b) $\frac{1}{64}$;   (c) $\frac{1}{125}$;
   (d) $\frac{1}{8}$;   (e) 1;   (f) -1;
   (g) 1 000 000;   (h) $(b + 3)^{-3}$;   (i) a^{-3}.

9 (a) $\sqrt{5}$;   (b) $\sqrt{7}$;   (c) 3;   (d) $\sqrt{3}$;   (e) 1.

EXERCISE 1B

1 $\{2, 3, 6, 11\}$.

2 $(-132, -34, -8, -6, 20\}$.

3 $-2 \leqslant \mathrm{h}(x) \leqslant 10$.

4 (a)   (b) $1 \leqslant \mathrm{f}(x) \leqslant 5$.

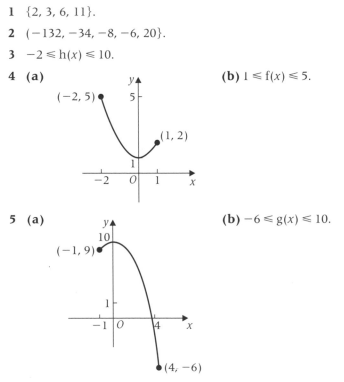

5 (a)   (b) $-6 \leqslant \mathrm{g}(x) \leqslant 10$.

6 (a) (i) (0, 1), (−0.5, 0), (0.5, 0),
  (ii)   **(b)** f(x) ≤ 1.

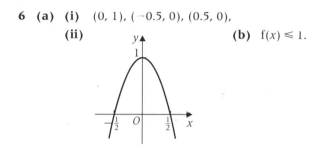

EXERCISE 1C

1

2

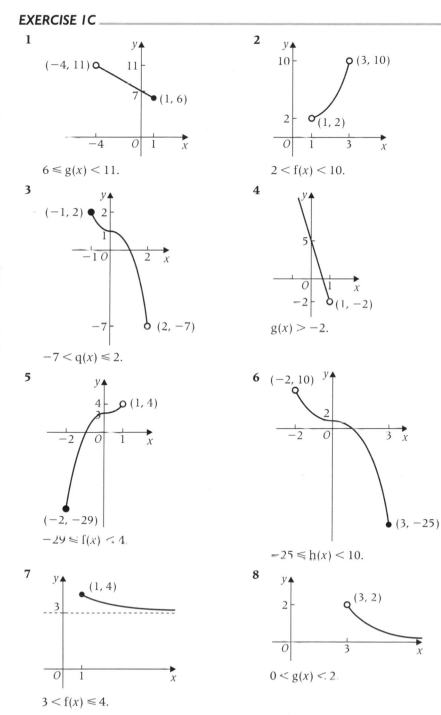

6 ≤ g(x) < 11.

2 < f(x) < 10.

3

4

−7 < q(x) ≤ 2.

g(x) > −2.

5

6

−29 ≤ f(x) < 4.

−25 ≤ h(x) < 10.

7

8

3 < f(x) ≤ 4.

0 < g(x) < 2.

9

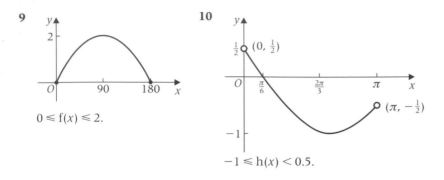

$0 \leqslant f(x) \leqslant 2.$

10

$-1 \leqslant h(x) < 0.5.$

EXERCISE 1D

1 $x \in \mathbb{R}, x \neq 1.$ **2** $x \geqslant -3.$

3 $x \in \mathbb{R}, x \neq 2.$ **4** $x \in \mathbb{R}, x \neq -1.$

5 $\mathbb{R}.$ **6** $x > 2.$

7 $x \in \mathbb{R}, x \neq 1, x \neq 2.$ **8** $x < 4, x \neq 0.$

EXERCISE 1E

1 Many-one function.   **2** Not.

3 One-one function.   **4** One-one function.

5 Many-one function.   **6** Not.

EXERCISE 1F

1 **(a)** $\text{fg}(x) = 4 - 2x;$ **(b)** $\text{fg}(x) = (2 + x)^2 - 3;$
     (c) $\text{fg}(x) = (3x - 1)^3 + 1;$ **(d)** $\text{fg}(x) = (x + 1)^8 - 2.$

2 **(a)** $\text{gf}(x) = 7 - 2x;$ **(b)** $\text{gf}(x) = x^2 - 1;$
     (c) $\text{gf}(x) = 3x^3 + 2;$ **(d)** $\text{gf}(x) = (x^4 - 1)^2.$

3 **(a)** $9x^2;$ **(b)** $6x - 3x^2 - 2.$

4 **(a)** $-1;$ **(b)** $4a - 9: a = 3.$

5 **(a)** $4k - 2 - 3kx;$ **(b)** $10 - 3kx, k = 3.$

6 $\text{fg}(x) = (5 + x)^2,\ \text{gf}(x) = 5 + x^2,\ x = -2.$

7 **(a)** $g(x) \geqslant 1;$
     (b) $\text{fg}(x) = \dfrac{3}{2x^2 + 1},$ range $0 < \text{fg}(x) \leqslant 3.$

EXERCISE 1G

1 **(a)** $f^{-1}(x) = \dfrac{x - 7}{5};$ **(b)** $f^{-1}(x) = \sqrt[3]{x} - 2;$

     (c) $f^{-1}(x) = \dfrac{6x - 1}{2};$ **(d)** $f^{-1}(x) = \dfrac{x^3 + 1}{2};$

     (e) $f^{-1}(x) = \dfrac{5x - 3}{-2} = \dfrac{3 - 5x}{2};$ **(f)** $f^{-1}(x) = \dfrac{7 - 4x}{3}.$

2 (a)

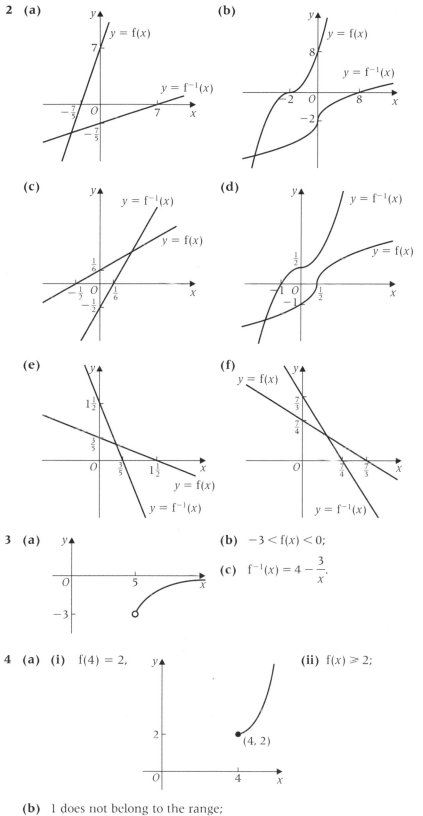

(b)

(c)

(d)

(e)

(f)

3 (a)

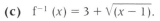

(b) $-3 < f(x) < 0$;

(c) $f^{-1}(x) = 4 - \dfrac{3}{x}$.

4 (a) (i) $f(4) = 2$,   **(ii)** $f(x) \geqslant 2$;

(b) 1 does not belong to the range;

(c) $f^{-1}(x) = 3 + \sqrt{(x - 1)}$.

5 (a) (i) $f(x) \geqslant -64$, **(ii)** $f^{-1}(x) = \dfrac{\sqrt[3]{x} + 1}{3}$, $x \geqslant -64$,

(iii) $f^{-1}(x) \geqslant -1$;

(b) (i) $-1 \leqslant g(x) < 1$, **(ii)** $g^{-1}(x) = \dfrac{2}{1 - x}$, $-1 \leqslant x < 1$,

(iii) $g^{-1}(x) \geqslant 1$;

(c) (i) $h(x) > 243$, **(ii)** $h^{-1}(x) = \dfrac{\sqrt[5]{x} - 3}{2}$, $x > 243$,

(iii) $h^{-1}(x) > 0$;

(d) (i) $0 < q(x) < 1$, **(ii)** $q^{-1}(x) = \dfrac{5}{x - 1}$, $0 < x < 1$,

(iii) $q^{-1}(x) < -5$;

(e) (i) $-2 \leqslant r(x) < 0$, **(ii)** $r^{-1}(x) = 3 - \dfrac{4}{x}$, $-2 \leqslant x < 0$,

(iii) $r^{-1}(x) \geqslant 5$.

6 (a) (i) $f(x) \in \mathbb{R}$, $f(x) \neq 2$, **(ii)** $f^{-1}(x) = \dfrac{3x + 5}{x - 2}$, $x \in \mathbb{R}$, $x \neq 2$,

(iii) $f^{-1}(x) \in \mathbb{R}$, $f^{-1}(x) \neq 3$;

(b) (i) $g(x) \in \mathbb{R}$, $f(x) \neq 2.5$, **(ii)** $g^{-1}(x) = \dfrac{4 + x}{5 - 2x}$, $x \in \mathbb{R}$, $x \neq 2.5$,

(iii) $g^{-1}(x) \in \mathbb{R}$, $g^{-1}(x) \neq -0.5$.

7 (a)

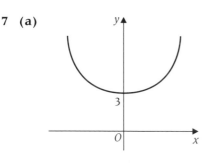

f is many-one, e.g. $f(-2) = f(2)$;

(b)

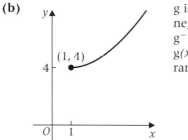

g is one-one – domain excludes negative values
$g^{-1}(x) = \sqrt{x - 3}$; since range of g is $g(x) \geqslant 4$, domain of g^{-1} is $x \geqslant 4$ and range is $g^{-1}(x) \geqslant 1$.

8 (a) Sketch of $y = h(x)$

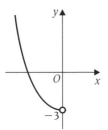

h is one-one since domain has only negative values of x;

(b) $h^{-1}(x) = -\sqrt{x + 3}$; domain of h^{-1} is $x > -3$; range is $h^{-1}(x) < 0$.

9 Functions f and h are self-inverse.

10 (a) 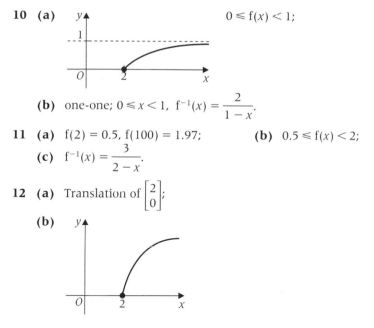 $0 \leqslant f(x) < 1$;

(b) one-one; $0 \leqslant x < 1$, $f^{-1}(x) = \dfrac{2}{1-x}$.

11 (a) $f(2) = 0.5$, $f(100) = 1.97$;   **(b)** $0.5 \leqslant f(x) < 2$;

(c) $f^{-1}(x) = \dfrac{3}{2-x}$.

12 (a) Translation of $\begin{bmatrix} 2 \\ 0 \end{bmatrix}$;

(b)

(c) f is a one-one function, domain of f^{-1} is $x \geqslant 0$, $f^{-1}(x) = x^2 + 2$.

MIXED EXERCISE

1 (a) $f(-1) = 6$, $f(2) = 9$;   **(b)**

(c) range of f is $5 \leqslant f(x) \leqslant 9$;

(d) not one-one so no inverse function; $(-1, 6)$

(e) $x^4 + 10x^2 + 30$.

2 (a) $g(x) \geqslant 2$;   **(b)** $fg(x) = \dfrac{6}{2x^2 + 3}$;

(c) $f^{-1}(x) = \dfrac{6+x}{2x}$;   **(d)** $(2, 2)$ and $(-1.5, -1.5)$.

3 (a) $fg(x) = \dfrac{2}{6 - x^2}$;

(b) (i) 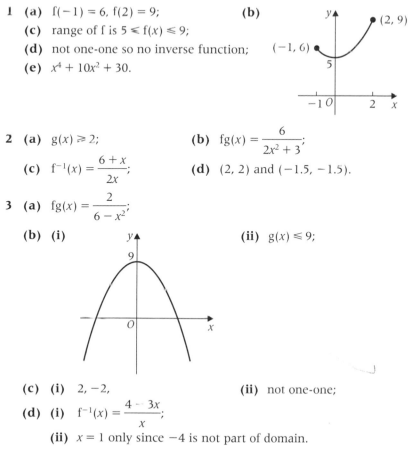   **(ii)** $g(x) \leqslant 9$;

(c) (i) $2, -2,$ **(ii)** not one-one;

(d) (i) $f^{-1}(x) = \dfrac{4 - 3x}{x}$;

(ii) $x = 1$ only since -4 is not part of domain.

4 (a)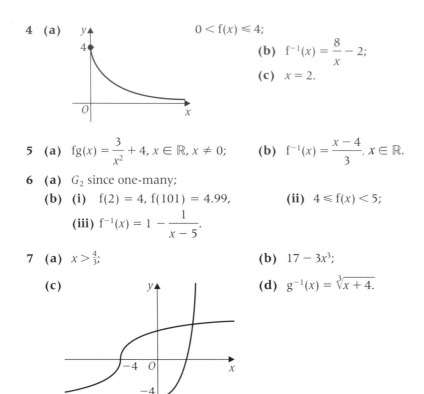
$0 < f(x) \leqslant 4$;

(b) $f^{-1}(x) = \dfrac{8}{x} - 2$;

(c) $x = 2$.

5 (a) $fg(x) = \dfrac{3}{x^2} + 4$, $x \in \mathbb{R}$, $x \neq 0$;

(b) $f^{-1}(x) = \dfrac{x - 4}{3}$, $x \in \mathbb{R}$.

6 (a) G_2 since one-many;

(b) (i) $f(2) = 4$, $f(101) = 4.99$,

(ii) $4 \leqslant f(x) < 5$;

(iii) $f^{-1}(x) = 1 - \dfrac{1}{x - 5}$.

7 (a) $x > \frac{4}{3}$;

(b) $17 - 3x^3$;

(c)

(d) $g^{-1}(x) = \sqrt[3]{x + 4}$.

2 Transformations of graphs and the modulus function

EXERCISE 2A

1 (a) $y = (x - 2)^3 - 1$;

(b) $y = x^2$;

(c) $y = \tan(x - 2) - 1$;

(d) $y = 3^{x - 2} - 1$.

2 (a) $y = 8x^3$;

(b) $y = 4x^2 + 8x + 5$;

(c) $y = \tan 2x$;

(d) $y = 3^{2x}$ or $y = 9^x$.

3 (a) translation $\begin{bmatrix} 0 \\ -1 \end{bmatrix}$;

(b) reflection in y-axis;

(c) translation $\begin{bmatrix} -2 \\ 4 \end{bmatrix}$;

(d) stretch in y-direction SF 4;

(e) reflection in x-axis;

(f) stretch in x-direction SF 0.5.

4 (a) stretch in x-direction SF 0.2;

(b) translation $\begin{bmatrix} 3 \\ 0 \end{bmatrix}$;

(c) stretch in x-direction SF 3;

(d) translation $\begin{bmatrix} -7 \\ 0 \end{bmatrix}$;

(e) reflection in x-axis.

5 (a) reflection in y-axis;

(b) translation $\begin{bmatrix} 4 \\ 0 \end{bmatrix}$;

(c) stretch in x-direction SF 0.4;

(d) stretch in y-direction SF 2;

(e) translation $\begin{bmatrix} 0 \\ 4 \end{bmatrix}$;

(f) stretch in x-direction SF 0.5.

EXERCISE 2B

1   translation $\begin{bmatrix} 5 \\ 0 \end{bmatrix}$, stretch in y-direction SF 3.

2   $y = 2 - \cos\left(\dfrac{x}{3}\right)$.

3   $y = 5 + 3^{1-x}$.

4   $y = -4 \sin(x - \pi)$ or $y = 4 \sin x$.

5   $2(x-3)^2 + 1$, translation $\begin{bmatrix} 3 \\ 0 \end{bmatrix}$, stretch in y-direction SF 2, translation $\begin{bmatrix} 0 \\ 1 \end{bmatrix}$.

6   **(a)**   translation $\begin{bmatrix} 2 \\ 0 \end{bmatrix}$, stretch in y-direction SF 4;

    (b)   translation $\begin{bmatrix} -1 \\ 0 \end{bmatrix}$, stretch in y-direction SF 3, translation $\begin{bmatrix} 0 \\ 4 \end{bmatrix}$.

    (c)   stretch in x-direction SF 0.5, translation $\begin{bmatrix} 1 \\ 0 \end{bmatrix}$;

    (d)   translation $\begin{bmatrix} 3 \\ 0 \end{bmatrix}$, reflection in x-axis;

    (e)   stretch in x-direction SF 1/3, translation $\begin{bmatrix} -5 \\ 0 \end{bmatrix}$;

    (f)   stretch in x-direction SF 3, translation $\begin{bmatrix} 2 \\ 0 \end{bmatrix}$, stretch in y-direction SF 4.

7   translation $\begin{bmatrix} 0 \\ 5 \end{bmatrix}$, stretch in y-direction SF 1/3
    (or stretch in x-direction SF$\sqrt{3}$).

8   translation $\begin{bmatrix} -4 \\ 0 \end{bmatrix}$, stretch in y-direction SF $\frac{1}{5}$.

9   stretch in x-direction SF 2, and in y-direction SF 5, then translation $\begin{bmatrix} 0 \\ -3 \end{bmatrix}$.

10   stretch in x-direction SF 2, stretch in y-direction SF $\dfrac{1}{320}$.

11   **(a)**   reflection in $y = x$;
     (b)   **(i)** $x = 3y + 2$,   **(ii)** $x = y^2$,   **(iii)** $(y-1)^2 + x^2 = 4$.

12   **(a)** $g(x) = -f(-x)$;   **(b)**   half turn about the origin.

13   **(a)** $y = 2 - x^2$;   **(b)**   reflection in x-axis then translation $\begin{bmatrix} 0 \\ 2 \end{bmatrix}$.

14   **(a)** $y = 2^{10-x}$;   **(b)**   reflection in y-axis then translation $\begin{bmatrix} 10 \\ 0 \end{bmatrix}$.

15   **(a)**   reflection in x-axis then translation $\begin{bmatrix} 0 \\ 6 \end{bmatrix}$;
     (b)   reflection in $y = 3$;
     (c)   reflect then in $y = p$;

16   **(a)**   reflection in y-axis then translation $\begin{bmatrix} 4 \\ 0 \end{bmatrix}$;
     (b)   reflection in $x = 2$;
     (c)   reflection in $x = q$.

EXERCISE 2C _____

1
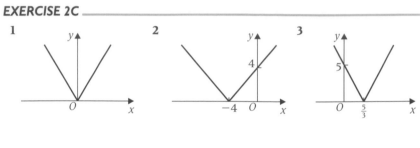

2

3

4
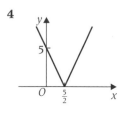

5

6

7
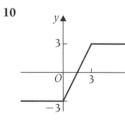

8

9

10
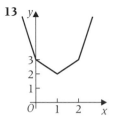

11
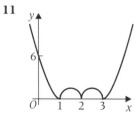

12
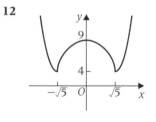

13
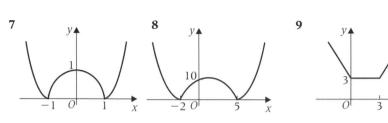

14

15

16
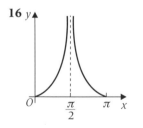

17
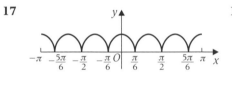

18
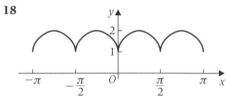

EXERCISE 2D

1 $x = 1, 5$.

2 $x = -1, 9$.

3 $x = 5, -9$.

4 $x = \frac{1}{2}$.

5 $x = \frac{1}{2}$.

6 No solution.

7 No solution.

8 $x = 1, 2$.

9 $x = 1, 2$.

10 No solutions.

11 $x = 1, \frac{4}{3}$.

12 $x = 2, \dfrac{\sqrt{17} - 5}{2}$.

13 No solution.

14 No solution.

15 $x = \frac{3}{4}, 1$.

EXERCISE 2E

1 $x < 3, x > 7$.

2 $3 \leqslant x \leqslant 5$.

3 $-10 < x < 4$.

4 $x \geqslant 1$.

5 $x < -\frac{1}{2}$.

6 No solution.

7 $x \in \mathbb{R}$.

8 $\frac{1}{3} < x < \frac{1}{2}$.

9 $x \leqslant 1, x \geqslant 4$.

10 $x < -\frac{1}{2}, x > 1$.

11 $\frac{1}{3}(\sqrt{19} - 1) \leqslant x \leqslant 2$.

12 $x \leqslant \frac{1}{2}(\sqrt{57} - 9), x \geqslant 6$.

MIXED EXERCISE

1 **(a)** **(b)** $f(x) \geqslant -1$;

(c) $x = \frac{2}{3}, 4$.

2 $x \geqslant \frac{1}{2}$.

3 **(a)** **(b)** $2 < x < 6$.

4 **(a)** $x = \pm 2, \pm 4$; **(b)** $x = 3, 4$.

5 **(a)** $x = -2, \frac{8}{3}$;

(b) Lots of possibilities such as

6 **(a)** **(i)** **(ii)**

(b) $x = 2, 4, 3 + \sqrt{7}, 3 - \sqrt{7}$;

(c) $3 - \sqrt{7} \leqslant x \leqslant 2, 4 \leqslant x \leqslant 3 + \sqrt{7}$.

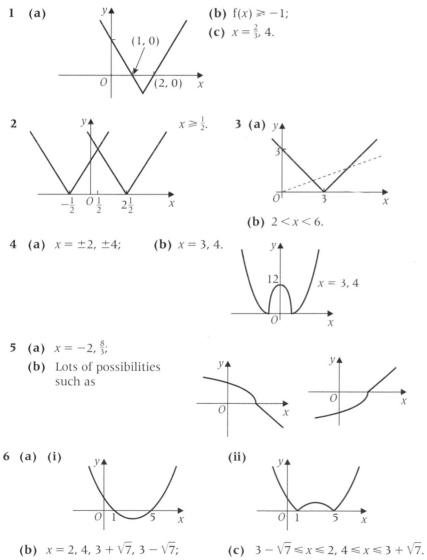

7 (a) 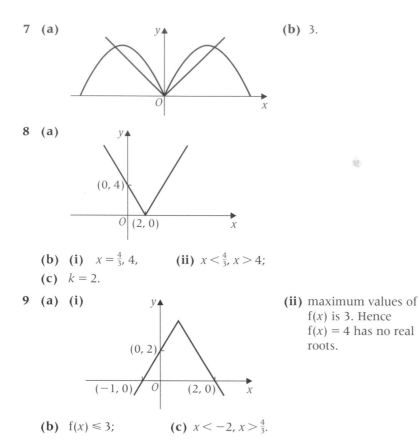   **(b)** 3.

8 (a)

(b) **(i)** $x = \frac{4}{3}, 4,$ **(ii)** $x < \frac{4}{3}, x > 4;$

(c) $k = 2.$

9 (a) (i)

(ii) maximum values of f(x) is 3. Hence f(x) = 4 has no real roots.

(b) f(x) $\leqslant$ 3;   **(c)** $x < -2, x > \frac{4}{3}.$

3 Inverse trigonometric functions and secant, cosecant and cotangent

EXERCISE 3A

1 (a) $\frac{\pi}{3}$;   **(b)** 0;   **(c)** $\frac{2\pi}{3}$;

(d) $\frac{2\pi}{3}$;   **(e)** $\frac{\pi}{6}$;   **(f)** $\frac{5\pi}{6}$.

2 (a) $\frac{\pi}{6}$;   **(b)** -1;   **(c)** $-\frac{\pi}{4}$;

(d) $\sqrt{3}$;   **(e)** $2\sqrt{2}$;   **(f)** $\frac{1}{\sqrt{5}}$.

3 (a) $-\frac{\pi}{4}$;   **(b)** $\frac{5\pi}{4}$.

4 1.

5 0.

6 3.

EXERCISE 3B

1 (a) 2.9238;   **(b)** 1.0642;   **(c)** 2.7475;

(d) 2.9238;   **(e)** 1;   **(f)** 4.3640;

(g) $\frac{1}{3}$;   **(h)** 0.4.

2 (a) -2.4030; (b) 1.5523; (c) 1.8305;
 (d) 1.8508; (e) 2.6131; (f) 6.4142;
 (g) 0.4875; (h) 0.1356.

3 (a) 2; (b) $\dfrac{2\sqrt{3}}{3}$; (c) $\sqrt{3}$;

 (d) -1; (e) $\sqrt{2}$; (f) $\dfrac{3+\sqrt{3}}{3}$;

 (g) $\sqrt{3}$; (h) 0.5.

4 (a) $\sqrt{2}$; (b) 3;
 (c) $\frac{1}{3}$; (d) $6+4\sqrt{3}$.

5 (a) $56.3°$, $303.7°$;
 (b) $206.4°$, $333.6°$;
 (c) $18.4°$, $198.4°$;
 (d) $140.3°$, $219.7°$;
 (e) $19.5°$, $160.5°$;
 (f) $157.4°$, $337.4°$;
 (g) $62.4°$, $117.6°$, $242.4°$, $297.6°$;
 (h) $55.9°$, $145.9°$, $235.9°$, $325.9°$.

6 (a) 1.05, 5.24;
 (b) 3.67, 5.76;
 (c) 0.393, 1.96, 3.53, 5.11;
 (d) 0.628, 1.88, 3.14, 4.40, 5.65;
 (e) 0.349, 0.698, 2.44, 2.79, 4.54, 4.89;
 (f) 1.31, 2.88, 4.45, 6.02;
 (g) 0, 2.09, 4.19, 6.28;
 (h) 0.349, 1.40, 2.44, 3.49, 4.54, 5.59.

EXERCISE 3C

1 (a) horizontal stretch, scale factor $\frac{1}{4}$;
 (b) $\dfrac{\pi}{2}$.

2 (a) horizontal stretch, scale factor 4;
 (b) 4π.

3
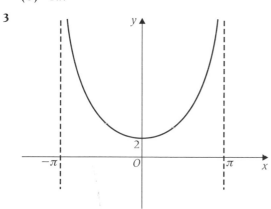

4 (a)

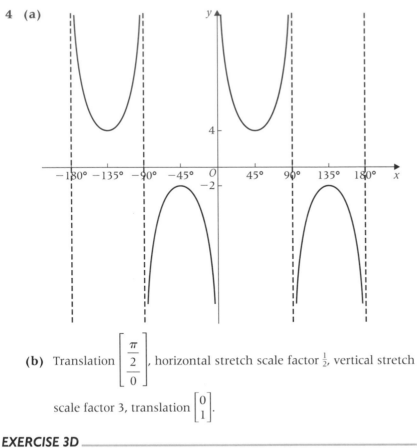

(b) Translation $\begin{bmatrix} \dfrac{\pi}{2} \\ 0 \end{bmatrix}$, horizontal stretch scale factor $\frac{1}{2}$, vertical stretch

scale factor 3, translation $\begin{bmatrix} 0 \\ 1 \end{bmatrix}$.

EXERCISE 3D

4 $\pm\sqrt{2}$.

5 $\pm\dfrac{1}{\sqrt{3}}$.

9 $x^2 = 16y^2 + 1$.

10 $(x - 2)^2 = 1 + \dfrac{16y^2}{16y^2}$.

11 $63.4°$, $135°$, $243.4°$, $315°$.

12 $45°$, $146.3°$, $225°$, $326.3°$.

13 $45°$, $63.4°$, $225°$, $243.4°$.

14 $0°$, $\pm120°$.

15 $19.5°$, $160.5°$, $194.5°$, $345.5°$.

16 $108°$, $252°$.

17 0.524, 2.62, 3.67, 5.76 (to 3 s.f.)

18 0, 1.05, 5.24 (to 3 s.f.).

19 3.318, 6.107 (to 3 d.p.).

20 -2.588, -1.017, 0.554, 2.124, -1.963, -0.393, 1.178, 2.749 (to 3 d.p.)

4 The number e and calculus

EXERCISE 4A

1 (a) $-e^{-x}$;

(b) $3e^{3x}$;

(c) $7e^{7x}$;

(d) $-2e^{-2x}$;

(e) $4e^{4x}$;

(f) $-6e^{-6x}$.

2 (a) -2;

(b) 9;

(c) 2.

3 (a) $3 + 6e^{-3}$;

(b) $20 + 3e^{-1}$;

(c) $10e^2 - 3$.

4 (a) e^{x+4};

(b) $-e^{2-x}$;

(c) $3e^{3x+1}$;

(d) $7e^{7x-5}$;

(e) ae^{ax+b}.

5 6.

6 (a) $5e^{5x} - 12e^{3x} + 4e^{2x}$; **(b)** $2e^{2x} - 7e^{-x} - 3e^{3x}$; **(c)** $4e^{-x} - 2e^{-2x}$.

7 0, 4, Stationary point at P which is a minimum.

EXERCISE 4B

1 (a) $-e^{-x} + c$; **(b)** $\dfrac{1}{3}e^{3x} + c$; **(c)** $\dfrac{1}{7}e^{7x} + c$;

   (d) $-\dfrac{1}{2}e^{-2x} + c$; **(e)** $\dfrac{1}{4}e^{4x} + c$; **(f)** $-\dfrac{1}{6}e^{-6x} + c$.

2 (a) $2e^{\frac{1}{2}x} + c$; **(b)** $\dfrac{x^3}{3} - \dfrac{3e^{2x}}{2} + c$; **(c)** $\dfrac{2x\sqrt{x}}{3} + 2e^{3x} + c$.

3 (a) $\dfrac{1 - e^{-3}}{3}$; **(b)** $2(e - 1)$; **(c)** $5e^{-1} - 3$.

4 $3 + \frac{1}{3}(e^6 - e^3)$.

5 $3(4 + e)$.

6 $3(e^{-1} - e^{-2}) + \frac{1}{2}(e^4 - e^2)$.

EXERCISE 4C

1 (a) 3; **(b)** -2; **(c)** $1 + \ln 2$; **(d)** 8; **(e)** $\frac{1}{5}$.

2 (a) $\sin x$; **(b)** $(1 + x)^2$; **(c)** $\dfrac{1}{5 - x}$.

3 (a) $\ln 5$; **(b)** $\ln 7$; **(c)** $-\ln 2$; **(d)** $\ln 2$.

4 (a) $\ln 2$; **(b)** $\frac{1}{2}\ln 5$; **(c)** $\ln 3$; **(d)** $-\ln 3$.

5 (a) e^2; **(b)** $\frac{1}{2}e^3$; **(c)** $3e^4$; **(d)** $2c^{-5}$.

6 (a) $\pm e^4$; **(b)** e^{-1}; **(c)** 3; **(d)** $\sqrt{5} - 1$; **(e)** $\dfrac{e - 2}{1 - e}$.

7 (a) One way stretch SF 2 parallel to y-axis.
   (b) Reflection in y-axis; **(c)** Translation $\begin{bmatrix} 0 \\ 2 \end{bmatrix}$;
   (d) Reflection in $x = 2$;
   (e) One way stretch SF 3 parallel to x-axis
   (f) Translation $\begin{bmatrix} 5 \\ 0 \end{bmatrix}$;
   (g) Enlargement SF 3;
   (h) Reflection in $y = x$.

8 (a) One way stretch SF 3 parallel to y-axis;
   (b) Translation $\begin{bmatrix} 0 \\ 3 \end{bmatrix}$;
   (c) Reflection in $y = \frac{3}{2}$;
   (d) Two way stretch SF 3 parallel to y-axis, SF $\frac{1}{3}$ parallel to x-axis.

9 (a)　　　　　　　　　**(b)**

EXERCISE 4D

1. $(\ln 5, 5 - 5 \ln 5)$, second derivative $= 5$ which is positive.

2. (a) $2e^{2x} - 6, 4e^{2x}$;   (b) $(\frac{1}{2} \ln 3, 6 - 3 \ln 3)$, minimum.

3. (a) $10 - 3e^{3x}, -9e^{3x}$;   (b) $(\frac{1}{3} \ln \frac{10}{3}, \frac{10}{3}\{\ln \frac{10}{3} - 1\})$, maximum.

4. (a) $4 - e^{-x}, e^{-x}$;   (b) $(-2 \ln 2, 9 - 8 \ln 2)$, minimum.

5. $-\frac{1}{2} \ln 2$.   6 $\frac{19}{3}$.   7 $\frac{1}{6}$.   8 $\frac{5}{72}$.   9 5.

EXERCISE 4E

1. (a) $\frac{1}{x}$;   (b) $2x - \frac{1}{x}$;   (c) $\frac{4}{x}$;   (d) $\frac{1}{2x}$.

2. (a) $2 \ln x + c$;   (b) $3 \ln 2$;   (c) $5 \ln 3 - 4$.

3. $14 + 5 \ln 2$.   4 $b = e^3$.

5. (a) $2x - \frac{2}{x^2}$; $M(1, 8)$ (b) 6;   (c) $\frac{22}{3} + 2 \ln 2$.

6. (a) (i)   (ii)

     (b) (i) $\frac{1}{2}e^{2x} - 2x + c$.

7. (a) (i) $\frac{3}{4}x^2 - \frac{6}{x}$;

     (b) (i) $x = 2$,   (ii) $\frac{3}{2}x + \frac{6}{x^2}$,   (iii) minimum point;
     (c) $28 - \frac{3}{2} \ln 2$.

8. (a) (i) $-2e^{-2x} - \frac{3}{x^2}$;   (b) $f(x) > 3$.

9. (a) (i) $\frac{dy}{dx} = 2x - 3 + \frac{1}{x}$;

     (b) (ii) $x = \frac{1}{2}, 1$.   (iii) $2 - \frac{1}{x^2}$.   (iv) -2 and 1.

MIXED EXERCISE

1. (a) $\frac{1}{3} \ln 2$;   (b) No solution;   (c) $\frac{e^2}{5}$;   (d) $\frac{e^{-5}}{3}$;

     (e) $5 - e^e$;   (f) $\frac{2 - e}{3}$;   (g) $\pm\frac{e^2}{3}$.

2. (a) $\ln 3$;   (b) $\ln 4$;   (c) $0, \ln 4$;
     (d) No solutions;   (e) $\ln(2 - \sqrt{2}), \ln(2 + \sqrt{2})$.

3. (a) $\frac{e^4}{2}, \frac{e^{-4}}{2}$;   (b) e^2, e^3;   (c) e^5, e^{-4};
     (d) $e^{2+\sqrt{5}}, e^{2-\sqrt{5}}$.

4 **(a)** 0; **(b)** 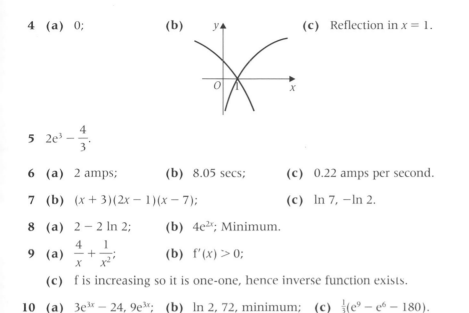 **(c)** Reflection in $x = 1$.

5 $2e^3 - \dfrac{4}{3}$.

6 **(a)** 2 amps; **(b)** 8.05 secs; **(c)** 0.22 amps per second.

7 **(b)** $(x + 3)(2x - 1)(x - 7)$; **(c)** $\ln 7, -\ln 2$.

8 **(a)** $2 - 2\ln 2$; **(b)** $4e^{2x}$; Minimum.

9 **(a)** $\dfrac{4}{x} + \dfrac{1}{x^2}$; **(b)** $f'(x) > 0$;

   (c) f is increasing so it is one-one, hence inverse function exists.

10 **(a)** $3e^{3x} - 24, 9e^{3x}$; **(b)** $\ln 2, 72$, minimum; **(c)** $\frac{1}{3}(e^9 - e^6 - 180)$.

5 Further differentiation and the chain rule

EXERCISE 5A

1 **(a)** Tangent $y = 6x + 6$, Normal $6y + x = 36$.

3 **(b)** **(i)** $y = 7x - 2$, **(ii)** $y = 9x - 2 - 2\ln 2$.

5 **(a)** **(i)** $-2e^{-2x} - \dfrac{3}{x^2}$; **(b)** $f(x) > 3$.

6 **(a)** $(0, -2)$; **(b)** $\frac{1}{2}\ln 3$.

7 **(a)** $(3, 6 + 2\ln 3)$; **(b)** minimum point; **(c)** $4y - x = 39$.

EXERCISE 5B

1 **(a)** $5\cos x + 3x^2$; **(b)** $8\cos x + 6e^{3x}$;

   (c) $5 - 2\sin x$; **(d)** $\dfrac{1}{x} + 7\sin x$;

   (e) $4\cos x - 5\sin x$; **(f)** $6\cos x + 8\sin x$.

2 **(a)** $\cos^{-1}\left(\frac{3}{4}\right)$;

   (b) **(i)** $-4\sin x$, **(ii)** maximum point.

3 **(a)** $\sin^{-1}\left(\frac{2}{5}\right)$;

   (b) **(i)** $-5\cos x$, **(ii)** maximum point;

   (c) tangent $y = 2x + 9$.

4 **(a)** $\dfrac{2\pi}{3}$;

   (b) **(i)** $6\sin x$; **(ii)** minimum point;

   (c) normal $y + 1 + \dfrac{3\pi}{2} = \dfrac{1}{3}\left(x - \dfrac{\pi}{2}\right)$ or $x - 3y = 3 + 5\pi$.

6 **(b)** $x + 9y + 63 = 0$.

EXERCISE 5C

1 (a) $32(8x + 3)^3$; (b) $-24(3 - 4x)^5$;
   (c) $150(3x - 2)^9$; (d) $-56(7 - x)^7$.

2 (a) $5e^{5x - 2}$; (b) $18 \cos (6x + 2)$; (c) $-16 \sin 8x$;
   (d) $\dfrac{45}{9x + 1}$; (e) $-\dfrac{2}{(2x + 7)^2}$.

3 (a) $\dfrac{9(2x + 1)}{3x^2 + 3x - 1}$; (b) $\dfrac{8}{(3 - 2x)^2}$;
   (c) $-\dfrac{24}{(6x + 1)^3}$; (d) $-\dfrac{e^x}{(e^x + 4)^2}$; (e) $2xe^{x^2}$.

4 (a) $48(4x - 1)^5$; (b) $-(8 - 3x)^4$;
   (c) $\frac{1}{2} \cos (6x + 1)$; (d) $\dfrac{4}{x - 3}$;
   (e) $-6e^{9 - 2x}$; (f) $-3 \sin (5 - 3x)$;
   (g) $3e^{3x} + 12x$; (h) $18x^2 - \dfrac{2}{x}$;
   (i) $8 \cos (6x - 5)$; (j) $-3e^{4 - 3x} + 5 \cos 5x$.

5 (a) 648; (b) -8; (c) $-4e$;
   (d) -2; (e) -74.

6 $y + 8x = 9$.

7 (a) $\dfrac{8}{(2x - 1)^3} - 1$, $(\frac{3}{2},\ 2)$ is a stationary point;

   (b) $\dfrac{d^2y}{dx^2} = -\dfrac{48}{(2x - 1)^4}$, maximum point.

8 (a) $9(3x - 2)^2 - 10x$; (b) $162x - 118$;
   maximum $(0.435, -1.28)$, minimum $(1.02, -4.01)$.

EXERCISE 5D

1 $10 \sin 5x \cos 5x$. 2 $-36 \cos^3 9x \sin 9x$.

3 $4 \cot 4x$. 4 $-40 \cos^4 4x \sin 4x$.

5 $6e^{2x}(e^{2x} - 5)^2$. 6 $\dfrac{3 \cos \sqrt{3x + 1}}{2\sqrt{3x + 1}}$.

7 $\dfrac{\ln x}{2x}$. 8 $\dfrac{3e^{\sqrt{6x}}}{\sqrt{6x}}$.

9 $\dfrac{-\sin (\ln 8x)}{x}$. 10 $\dfrac{-5\sin(\sqrt{x}) \cos^4 (\sqrt{x})}{2\sqrt{x}}$.

MIXED EXERCISE

1 (a) $\dfrac{dy}{dx} = \dfrac{3}{x}$ so $\dfrac{dx}{dy} = \dfrac{x}{3}$;

   (b) $\dfrac{dy}{dx} = -10(2x + 4)^{-6}$ so $\dfrac{dx}{dy} = -\dfrac{(2x + 4)^6}{10}$;

   (c) $\dfrac{dy}{dx} = \dfrac{-8}{(2x - 3)^2}$ so $\dfrac{dx}{dy} = -\dfrac{(2x - 3)^2}{8}$;

(d) $\dfrac{dy}{dx} = -2 \sin 2x$ so $\dfrac{dx}{dy} = -\dfrac{1}{2} \operatorname{cosec} 2x;$

(e) $\dfrac{dy}{dx} = 20 \cos 5x$ so $\dfrac{dx}{dy} = \dfrac{1}{20} \sec 5x;$

(f) $\dfrac{dy}{dx} = 2e^{2x} - 15 \sin 3x$ so $\dfrac{dx}{dy} = \dfrac{1}{2e^{2x} - 15 \sin 3x}.$

2 $\dfrac{dy}{dx} = (2 \cos 2x) \, e^{\sin 2x} = 2y \cos 2x.$

3 (a) (i) $6 \cos 2x - 2 \sin 2x;$ **(b)** $y = 3 - 2x + \dfrac{\pi}{2}.$

4 (a) (i) $12 \cos 3x - 6 \sin 3x;$ **(iii)** $0.369, 1.416, 2.463;$

    (b) $y - 5 = \dfrac{1}{12}\left(x - \dfrac{\pi}{3}\right).$

5 $3 \cos y;$ **(a)** $\dfrac{1}{3 \cos y} = \tfrac{1}{3} \sec y;$ **(c)** $\sin^{-1}\left(\dfrac{x}{3}\right).$

6 $\dfrac{1}{5 + y};$ **(a)** $5 + y;$ **(b)** $\dfrac{dy}{dx} = e^x = 5 + y.$

7 (a) $2x \cos (x^2 + 3);$ **(b)** $\tfrac{1}{2} \sin (x^2 + 3) + \text{constant}.$

8 (a) greatest depth 25 m, least depth 15 m;

    (b)

    (c) $k = 0.559$ (3 s.f.);

    (d) $-2.51 \, \text{m h}^{-1}$, minus sign $\Rightarrow$ depth is decreasing.

6 Differentiation using the product rule and the quotient rule

EXERCISE 6A

1 (a) $x \cos x + \sin x;$ **(b)** $x^2(3 \cos x - x \sin x);$

    (c) $4(\ln x + 1);$ **(d)** $\sqrt{x} \cos x + \dfrac{1}{2\sqrt{x}} \sin x;$

    (e) $5x^2 e^x(x + 3);$ **(f)** $\cos^2 x - \sin^2 x = \cos 2x;$

    (g) $e^x \left(\dfrac{1}{x} + \ln x\right);$ **(h)** $e^x(\cos x + \sin x);$

    (i) $\dfrac{1}{\sqrt{x}}(2 + \ln x);$ **(j)** $3x^4 e^x(x + 5).$

2 (a) $x^2 e^{2x}(2x + 3);$ **(b)** $(2x - 1)^4(63 - 36x);$

    (c) $e^x(4 \cos 4x + \sin 4x);$ **(d)** $3 \cos 3x \cos 2x - 2 \sin 3x \sin 2x;$

(e) $\dfrac{\sqrt{x+1}}{x} + \dfrac{\ln x}{2\sqrt{x+1}}$;

(f) $6x(1 + 2\ln 5x)$;

(g) $e^{3x}(3\cos 2x - 2\sin 2x)$;

(h) $\dfrac{4x^3}{\sqrt{2x-1}} + 12x^2\sqrt{2x-1} = \dfrac{4x^2(7x-3)}{\sqrt{(2x-1)}}$;

(i) $\dfrac{12}{3x+2} + 4\ln(3x+2)$;

(j) $e^{5x}\cos x + 5e^{5x}\sin x + 12x$.

3 (a) $9e$;   **(b)** $\dfrac{\sqrt{3}\pi + 6}{12}$;   **(c)** 1.

4 $y + 27e^9 = 10e^9 x$.

5 (a) $\left(\dfrac{1}{e}, -\dfrac{1}{e}\right)$;   **(b)** $\dfrac{1}{x}$, minimum point.

6 (a) $x = -\tfrac{3}{5}$;   **(b)** -1, maximum point.

7 (a) $A(1, 0)$, 1;   **(b)** $B(e^{-\frac{1}{2}}, -\tfrac{1}{2}e^{-1})$, 2.

EXERCISE 6B

1 (a) $\dfrac{-x\sin x - \cos x}{x^2}$;   **(b)** $\dfrac{e^x(x-2)}{x^3}$;

(c) $\dfrac{2x\cos x - \sin x}{2x\sqrt{x}}$;   **(d)** $\dfrac{1 - \ln x}{x^2}$;

(e) $\dfrac{4x(2-x)}{e^x}$;   **(f)** $\dfrac{5(1 - 3\ln x)}{x^4}$;

(g) $\dfrac{e^x(x-2)}{3x^3}$;   **(h)** $\dfrac{\cos x - \sin x}{e^x}$;

(i) $\sec^2 x$;   **(j)** $-\operatorname{cosec}^2 x$.

2 (a) $\dfrac{2x\cos 2x - 3\sin 2x}{x^4}$;   **(b)** $\dfrac{-e^{-x}(\sin x + \cos x)}{\sin^2 x}$

(c) $\dfrac{4x^4(5 - 3x)}{e^{3x}}$;   **(d)** $\dfrac{12x^2 - 8x - 39}{(2x+3)^4} = \dfrac{6x - 13}{(2x+3)^3}$;

(e) $\dfrac{1 - \ln 5x}{x^2}$;   **(f)** $\dfrac{3x^2(\cos 3x + x\sin 3x)}{\cos^2 3x}$;

(g) $\dfrac{2(2x\cos 2x - 3\sin 2x)}{x^4}$ **(h)** $\dfrac{2e^{3x}(3x - 1)}{x^2}$;

(i) $\dfrac{7(1 - 2\ln 3x)}{x^3}$;   **(j)** $\dfrac{-e^{2-x}(\sin 3x + 3\cos 3x)}{\sin^2 3x}$.

4 $y = 5x + 2$, $(\tfrac{8}{5}, 10)$.

5 (a) $\dfrac{2 - x}{4x^{\frac{3}{4}}(x+2)^{\frac{3}{2}}}$, $x = 2$;   **(b)** $0.000\,481$ cm^3 s^{-1}.

6 (a) $x = 1$;   **(c)** $\left(e^{\frac{1}{2}}, \dfrac{1}{2e}\right)$, $-\dfrac{2}{e^2}$.

EXERCISE 6C

1 (a) $15x^2 + 4 \sec^2 4x$;   (b) $6 \sec(6x - 3) \tan(6x - 3)$;

    (c) $\mathrm{cosec}^2(1 - x)$;   (d) $-15 \, \mathrm{cosec} \, 3x \cot 3x$.

2 (a) $3t^3 \sec t(t \tan t + 4)$;   (b) $\dfrac{-x \, \mathrm{cosec}^2 x - 3 \cot x}{x^4}$;

    (c) $\tan t$;   (d) $\cos x + x^2 \sec^2 x + 2x \tan x$.

3 $y = 6x - \frac{3}{2}\pi - 1$;   **5** $e^{-x} \sec 2x(2 \tan 2x - 1)$.

8 $y = 12x + 2 - \pi$.   **9** -4.

10 (a) $f'(x) = 2 \sec 2x \tan 2x + 2 \sec^2 2x$
           $= 2 \sec^2 2x \, (\sin 2x + 1) > 0$;

    (b) $2y + x = 2$.

MIXED EXERCISE

1 (a) $2x - 1$;   (b) $15x^2 - 2 + 2e^{2x}$;   (c) $8x + \dfrac{2}{x^2}$;

    (d) $3x^2 + 6x - 10$;   (e) $e^{-x}(\cos x - \sin x)$;   (f) $15 \sec^2 5x$;

    (g) $\dfrac{2(1 - 3 \ln x)}{x^4}$;   (h) $12 \, \mathrm{cosec}^2 (2 - 3x)$.

2 (a) 5;   (b) $2e$;   (c) $\frac{14}{3}$.

3 $4y + 28e^4 = 9e^4 x$.

4 $y = 8x + 2 - \pi$.

5 $(0, 0)$ minimum, $(2, 4e^{-2})$ maximum.

6 $9 - \ln 2$.

7 (a) $\frac{1}{3}$ s;   (b) $\frac{7}{25}$ cm s^{-1};   (c) 4 cm.

8 (a) $-4e^{-3}$;   (b) $y + 4e^{-3}x = 17e^{-3}$; $x = 4\frac{1}{4}$;   (c) $(-1, e)$.

9 (a) $x = 2 \pm \sqrt{3}$;   (b) $(1, -2e^{-1})$, $(5, 6e^{-5})$.

10 (a) -2;

    (b) (i) $-1.6 \, \mathrm{m \, s^{-1}}$,   (ii) plane losing height.

7 Numerical solution of equations and iterative methods

EXERCISE 7A

5 Between -2 and -1.

6 Between 2 and 3.

7 Between 1.80 and 1.85.

8 (b) (i) No, there may be an even number of roots in this interval.
      (ii) $g(-\frac{1}{2}) = -2.531\,25$ and $g(1\frac{1}{2}) = -12.906\,25$. At least one root lies
         between -1 and -0.5 and between 1.5 and 2.

9 (a) $\sqrt[3]{16} - \sqrt[3]{4}$;   (b) $\sqrt[3]{4} - \sqrt[3]{2}$;   (c) $(14\sqrt{2} + 20)^{\frac{1}{3}} + (20 - 14\sqrt{2})^{\frac{1}{3}} = 4$.

10 $f(0) = -1$ and $f(1) = 1$. No, the function has a discontinuity when $x = \frac{1}{2}$ and is
therefore not continuous between 0 and 1. No conclusion can be made about
roots in this interval.

EXERCISE 7B

8 (a) $-0.58 < a < -0.57$;   **(b)** $1.36 < c < 1.37$.

EXERCISE 7C

1 $x_2 = 5, x_3 = -\frac{5}{7}, x_4 = \frac{15}{11}$.

2 $u_2 = 1, u_3 = \frac{1}{2}, u_4 = \frac{5}{14}$.

3 (a) 3.166 67, 3.162 28, 3.162 28;   **(b)** 3.666 67, 3.606 06, 3.605 55.

4 (a) 3.111 11, 3.107 237, 3.107 23;   **(b)** 4.666 67, 4.641 72, 4.641 59.

5 (a) 0.13, 0.1417, 0.142 847 77;   **(b)** 0.0188, 0.018 867 68, 0.018 867 924 5.

6 (a) $-1, -1, -1$, converges;   **(b)** $-6, -26, -126$, diverges;

   (c) 4, 1, 2.5, converges;

   (d) 1.703 125, 1.704 878… , 1.705 860… , converges.

EXERCISE 7D

1 (a)

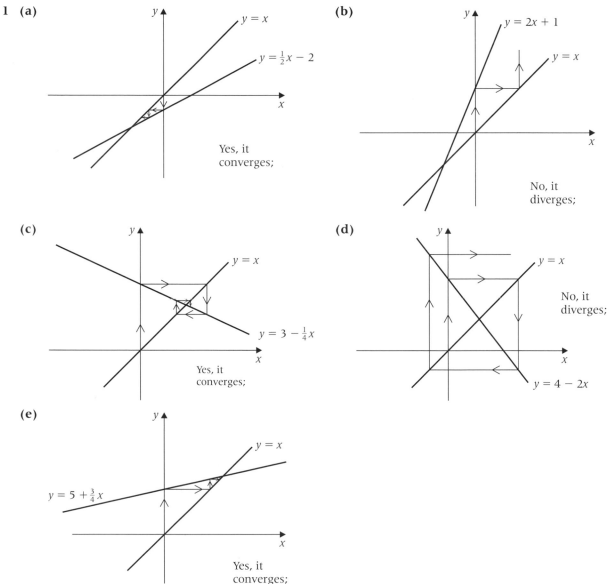

$y = x$

$y = \frac{1}{2}x - 2$

Yes, it converges;

(b)

$y = 2x + 1$

$y = x$

No, it diverges;

(c)

$y = x$

$y = 3 - \frac{1}{4}x$

Yes, it converges;

(d)

$y = x$

No, it diverges;

$y = 4 - 2x$

(e)

$y = x$

$y = 5 + \frac{3}{4}x$

Yes, it converges;

2 $-1 < k < 1$.

3 **(a)** 0.9093, 0.9695;

(b) cobweb convergence

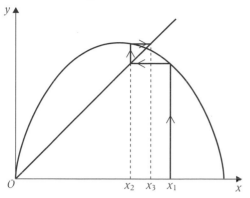

4 1.9463, 1.9303, 1.9361, 1.9340

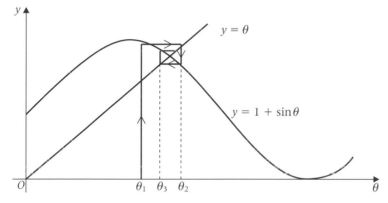

5 **(a)** between -2 and -1;

(b) $x_{n+1} = \sqrt[5]{(7 - 3x)}$, yes; $x_{n+1} = \frac{1}{3}(x_n^5 + 7)$, no; etc.;

6 4.224.

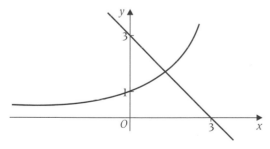

7 **(b)** $x = \ln(3 - x)$; 0.788, 0.794, 0.791, cobweb convergence

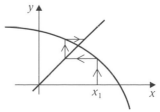

8 (a)

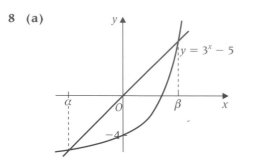

(b) $x_1 < \alpha$ converges to α; $\alpha < x_1 < \beta$, converges to α; $x_1 > \beta$ diverges;

(c) -4.996; **(d)** $x = \dfrac{\ln (x + 5)}{\ln 3}$

9 $x^3 - 3x^2 - 2 = 0$, 3.1958, cobweb convergence

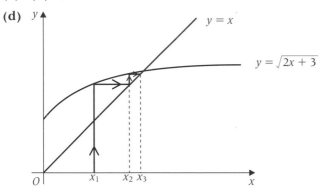

10 (a) $2.6457\ldots, 2.879\ldots, 2.959\ldots$;

(b) x_1 is positive and hence $x_2, x_3, x_4, \ldots$ are all positive. Always taking the square root of a positive number;

(c) (ii) 3;

(d)

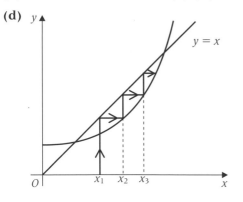

11 (a) $2.45, 2.4005, 2.352\ldots$; **(c) (ii)** 2;

(d)

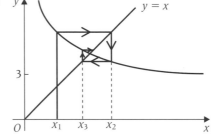

12 **(a)** $-2.6457\ldots$, $-1.436\ldots$, $-2.3856\ldots$;

(b) take the positive square root of a positive number and multiply the answer by -1;

(c) **(ii)** -2 **(d)**

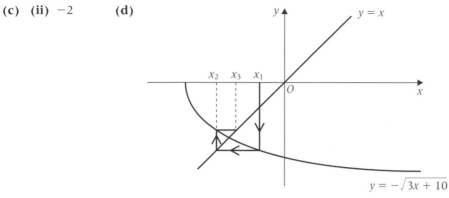

13 **(a)** $x_2 = 3.742$, $x_3 = 3.968$, $x_4 = 3.996$;

(b) **(ii)** $L = 4$; **(c)**

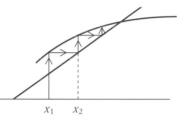

8 Integration by inspection and substitution

EXERCISE 8A

1 $2x^3 + 2\ln x + 7x + c.$

2 $\frac{2}{3}\sqrt{x^3} + \frac{1}{2}e^{2x} + c.$

3 $\frac{1}{2}\ln x + \frac{1}{2e^{4x}} + c.$

4 $\frac{1}{3}e^{3x} + \ln x - \frac{3}{x} + c.$

5 $\frac{1}{4}e^{4x} - e^{2x} + x + c.$

6 $24.2.$

EXERCISE 8B

1 **(a)** $\frac{1}{7}(x + 2)^7 + c;$ **(b)** $\frac{1}{45}(5x + 1)^9 + c;$

(c) $-\frac{1}{12}(3 - 2x)^6 + c;$ **(d)** $4x - \frac{1}{14}(1 + 2x)^7 + c;$

(e) $\frac{1}{6}\sqrt{(4x + 1^3)} + c;$ **(f)** $\frac{3}{32}\sqrt[3]{(8x - 27)^4} + c;$

(g) $-\frac{1}{x - 1} + c;$ **(h)** $\frac{1}{10(2 - 5x)^2} + c;$

(i) $-\frac{1}{24(5 + 4x)^3} + c;$ **(j)** $2\sqrt{2x - 1} + c;$

(k) $\frac{1}{2}e^{2x} + c;$ **(l)** $-e^{-3x} + c;$

(m) $\frac{1}{5}e^{5x + 6} - x + c;$ **(n)** $2e^{2x - 3} + c.$

2 **(a)** $24.2;$ **(b)** $6532.5;$ **(c)** $\frac{1}{3};$

(d) $\frac{9}{14};$ **(e)** $\frac{e^3 - 1}{2};$ **(f)** $\frac{e - 1}{8}.$

3 **(a)** $(x - 1)^2 + 2;$ **(c)** $2.$

4 $\frac{1}{2}.$

EXERCISE 8C

1 (a) $-\cos x + c$.

2 (a) 1;   (b) 1.

3 (a) $\frac{1}{2} \sin 2x + c$;   (b) $-\frac{1}{3} \cos 3x + c$;

  (c) $2 \tan \dfrac{\theta}{2} + c$;   (d) $x - \frac{1}{2} \sin(2x + 4) + c$;

  (e) $\frac{1}{2}x^2 + \frac{1}{4} \cos(4x - 1) + c$;   (f) $-\frac{1}{3} \tan(1 - 3\theta) + c$;

  (g) $-\frac{1}{4} \cot 4x + c$;   (h) $\frac{1}{3} \sec 3x + c$;

  (i) $-\frac{1}{2} \operatorname{cosec} 2x + c$.

4 $\tan x - x + c$.

5 (a) $\frac{1}{2} \tan 2x - x + c$;   (b) $\dfrac{4 - \pi}{8}$.

6 (a) $\frac{1}{3}$;   (b) 0.25;   (c) 0.75;

  (d) 0.25;   (e) 0;   (f) $\dfrac{\pi + 2}{4}$;

  (g) $\dfrac{\pi + 2}{4}$;   (h) 0.577 (to 3 s.f.);   (i) $\frac{1}{3}$.

7 $-\cot x - x + c$.   8 $\dfrac{4 - \pi}{8}$.

EXERCISE 8D

1 (a) $\ln|3x + 5| + c$;   (b) $\ln(3x^2 + 5) + c$;

  (c) $\ln|x^3 + 8| + c$;   (d) $\frac{1}{4} \ln(2x^2 + 3) + c$;

  (e) $\ln|x| - 2 \ln|x + 3| + c$;   (f) $\frac{3}{2} \ln(2x^2 + 5) + c$;

  (g) $-\frac{1}{2} \ln|4 - x^2| + c$;   (h) $-\frac{1}{3} \ln|4 - e^{3x}| + c$;

  (i) $\frac{1}{2} \ln|1 + \sin 2x| + c$;   (j) $\tan x + 2 \ln|\sec x| + c$;

  (k) $3 \ln|\ln x| + c$.

3 $\frac{1}{4} \ln|\sec 4x + \tan 4x| + c$.

4 $-\frac{1}{2} \ln|\operatorname{cosec} 2x + \cot 2x| + c$.

5 (a) $\ln \frac{5}{3}$;   (b) $\frac{1}{2} \ln 10$;

  (c) $\ln 2$;   (d) $\frac{1}{12}(7 \ln 2 - 4 \ln 9)$;

  (e) $\ln \frac{3}{2}$;   (f) $2 \ln \frac{8}{3}$;

  (g) $\frac{1}{2} \ln 2$;   (h) $\ln\left(\dfrac{e^2 + 1}{2}\right)$.

EXERCISE 8E

1 (a) $\frac{1}{36}(4x + 1)^9 + c$;   (b) $\frac{1}{5} \sin\left(5x + \dfrac{\pi}{3}\right) + c$;

  (c) $\dfrac{(2x + 1)^7}{7} - \dfrac{(2x + 1)^6}{6} + c$;   (d) $\frac{1}{2} \ln(5x^2 + 7) + c$;

  (e) $\dfrac{1}{8}\left[\dfrac{5}{2(4x + 5)^2} - \dfrac{1}{4x + 5}\right] + c$;   (f) $-\dfrac{1}{(x^2 + 3)^2} + c$;

  (g) $\frac{1}{2}(x + 3)^2 - 6(x + 3) + 9 \ln|x + 3| + c$;

  (h) $\sin x - \frac{1}{3} \sin^3 x + c$;   (i) $\frac{1}{5} \cos^5 x - \frac{1}{3} \cos^3 x + c$;

  (j) $\frac{1}{4} \cos^2 2x - \frac{1}{2} \ln \cos 2x + c$;   (k) $4\sqrt{x^2 - 1} + c$;

  (l) $2e^{x^2} + c$;   (m) $\frac{1}{2} x - \frac{1}{4} \ln(e^{2x} + 4) + c$;

  (n) $-2 \cos \sqrt{x} + c$.

2 (a) $\frac{11}{30}$; (b) $\frac{1}{7}$; (c) $\frac{1}{2}\ln 3$; (d) $\frac{1}{8}\ln 13 - \frac{3}{26}$;

(e) $-\frac{1}{12}$; (f) $\frac{15}{2} + 9\ln 2$; (g) $\frac{8}{15}$; (h) $\frac{31}{160}$;

(i) $\frac{1}{15}$; (j) 8; (k) $2(e-1)$; (l) $2\ln\frac{3}{2}$.

3 (a) $\frac{1}{9}(x^2+1)^9 + c$; (b) $4\sqrt{x^2+1} + c$;

(c) $\frac{1}{2}\sin 2x - \frac{1}{6}\sin^3 2x + c$.

4 (a) $\frac{1}{2}$; (b) $\frac{7}{3}$; (c) $\frac{1}{6}$.

5 $28\frac{8}{15}$.

MIXED EXERCISE

1 (a) $\frac{3}{4}\ln|x| + \frac{1}{5}e^{-5x} + c$; (b) $-\frac{2}{9}$; (c) $\dfrac{e^7 - 1}{4}$;

(d) $\frac{1}{2}\sin 2x - \frac{1}{4}\cos 4x + c$; (e) $\ln|4x+5| + c$;

(f) $-\frac{1}{2}\ln|6 - e^{2x}| + c$; (g) $\ln 2$.

3 $\frac{1}{24}$. **4** $\ln\frac{4}{3}$.

5 $a = 1$, $b = 2$, $c = 2$; $\frac{3}{2} - 2\ln 2$. **6** $\frac{232}{5}$.

9 Integration by parts and standard integrals

EXERCISE 9A

1 (a) $\frac{1}{2}xe^{2x} - \frac{1}{4}e^{2x} + c$; (b) $2xe^{\frac{x}{2}} - 4e^{\frac{x}{2}} + c$;

(c) $-xe^{-x} - e^{-x} + c$; (d) $-\theta\cos\theta + \sin\theta + c$;

(e) $\frac{1}{4}x\sin 4x + \frac{1}{16}\cos 4x + c$; (f) $\frac{1}{3}x\sin 3x + \frac{1}{9}\cos 3x + c$;

(g) $\frac{1}{2}x^2\ln x - \frac{1}{4}x^2 + c$; (h) $\frac{1}{7}x^7\ln x - \frac{1}{49}x^7 + c$;

(i) $-\dfrac{\ln 4x}{2x^2} - \dfrac{1}{4x^2} + c$.

2 (a) $e^2 + 1$; (b) $\dfrac{3e^4 + 1}{16}$; (c) $\dfrac{1 - 9e^{-8}}{4}$;

(d) 4; (e) $\dfrac{\pi}{2} - 1$; (f) $\dfrac{\pi^2}{4} - 2$;

(g) $4\ln 2 - \frac{15}{16}$; (h) 2; (i) $\dfrac{3\ln 2 + 1 - 32e^{-3}}{9}$.

3 (a) $(x+2)e^x + c$; (b) $(x-1)e^{3x} + c$;

(c) $\frac{1}{2}x^2 + \frac{1}{2}x\sin 2x + \frac{1}{4}\cos 2x + c$;

(d) $(3x + x^2)\ln x - 3x - \frac{1}{2}x^2 + c$.

4 (a) e^2; (b) 1; (c) $2\ln 2 - \frac{7}{8}$.

7 $\theta\tan\theta - \ln|\sec\theta| + c$.

EXERCISE 9B

1 (a) $\sin^{-1}\left(\dfrac{x}{5}\right) + c$; (b) $\dfrac{1}{5}\tan^{-1}\left(\dfrac{x}{5}\right) + c$;

(c) $\sin^{-1}(2x) + c$; (d) $10\tan^{-1}(10x) + c$.

2 $\frac{1}{2}\tan^{-1}(2x) + c$. **3** $\frac{1}{3}\sin^{-1}\left(\dfrac{3x}{2}\right) + c$.

4 **(a)** $\frac{1}{2}\sin^{-1}2x + c;$ **(b)** $\frac{1}{6}\tan^{-1}\left(\frac{2x}{3}\right) + c;$

   (c) $\frac{1}{4}\sin^{-1}(8x) + c;$ **(d)** $\frac{1}{12}\tan^{-1}\left(\frac{3x}{4}\right) + c.$

5 **(a)** 0.322;   **(b)** 0.464;   **(c)** 0.726;
   (d) 0.161;   **(e)** 0.0496;   **(f)** 0.258.

6 **(a)** $\ln(4 + x^2) + \frac{1}{2}\tan^{-1}\left(\frac{x}{2}\right) + c;$ **(b)** $\frac{1}{2}\ln(9 + x^2) + \frac{2}{3}\tan^{-1}\left(\frac{x}{3}\right) + c;$

   (c) $\sin^{-1}\left(\frac{x}{5}\right) + 2\sqrt{25 - x^2} + c;$ **(d)** $3\sin^{-1}\left(\frac{x}{5}\right) - \sqrt{25 - x^2} + c.$

7 **(a)** 1.87   **(b)** $-1.36.$

MIXED EXERCISE

1 0.25.   **2** 0.381 77.

4 **(a)** $\dfrac{2x}{x^2 + 9}.$ **5** $\frac{1}{2}te^{2t} - \frac{1}{4}e^{2t} + c, \frac{3}{2}e^4 + \frac{1}{2}.$

10 Volume of revolution and numerical integration

EXERCISE 10A

1 **(a)** $\frac{3}{2}\pi;$ **(b)** $\frac{62}{3}\pi;$ **(c)** $\frac{28}{15}\pi;$
   (d) $\pi;$ **(e)** $\pi;$ **(f)** $9\pi \ln 9.$

2 **(a)** $\dfrac{\pi}{2};$ **(b)** $\frac{31}{5}\pi;$
   (c) $\frac{9}{2}\pi;$ **(d)** $\pi\left(\frac{7}{2} + \ln 2\right).$

3 **(b)** $\frac{1}{10}\pi.$ **4** $\dfrac{\pi}{2}(e^2 - 1).$

5 $\frac{635}{7}\pi.$ **6** **(a)** $x^2 + y^2 = r^2.$

7 **(a)**   **(b)** $\frac{16}{3}\pi.$

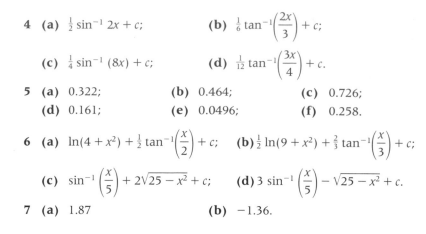

8 **(a)** $\frac{15}{2}\pi.$ **9** **(a)** $\dfrac{65\pi}{2}.$

EXERCISE 10B

1 **(a)** 1;   **(b)** 419.

2 **(a)** 0.5;   **(b)** 12.572 66... .

3 **(a)** 0.2;   **(b)** 0.551 58... .

EXERCISE 10C

1 9.384.

2 3.9974.

3 0.7960.

4 0.1348.

5 (a) 1.311;

(b) 2.6;

(c) 2.8;

(d) same to 1 s.f.

6 (a) (i) 0.609 16,

(ii) 0.763 34;

(iii) 0.971 46.

EXERCISE 10D

1 (a) 9.2522;

(b) Increase the number of ordinates (strips).

2 1.9248.

3 (a) 4.0071;

(b) 4.0070;

(c) 4.0070.

4 (a) 0.7709;

(b) 0.7775;

(c) 0.7803.

5 (a) 0.132 733;

(b) 0.132 728;

(c) 0.132 727.

6 (a) 0.795 581 8;

(b) 0.795 586 1;

(c) 0.795 586 8.

7 (a) 2.0510;

(b) 2.0506.

8 (a) (i) 0.7635,

(ii) 0.7635,

(iii) 0.7635;

(b) (i) 0.9624,

(ii) 0.9649,

(iii) 0.9660.

9 (a) 1.2953;

(b) exact answer, $3\ln 3 - 2$, is greater than answer to (a).

10 (a) 0.523 599 96;

(b) 3.141 599 8.

MIXED EXERCISE

1 $\dfrac{5\pi}{6}$.

2 0.565.

3 0.381 82.

4 $\dfrac{6\pi}{5}$.

5 0.6334.

6 π.

7 1.434.

8 200.907.

C3 Exam style practice paper

1 (a) $\tan x + x \sec^2 x$;

(b) $\dfrac{\cos x}{1 + \sin x}$.

2 (a) $0 < g(x) \le 1$;

(b) $(0, 1)$;

(c) $-3, -1$.

3 (a) $\frac{1}{4}$;

(b) -0.6869.

4 (a) One-way stretch in x-direction scale factor $\frac{1}{3}$, translation of $\begin{bmatrix} 0 \\ 2 \end{bmatrix}$;

(c) (iii) 0.732.

5 $\frac{1}{15}(3x + 1)(2x - 1)^{\frac{3}{2}} + c$.

6 (a)

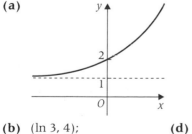

(b) $(\ln 3, 4)$;

(d) $\ln(x - 1)$.

7 (a)

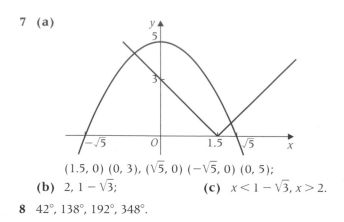

$(1.5, 0)$ $(0, 3)$, $(\sqrt{5}, 0)$ $(-\sqrt{5}, 0)$ $(0, 5)$;

(b) $2, 1 - \sqrt{3}$; **(c)** $x < 1 - \sqrt{3}, x > 2$.

8 $42°, 138°, 192°, 348°$.

Core maths 4

1 Binomial series expansion

EXERCISE 1A

1 $1 + 3x + 9x^2 + 27x^3$.

2 $1 - 4x + 16x^2 - 64x^3$.

3 (a) $1 + x + x^2 + x^3$, $|x| < 1$; **(b)** $1 + 4x + 16x^2 + 64x^3$, $|x| < \frac{1}{4}$;

   (c) $1 - 2x + 4x^2 - 8x^3$, $|x| < \frac{1}{2}$; **(d)** $1 - 3x + 9x^2 - 27x^3$, $|x| < \frac{1}{3}$.

4 $1 + \dfrac{x}{2} + \dfrac{x^2}{4} + \dfrac{x^3}{8}$.

5 (a) $1 + \dfrac{x}{5} + \dfrac{x^2}{25} + \dfrac{x^3}{125}$, $|x| < 5$; **(b)** $1 - \dfrac{x}{4} + \dfrac{x^2}{16} - \dfrac{x^3}{64}$, $|x| < 4$;

   (c) $1 + \dfrac{2x}{3} + \dfrac{4x^2}{9} + \dfrac{8x^3}{27}$, $|x| < \dfrac{3}{2}$; **(d)** $1 - \dfrac{3x}{4} + \dfrac{9x^2}{16} - \dfrac{27x^3}{64}$, $|x| < \dfrac{4}{3}$.

EXERCISE 1B

1 $1 + 7x + 21x^2 + 35x^3$. **2** $1 + 10x + 45x^2 + 120x^3$.

3 $1 - 4x + 10x^2 - 20x^3$. **4** $1 - 2x + 3x^2 - 4x^3$.

5 $1 - 5x + 15x^2 - 35x^3$. **6** $1 + \dfrac{x}{3} - \dfrac{x^2}{9} + \dfrac{5x^3}{81}$.

7 $1 - \dfrac{x}{2} + \dfrac{3x^2}{8} - \dfrac{5x^3}{16}$. **8** $1 - \dfrac{3x}{4} + \dfrac{21x^2}{32} - \dfrac{77x^3}{128}$.

EXERCISE 1C

1 $1 + x + x^2 + x^3$, $|x| < 1$. **2** $1 - 6x + 27x^2 - 108x^3$, $|x| < \frac{1}{3}$.

3 $1 - 8x + 40x^2 - 160x^3$, $|x| < \frac{1}{2}$. **4** $1 + 2x + 3x^2 + 4x^3$, $|x| < 1$.

5 $1 + 10x + 60x^2 + 280x^3$, $|x| < \frac{1}{2}$. **6** $1 - x + \frac{2}{3}x^2 - \frac{10}{27}x^3$, $|x| < 3$.

7 $1 - 6x + 30x^2 - 140x^3$, $|x| < \frac{1}{4}$. **8** $1 + 4x + 14x^2 + \frac{140}{3}x^3$, $|x| < \frac{1}{3}$.

9 $1 - 5x + \frac{45}{2}x^2 - \frac{195}{2}x^3$, $|x| < \frac{1}{4}$. **10** $1 - 2x + \frac{5}{2}x^2 - \frac{5}{2}x^3$, $|x| < 2$.

11 $1 + \dfrac{x}{6} - \dfrac{x^2}{72} + \dfrac{x^3}{432}$.

12 **(a)** $a = -12, b = 90, c = -540$; **(b)** $|x| < \frac{1}{3}$.

13 **(a)** $1 - 4x + 12x^2 - 32x^3$; **(b)** $|x| < \frac{1}{2}$.

EXERCISE 1D

1 $\dfrac{1}{3} - \dfrac{x}{9} + \dfrac{x^2}{27} - \dfrac{x^3}{81}$, $|x| < 3$.

2 $\dfrac{1}{4} - \dfrac{x}{4} + \dfrac{3x^2}{16} - \dfrac{x^3}{8}$, $|x| < 2$.

3 $\dfrac{1}{81} + \dfrac{4x}{243} + \dfrac{10x^2}{729} + \dfrac{20x^3}{2187}$, $|x| < 3$.

4 $\dfrac{1}{25} + \dfrac{2x}{125} + \dfrac{3x^2}{625} + \dfrac{4x^3}{3125}$, $|x| < 5$.

5 $\dfrac{1}{16} - \dfrac{x}{8} + \dfrac{5x^2}{32} - \dfrac{5x^3}{32}$, $|x| < 2$.

6 $\dfrac{1}{16} - \dfrac{3x}{32} + \dfrac{27x^2}{256} - \dfrac{27x^3}{256}$, $|x| < \dfrac{4}{3}$.

7 $\dfrac{1}{3} + \dfrac{2x}{9} + \dfrac{4x^2}{27} + \dfrac{8x^3}{81}$, $|x| < \dfrac{3}{2}$.

8 $\dfrac{1}{8} - \dfrac{3x}{64} + \dfrac{15x^2}{1024} - \dfrac{35x^3}{8192}$, $|x| < 4$.

10 **(a)** $1 - 2x + 3x^2 - 4x^3 + \ldots$; **(b)** $1 - 3x + 6x^2 - 10x^3 + \ldots$.

EXERCISE 1E

1 **(a)** $1 - y + y^2 - y^3 + y^4$, $1 - x^2 + x^4 - x^6 + x^8$;

 (b) $0.197\,396$;

 (c) series expansion not valid when $x = 5$.

2 **(a)** $1 - \dfrac{x}{3} - \dfrac{x^2}{9} - \dfrac{5x^3}{81} - \dfrac{10x^4}{243}$; **(b)** **(i)** $0.965\,49$, **(ii)** 9.6549.

3 **(b)** **(ii)** $1.414\,214$.

4 **(a)** $1 - 2x - 2x^2 - 4x^3$; **(b)** $0.979\,796$;

 (c) **(i)** $9.797\,96$, **(ii)** $2.449\,49$.

MIXED EXERCISE

1 $2x - 12x^2 + 48x^3$. **2** $a = \frac{1}{2}, n = 4$.

3 **(a)** **(i)** $1 + x + x^2 + x^3$, **(ii)** $1 + 2x + 4x^2 + 8x^3$;

 (c) $1 + 3x + 7x^2 + 15x^3$; **(d)** $2^{n+1} - 1$.

4 **(a)** $1 + \dfrac{3x}{10} + \dfrac{3x^2}{50} + \dfrac{x^3}{100}$; **(b)** $k = \frac{1}{2}$; **(c)** $1.003\,006\,010\,014\,95$.

5 $k = \dfrac{3}{256}$.

6 $1 + 3x + \dfrac{27}{2}x^2 + \dfrac{135}{2}x^3$, $\dfrac{351}{2}$.

7 (a) (i) $1 - x + \dfrac{3x^2}{2} - \dfrac{5x^3}{2},$ **(ii)** $1 + 4x - \dfrac{7x^2}{2} + 5x^3;$ **(b)** $0.577.$

8 (a) (i) $1 - x + x^2 - x^3,$ **(ii)** $1 + 2x + 4x^2 + 8x^3,$

(iii) $1 + 4x + 12x^2 + 32x^3;$ **(b) (i)** $6 + 7x + 31x^2 + 69x^3,$ **(ii)** $|x| < \frac{1}{2}.$

9 (a) $1 + \frac{1}{2}x - \frac{1}{8}x^2;$ **(b) (i)** $2 + \dfrac{x}{2} - \dfrac{x^2}{16},$ **(ii)** $|x| \le 2.$

10 (a) $4 + \dfrac{y}{8} - \dfrac{y^2}{512};$ **(b)** $|y| < 4.$

2 Rational functions and division of polynomials

EXERCISE 2A

1 $\dfrac{2x}{x+4}.$ **2** $\dfrac{x+3}{3(x+2)}.$ **3** $\dfrac{5+x}{4-9x}.$

4 $\dfrac{-(5+4x)}{2x+5}.$ **5** $\dfrac{4x-9}{2x-9}.$ **6** $x+2.$

7 $\dfrac{3(x-1)}{(x+1)(2x-5)}.$

EXERCISE 2B

1 $\dfrac{x+6}{2x}.$ **2** $\dfrac{1}{x+3}.$ **3** $\dfrac{3}{x-1}.$

4 $\dfrac{x-1}{x(2x-3)}.$ **5** $\dfrac{-(x+2)}{6}.$ **6** $\dfrac{x-1}{(x+2)(2x+3)}.$

EXERCISE 2C

1 (a) $\dfrac{4}{(x-2)(x+2)};$ **(b)** $\dfrac{11-4x}{(x-2)(x+2)};$

(c) $\dfrac{7x+18}{(x+3)(x+2)};$ **(d)** $\dfrac{2(x-1)}{(x-3)(x+2)};$

(e) $\dfrac{2}{x-1};$ **(f)** $\dfrac{2}{6-x};$

(g) $\dfrac{2}{x(x-3)};$ **(h)** $\dfrac{3-x}{x(x-1)(x+1)}.$

3 (a) $\dfrac{x-1}{(x+2)^2};$ **(b)** $\dfrac{x^2}{(x-1)^3}.$

EXERCISE 2D

1 (a) $3 + \dfrac{7}{x};$ **(b)** $3 + \dfrac{1}{x+2};$

(c) $\dfrac{3}{2} + \dfrac{13}{2x-4};$ **(d)** $x - 1 + \dfrac{1}{x+1};$

(e) $4x^2 - x - 3 - \dfrac{4}{x-1};$ **(f)** $2x^2 + 11x + 18 + \dfrac{41}{x-2};$

(g) $x^2 - \dfrac{5x}{3} + \dfrac{11}{9} - \dfrac{20}{9(3x+1)}$; **(h)** $3x^2 + \dfrac{5x}{2} - \dfrac{9}{4} + \dfrac{3}{4(2x-1)}$;

(i) $x^2 - \dfrac{4x}{5} + \dfrac{23}{25} - \dfrac{71}{25(5x+2)}$; **(j)** $x^2 + \dfrac{3}{5} + \dfrac{1}{5(5x-2)}$;

(k) $2x^2 - \dfrac{3x}{2} - \dfrac{11}{4} + \dfrac{15}{4(2x+1)}$; **(l)** $3x^2 + 2x + \dfrac{4}{3} + \dfrac{5}{3(3x-2)}$.

2 **(a)** **(i)** $2x - 1$, **(ii)** -2;

(b) **(i)** $3x^2 - \dfrac{3x}{2} - \dfrac{1}{4}$, **(ii)** $7\frac{1}{4}$;

(c) **(i)** $x^3 - \frac{1}{2}x^2 + \frac{1}{4}x - \frac{1}{8}$, **(ii)** $-\frac{7}{8}$;

(d) **(i)** $x^2 + \dfrac{7x}{2} - \dfrac{7}{4}$, **(ii)** $-7\frac{1}{4}$.

3 $k = 2$, quotient $4x^2 + x + 1$.

EXERCISE 2E

1 $Q(x) = (3x + 1)(6x - 5)$.

2 $f(x) = (2x + 5)(3 - 4x)$.

3 $P\left(\frac{1}{2}\right) = -\frac{1}{2}$, $P\left(\frac{1}{3}\right) = 0$; $(3x - 1)$.

4 $(4x - 3)(3x - 2)(3x + 2)$.

5 $a = -20$, $b = 6$; $(2x - 3)$.

6 $k = -4$ $f(x) = (3x + 1)(2x^2 - 2x + 1)$
$x = -\frac{1}{3}$ $(2x^2 - 2x + 1 = 0$ has no real solutions).

7 $Q(x) = x^2 - 2x - 1$; $x = -\frac{1}{2}$, $1 + \sqrt{2}$, $1 - \sqrt{2}$.

8 **(b)** $f(x) = (x - 3)(4x - 1)(x - 1)$.

9 **(a)** $(x + 2)(x + 3)(2x - 1)$;
(b) $(x - 4)(2x + 1)(2x + 3)$;
(c) $(x - 5)(2x - 1)(3x + 4)$;
(d) $(x - 2)(3x - 1)(3x + 4)$.

EXERCISE 2F

1 **(a)** $-\frac{11}{4}$; **(b)** $\frac{79}{4}$; **(c)** $-\frac{13}{4}$.

2 **(a)** -2; **(b)** $\frac{5}{4}$; **(c)** $\frac{7}{8}$; **(d)** $\frac{13}{3}$; **(e)** 5.

4 34. **5** $\frac{270}{8}$. **6** $-\frac{19}{8}$. **7** -7.

8 **(a)** $a = 24$, $b = 26$.

9 **(a)** $k = 4$; **(c)** $-\frac{1}{3}$ and 2.

10 $p = 5$, $R = 15$.

11 **(a)** -108; **(b)** **(ii)** $\dfrac{x + 3}{2x^2 - x + 6}$.

3 Partial fractions and applications

EXERCISE 3A

1 $A = 3, B = -2.$

2 $A = 13, B = -10.$

3 $A = \frac{1}{3}, B = -\frac{2}{3}.$

4 $A = -1, B = 1.$

5 $A = 1, B = -\frac{1}{2}, C = \frac{1}{2}.$

6 $A = 1, B = \frac{1}{2}, C = -\frac{1}{2}.$

7 $\dfrac{6}{x+2} - \dfrac{3}{x+1}.$

8 $\dfrac{4}{x+1} - \dfrac{1}{x-1}.$

9 $\dfrac{1}{x} - \dfrac{2}{(x+1)^2}.$

10 $\dfrac{2}{x+2} - \dfrac{2}{x-1} + \dfrac{6}{(x-1)^2}.$

11 $-1 + \dfrac{1}{2+x} + \dfrac{1}{2-x}.$

12 $\dfrac{6}{x+1} - \dfrac{3}{(x+1)^2} - \dfrac{6}{x+2}.$

13 $\dfrac{1}{x-1} + \dfrac{2}{(x-1)^2} + \dfrac{3}{(x-1)^3}.$

14 $\dfrac{1}{2(x+3)} - \dfrac{2}{x+2} + \dfrac{3}{2(x+1)}.$

EXERCISE 3B

1 (a) $1 + \ln 2;$ (b) $\frac{1}{2} + 2\ln\frac{2}{3};$ (c) $2\frac{1}{2} + \ln\frac{2}{3}.$

2 (a) $2\ln\frac{4}{3};$ (b) $2\ln\frac{3}{2};$ (c) $\ln\frac{4}{3};$

   (d) $\ln\frac{9}{4};$ (e) $3 - 2\ln\frac{8}{5};$ (f) $\frac{1}{4}(\ln\frac{3}{2} - \frac{1}{6}).$

3 (a) $\dfrac{3}{1+x} - \dfrac{2}{2-x} - \dfrac{1}{(1+x)^2}.$

4 (a) $\ln|x| - \frac{1}{2}\ln|2x+1| + c;$ (b) $x^2 - x + \ln|x+1| + c.$

5 (a) $A = 1, \ B = 4, \ C = -3;$ (b) $p = 4, q = -3, r = 4.$

6 (a) $\dfrac{1}{x-2} + \dfrac{3}{5-x};$ (b) $4\ln 2.$

7 (a) $A = 2, B = -6;$ (b) $p = -1, q = 4.$

EXERCISE 3C

1 (a) (i) $1 - x + x^2 - x^3,$ (ii) $1 - 2x + 4x^2 - 8x^3;$

   (b) $A = 2, B = -1;$ (c) $1 - 3x + 7x^2 - 15x^3, \ |x| < \frac{1}{2}.$

2 (a) (i) $1 - 2x + 3x^2,$ (ii) $\frac{1}{2} - \frac{1}{4}x + \frac{1}{8}x^2;$

   (b) $A = -1, B = 1, C = 1;$ (c) $\frac{1}{2} - \frac{5}{4}x + \frac{17}{8}x^2, \ |x| < 1.$

3 (a) (i) $1 + x + x^2 + x^3,$ (ii) $1 + 2x + 4x^2 + 8x^3;$

   (b) $3x + 9x^2 + 21x^3.$

4 (a) $A = -4, B = 3, C = 2;$ (b) $1 - 6x + 22x^2, \ |x| < \frac{1}{2}.$

5 $3 - 3x + 9x^2 - 15x^3, \ |x| < \frac{1}{2}.$

6 $\frac{3}{2} - \frac{5}{4}x + \frac{9}{8}x^2 - \frac{17}{16}x^3, \ |x| < 1.$

7 $1 + x - x^2 - 11x^3, \ |x| < \frac{1}{3}.$

8 $\frac{1}{2}x - \frac{9}{4}x^2 + \frac{57}{8}x^3, \ |x| < \frac{1}{2}.$

9 $3 - x + \frac{19}{3}x^2, \ |x| < 1.$

10 $\frac{1}{3} - \frac{4}{9}x + \frac{11}{9}x^2, \ |x| < \frac{1}{2}.$

MIXED EXERCISE

1 (a) $A = 1. B = 2$;   (b) (ii) $1 + x + x^2$;

  (c) $2 + \frac{1}{2}x + \frac{5}{4}x^2$.

2 (a) (ii) $A = 6, B = -3, C = 8$;   (b) $\frac{4}{3} + 3\ln 2$.

3 (a) $\dfrac{3}{x-2} - \dfrac{2}{x+3}$;   (b) (i) $\dfrac{1}{6}$, (ii) $x - 6y = 11$.

4 (a) $\dfrac{1}{1-x} - \dfrac{1}{2-x}$;   (b) $\dfrac{1}{2} + \dfrac{3}{4}x + \dfrac{7}{8}x^2$;   (c) $|x| < 1$.

5 (a) $\dfrac{1}{x-1} + \dfrac{4}{x+2}$;   (b) $4\ln 5 - 7\ln 2$.

6 (a) $A = 8, B = 2, C = -1$;   (b) $5 + 2x + \frac{5}{2}x^3$.

7 (a) $A - 1, B = 4, C = -3$;   (b) $4 - 3\ln 2 + 4\ln 3$.

8 (a) $\dfrac{2}{1+2x} + \dfrac{1}{4-x}$;

  (b) (ii) $1 - 2x + 4x^2$, (iii) $\frac{9}{4} - \frac{63}{16}x + \frac{513}{64}x^2$, (iv) $|x| < \frac{1}{2}$;

  (c) (i) $\ln|1+2x| - \ln|4-x| + c$, (ii) 0.011.

4 Implicit differentiation and applications

EXERCISE 4A

3 (a) $2\dfrac{dy}{dx}$;   (b) $6y^5\dfrac{dy}{dx}$;   (c) $-\dfrac{2}{y^3}\dfrac{dy}{dx}$;

  (d) $2x + 2y\dfrac{dy}{dx}$;   (e) $\sec^2 y\dfrac{dy}{dx}$;   (f) $\dfrac{1}{y}\dfrac{dy}{dx}$;

  (g) $2y\dfrac{dy}{dx} + 6x + 4\dfrac{dy}{dx}$; (h) $-3\sin 3y\dfrac{dy}{dx}$;   (i) $2e^{2y}\dfrac{dy}{dx} + 2\dfrac{dy}{dx}$;

  (j) $8(2y-1)^3\dfrac{dy}{dx}$.

4 (a) $y + x\dfrac{dy}{dx}$;   (b) $2xy^2 + 2yx^2\dfrac{dy}{dx}$;

  (c) $e^y + xe^y\dfrac{dy}{dx}$;   (d) $y\cos x + \dfrac{dy}{dx}\sin x$;

  (e) $e^{2x+y}\left(2 + \dfrac{dy}{dx}\right)$;   (f) $\dfrac{dy}{dx}(\sin y + y\cos y)$;

  (g) $\ln y + \dfrac{x}{y}\dfrac{dy}{dx}$;   (h) $2x^2e^{2y}\dfrac{dy}{dx} + 2xe^{2y}$.

5 (a) $\dfrac{x\dfrac{dy}{dx} - y}{x^2}$;   (b) $\dfrac{y - (x+1)\dfrac{dy}{dx}}{y^2}$;

  (c) $\dfrac{\sin x\dfrac{dy}{dx} - y\cos x}{\sin^2 x}$; (d) $\dfrac{2(x+y) - 2x\left(1 + \dfrac{dy}{dx}\right)}{(x+y)^2}$;

(e) $\dfrac{\cos x(x+\cos y)-\sin x\left(1-\sin y\dfrac{dy}{dx}\right)}{(x+\cos y)^2}$;

(f) $\dfrac{(x+e^y)2e^{2y}\dfrac{dy}{dx}-e^{2y}\left(1+e^y\dfrac{dy}{dx}\right)}{(x+e^y)^2}$.

7 $\frac{1}{3}$. **8** 1.

EXERCISE 4B

1 $y=x$.

2 $y+x=3$.

3 **(b)** $5y+4x=40$.

4 $3y=2x+5$.

5 **(a)** $y=2x-4$;

(b) $\left(\frac{19}{3},\frac{1}{3}\right)$.

7 $y+x=1$.

8 $3y=x+28,\ 3y=x-28$.

MIXED EXERCISE

2 **(b)** $4y+5x=16$.

4 **(b)** $3y+4x=7$. **5** 0.

6 **(a)** $\left(3,\frac{7}{3}\right),\left(3,-\frac{1}{3}\right)$;

(b) $1,-1$.

7 $\pm\dfrac{\sqrt{13}}{13}$.

8 -3.

9 **(a)** $B(1,-1)$;

(b) $2y\dfrac{dy}{dx}+x\ln x\dfrac{dy}{dx}+y\ln x+y=1$;

(c) **(ii)** -1.

10 **(b)** $(0,1)$.

5 Parametric equations

EXERCISE 5A

1 **(a)** $(8,-1),(2,0),(0,1),(2,2),(8,3)$; **(b)** $(2,0),(0,1)$;

(c)

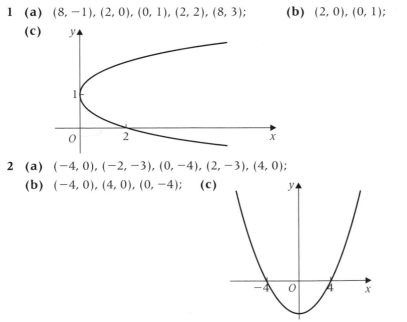

2 **(a)** $(-4,0),(-2,-3),(0,-4),(2,-3),(4,0)$;

  (b) $(-4,0),(4,0),(0,-4)$; **(c)**

3 (a) (4, 18), (1, 15), (0, 12), (1, 9), (4, 6);

(b) (0, 12), (16, 0); **(c)**

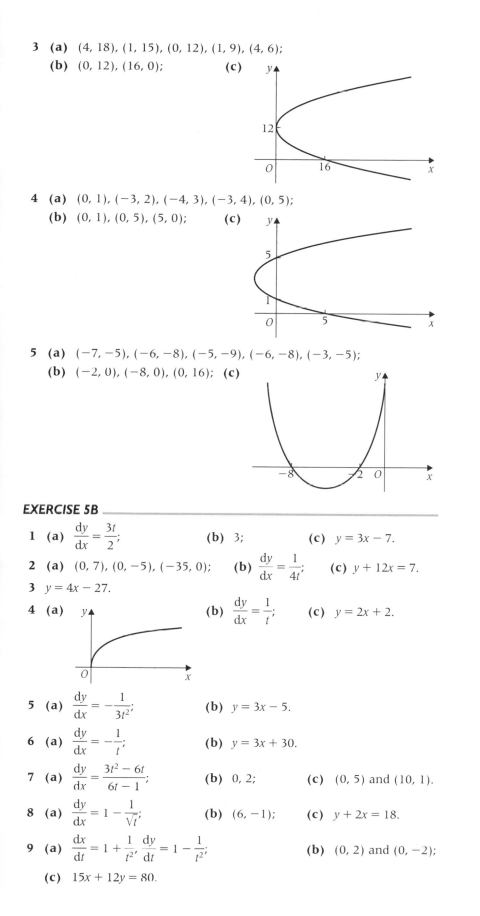

4 (a) (0, 1), (−3, 2), (−4, 3), (−3, 4), (0, 5);

(b) (0, 1), (0, 5), (5, 0); **(c)**

5 (a) (−7, −5), (−6, −8), (−5, −9), (−6, −8), (−3, −5);

(b) (−2, 0), (−8, 0), (0, 16); **(c)**

EXERCISE 5B

1 (a) $\dfrac{dy}{dx} = \dfrac{3t}{2}$; **(b)** 3; **(c)** $y = 3x − 7$.

2 (a) (0, 7), (0, −5), (−35, 0); **(b)** $\dfrac{dy}{dx} = \dfrac{1}{4t}$; **(c)** $y + 12x = 7$.

3 $y = 4x − 27$.

4 (a) **(b)** $\dfrac{dy}{dx} = \dfrac{1}{t}$; **(c)** $y = 2x + 2$.

5 (a) $\dfrac{dy}{dx} = -\dfrac{1}{3t^2}$; **(b)** $y = 3x − 5$.

6 (a) $\dfrac{dy}{dx} = -\dfrac{1}{t}$; **(b)** $y = 3x + 30$.

7 (a) $\dfrac{dy}{dx} = \dfrac{3t^2 − 6t}{6t − 1}$; **(b)** 0, 2; **(c)** (0, 5) and (10, 1).

8 (a) $\dfrac{dy}{dx} = 1 − \dfrac{1}{\sqrt{t}}$; **(b)** (6, −1); **(c)** $y + 2x = 18$.

9 (a) $\dfrac{dx}{dt} = 1 + \dfrac{1}{t^2}, \dfrac{dy}{dt} = 1 − \dfrac{1}{t^2}$; **(b)** (0, 2) and (0, −2);

(c) $15x + 12y = 80$.

EXERCISE 5C

1 (b) $y = -\frac{1}{3}x + 1.414$. **2 (b)** $y = 4x - 8$.

3 (a) $\dfrac{dy}{dx} = \dfrac{\sin 2t}{2 \cos 4t}$; **(b)** $-\dfrac{1}{2}$; **(c)** $y = 2x + 3$ cuts at $(-1.5, 0)$.

4 (b) $\dfrac{\pi}{6}, \dfrac{5\pi}{6}$; **(c)** $x + y = 7$.

5 (b) $y = -2x + 8$.

EXERCISE 5D

1 (a) $y^2 = 4x$; **(b)** $3y^2 = 4x^3$; **(c)** $(x + 5)^2 = (y - 2)^3$.

2 (a) $t = \dfrac{y - 2}{3}$; **(b)** $x = \left(\dfrac{y - 2}{3}\right)^2 - 1$.

3 $y^2 = 2x - 10$. **4** $y = 2x^2 - 9x + 10$.

5 $y^2 + 5 = 4x + 6y$. **6** $xy = 6$.

7 $\left(\dfrac{x}{2}\right)^2 + \left(\dfrac{y}{5}\right)^2 = 1$. **8** $(x - 3)(5 - y) = 8$.

9 $\left(\dfrac{y + 1}{2}\right)^2 = 1 + (x - 4)^2$.

10 (a) $\dfrac{1}{x} = p - 2, \dfrac{1}{y} = p + 4$; **(b)** $\dfrac{1}{y} - \dfrac{1}{x} = 6$.

11 $\left(\dfrac{x - 3}{2}\right)^2 = 1 + \left(\dfrac{2 - y}{3}\right)^2$. **12** $(x - 1)^2 + \left(\dfrac{1}{2 - y}\right)^2 = 1$.

13 (a) $x + y = 2^{p + 1}, x - y = 2^{1 - p}$; **(b)** $x^2 - y^2 = 4$.

MIXED EXERCISE

1 (a) $\dfrac{dy}{dx} = 12t^2 - 4t$; **(b)** $x + 8y = 35$; **(c)** $y = (x - 1)^3 - (x - 1)^2$.

2 (b) $5x + 3y = 36$; **(c)** $(x - y)^2 = 2(x + y)$.

3 (a) $2e^p$; **(b)** $y = 2x$; **(c)** $y = x^2 - 4x + 9$.

4 (a) (i) $xy = \sec^2 \theta - \tan^2 \theta$, **(ii)** $xy = 1$; **(b) (ii)** $x + y = 2$.

5 (b) $\dfrac{dy}{dx} = \dfrac{2}{t(t^2 + 3)}$.

6 (a) $x^{\frac{2}{3}} = \cos^2 \theta, x^{\frac{2}{3}} + y^{\frac{2}{3}} = 1$; **(b)**

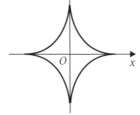

7 (a) $\dfrac{1}{(2 - x)^2} = 1 + \dfrac{1}{(1 - y)^2}$; **(b)** $n = -3$.

8 (a) $x^2 + 2xy = 12y$; **(b)** $y = 4x - 12$.

9 (b) $x + y = 1$.

10 (a) -1, straight line; **(b)** $x + y = 1$; **(c)** $(1, 0)$.

6 Further trigonometry with integration

EXERCISE 6A

2 (a) $\sin 75°$; (b) $\cos 15°$

3 $\tan 0.5^c$.

4 (a) $\sin 60°$; (b) $\sin 20°$; (c) $\sin 80°$; (d) $\sin 20°$.

8 $\tan y = \dfrac{1-k}{1+k}$. **9** 1.84, 5.49. **10** $0, \frac{1}{2}, x = 75°, y = 15°$.

12 240°, 300°. **13** translation $\begin{bmatrix} 60° \\ 0 \end{bmatrix}$. **14** translation $\begin{bmatrix} \frac{\pi}{3} \\ 0 \end{bmatrix}$.

EXERCISE 6B

3 0°, 75.5°, 180°, 284.5°. **4** 0.167, 1.57, 2.97, 4.71.

5 1.05, 3.14, 5.24. **9** 0.524, 2.62.

10 0, 0.524, 2.62, 3.14, 3.67, 5.76, 6.28. **11** 22.5°, 112.5°, 202.5°, 292.5°.

12 0°, 60°, 120°, 180°, 240°, 300°. **13** $\dfrac{2\tan x}{3\tan^2 x - 1}$.

15 ±2.09. **17** (b) 120°.

18 (b) 0.262, 1.31, 2.36. **19** 0, 1.05, 3.14, 5.24, 6.28.

EXERCISE 6C

1 0.5. **2** $\dfrac{\pi+2}{4}$. **3** 0.25. **4** $\dfrac{\pi-3}{3}$.

5 $2x - \tan x + c$. **7** $\dfrac{\pi+2}{4}$. **8** 0.351 **9** 0.181.

EXERCISE 6D

1 (a) $2\sqrt{3}\cos(x-30°)$; (b) $5\cos(x-53.1°)$;
   (c) $\sqrt{3}\cos(x-35.3°)$; (d) $2\cos(x-60°)$.

2 (a) (i) $2\sqrt{3}$, (ii) $-2\sqrt{3}$; (b) (i) 5, (ii) -5;
   (c) (i) $\sqrt{3}$, (ii) $-\sqrt{3}$; (d) (i) 2, (ii) -2.

3 (a) $2\sqrt{3}\sin(x+0.524^c)$; (b) $13\sin(x+1.18^c)$;
   (c) $2\sin(x+0.524^c)$; (d) $3\sin(x+1.23^c)$.

4 (a) $-2\sqrt{3} \le f(x) \le 2\sqrt{3}$; (b) $-13 \le f(x) \le 13$;
   (c) $-2 \le f(x) \le 2$; (d) $-3 \le f(x) \le 3$.

5 (a) $2\sqrt{3}\sin(x-30°)$; (b) $5\sin(x-53.1°)$;
   (c) $\sqrt{3}\sin(x-35.3°)$; (d) $2\sin(x-60°)$.

6 (a) 120°; (b) 143.1°; (c) 125.3°; (d) 150°.

7 (a) 300°; (b) 323.1°; (c) 305.3°; (d) 330°.

8 (a) $2\cos(2\theta + 0.524^c)$;
   (b) minimum point $(1.31, -2)$, maximum point $(2.88, 2)$.

10 $\sqrt{2}\ln(1+\sqrt{2}) = 1.25$ (to 3 sf).

11 1.54.

EXERCISE 6E

1 (a) $10 \sin(x + 0.927^c)$;   **(b)** 1.44, 6.13 (to 3 sf).

2 232.6°, 352.6°.   **3** 115.3°, 318.4°.

4 0, 1.287.   **5** 1.373, 3.055.

6 −2.721, 0.4205.   **7** 1.006, 2.779.

8 90°, 180°.   **9** −125.7°, 19.4°.

10 67.5°, 157.5°, 247.5°, 337.5°.

MIXED EXERCISE

1 (a) $5 \sin(\theta - 36.9°)$;   **(b)** 60.4°, 193.3°.

2 48°, 120°, 240°, 312°.

3 (a) $3 - 3\cos 2\theta$;   **(b)** $\dfrac{\pi - 3}{4}$;   **(c)** 0.766, 2.38.

4 (a) (i) $L = 2\sin\theta + 4\cos\theta$,   **(ii)** $\sqrt{20}\sin(\theta + 1.107)$;
   (b) (i) $\sqrt{20}$,   **(ii)** 0.46.

5 (b) 114°, 336°.

6 (a) 1.176;
   (b) $R = 26$, $\alpha = 1.176^c$;
   (c) (i) 26,   **(ii)** $\theta = 0.395^c$.

7 (b) (i) $\dfrac{\pi + 3}{24}$,   **(ii)** 0.245^c, 1.57^c, 3.39^c, 4.71^c.

7 Exponential growth and decay

EXERCISE 7A

1 (a) 6;   **(b)** 7;   **(c)** −3;   **(d)** 5.

2 (a) 5.95;   **(b)** 7.01;   **(c)** −2.94;   **(d)** 5.00.

3 (a) −0.399;   **(b)** 3.82;   **(c)** 2.68;   **(d)** 1.64.

5 (a) $\left(\dfrac{\ln 6}{\ln 2} - 4\right)$;   **(b)** $\dfrac{1}{3}\left(\dfrac{\ln 17}{\ln 3} + 1\right)$;

   (c) $\dfrac{1}{4}\left(1 - \dfrac{\ln 5}{\ln 2}\right)$;   **(d)** $\dfrac{1}{3}\left(\dfrac{\ln 31}{\ln 5} - 4\right)$.

6 (a) 3.39;   **(b)** 0.363;   **(c)** 1.25;   **(d)** 0.743.

7 (a) 8.303;   **(b)** 0.7731;
   (c) 5.116;   **(d)** 0.9882;
   (e) 1.504;   **(f)** 0.8244.

8 (a) −2.30;   **(b)** 0.519;
   (c) 0.683;   **(d)** −0.419;
   (e) 0.117;   **(f)** 0.862.

10 $b > 0$.

EXERCISE 7B

1

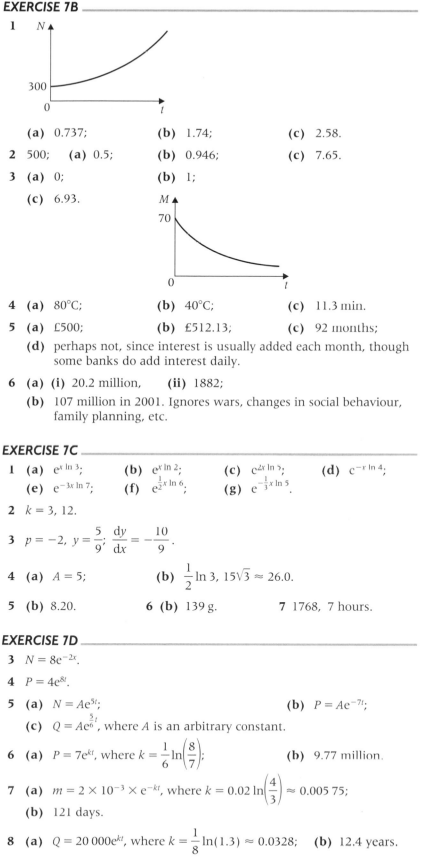

   (a) 0.737;   (b) 1.74;   (c) 2.58.

2 500;   (a) 0.5;   (b) 0.946;   (c) 7.65.

3 (a) 0;   (b) 1;

   (c) 6.93.

4 (a) 80°C;   (b) 40°C;   (c) 11.3 min.

5 (a) £500;   (b) £512.13;   (c) 92 months;

   (d) perhaps not, since interest is usually added each month, though some banks do add interest daily.

6 (a) (i) 20.2 million,   (ii) 1882;

   (b) 107 million in 2001. Ignores wars, changes in social behaviour, family planning, etc.

EXERCISE 7C

1 (a) $e^{x \ln 3}$;   (b) $e^{x \ln 2}$;   (c) $e^{2x \ln 5}$;   (d) $e^{-x \ln 4}$;

   (e) $e^{-3x \ln 7}$;   (f) $e^{\frac{1}{2}x \ln 6}$;   (g) $e^{-\frac{1}{3}x \ln 5}$.

2 $k = 3, 12$.

3 $p = -2$, $y = \dfrac{5}{9}$; $\dfrac{dy}{dx} = -\dfrac{10}{9}$.

4 (a) $A = 5$;   (b) $\dfrac{1}{2}\ln 3$, $15\sqrt{3} \approx 26.0$.

5 (b) 8.20.   6 (b) 139 g.   7 1768, 7 hours.

EXERCISE 7D

3 $N = 8e^{-2x}$.

4 $P = 4e^{8t}$.

5 (a) $N = Ae^{5t}$;   (b) $P = Ae^{-7t}$;

   (c) $Q = Ae^{\frac{5}{6}t}$, where A is an arbitrary constant.

6 (a) $P = 7e^{kt}$, where $k = \dfrac{1}{6}\ln\!\left(\dfrac{8}{7}\right)$;   (b) 9.77 million.

7 (a) $m = 2 \times 10^{-3} \times e^{-kt}$, where $k = 0.02 \ln\!\left(\dfrac{4}{3}\right) \approx 0.005\,75$;

   (b) 121 days.

8 (a) $Q = 20\,000e^{kt}$, where $k = \dfrac{1}{8}\ln(1.3) \approx 0.0328$;   (b) 12.4 years.

9 (a) $\frac{1}{10}\ln\left(\frac{3}{2}\right) \approx 0.040\,55$;   **(b)** $t = 27.1$.

10 (a) £100;   **(b)** £115.76;

     (c) $t = 8.31$.

11 (b) 11.9 grams.

12 (a) (i) $\dfrac{dt}{dN} = \dfrac{1}{4N}$;   **(ii)** $t = \dfrac{1}{4}(\ln N - \ln 200)$;

     (b) 19 minutes.

13 (a) 1000;

     (c) (i) $t = \dfrac{1}{1.0423}\ln\left(\dfrac{N}{1000}\right)$;   **(ii)** 7 minutes;

14 (a) (i) 15 000;   **(ii)** 0.155;

     (b) 2014.

8 Differential equations

EXERCISE 8A

1 (a) $\dfrac{dm}{dt} = -km$, where k is a positive constant;

   (b) $\dfrac{dP}{dt} = kP$, where k is a positive constant;

   (c) $\dfrac{dV}{dt} = -kV$, where k is a positive constant;

   (d) $\dfrac{dT}{dt} = -k(T - T_0)$, where k is a positive constant;   **(e)** $\dfrac{dy}{dx} = 2xy$.

2 (a) $\dfrac{dy}{dx} = \dfrac{y}{x}$;   **(b)** $\dfrac{dy}{dx} = \dfrac{y-2}{x}$;   **(c)** $\dfrac{dy}{dx} = \dfrac{y}{2x}$;

   (d) $\dfrac{dy}{dx} = 2y$;   **(e)** $\dfrac{dy}{dx} = y\cot x$;   **(f)** $\dfrac{dy}{dx} = \dfrac{y}{x-2}$;

   (g) $\dfrac{dy}{dx} = \dfrac{1 + y - \ln x}{x}$;   **(h)** $\dfrac{dy}{dx} = \dfrac{-x}{y}$;   **(i)** $\dfrac{dy}{dx} = \dfrac{3x^2 y}{x^3 - 2y^3}$.

3 (a) $\dfrac{dy}{dx} = 2x + 4y$.

EXERCISE 8B

1 $y = 2x^2 + A$ **2** $y = Ae^x$ **3** $y = Ae^x - 2$

4 $y = Ae^{\frac{x^2}{2}} - 1$ **5** $y = -\ln|A - x|$ **6** $y = \sin^{-1}(x + A)$

7 $x = Ae^t - 2$ **8** $y^2 = 4x^2 + A$ **9** $y^2 = x^2 + 8x + A$

10 $y = Ax$ **11** $y = A(x + 4)$ **12** $y = Ae^{\frac{1}{2}x} - 2$

13 $y^2 = Ax^{\frac{1}{4}} - 1$ **14** $x^2 = 2e^t + A$ **15** $x = A(t + 1) - 2$

16 $\tan x = A - \cos t$ **17** $x = -\ln|A + e^{-t}|$ **18** $x^2 = Ae^{\frac{1}{2}t^2} - 3$

19 $x = -\ln|A - e^t|$ **20** $y = \left(\dfrac{1}{2}x^{\frac{3}{2}} + 3x^{\frac{1}{2}} + A\right)^{\frac{2}{3}}$

EXERCISE 8C

1 $y = 4x^2 + 2x + 2$.

2 $y = e^{4(x+1)}$.

3 $N = 1 - \dfrac{1}{2}e^{-t}$.

4 $P = 200e^{2t}$.

5 $y = 2\ln x$.

6 $y = e^{(x+4)(x-1)} - 1$.

7 $y = \dfrac{1}{2}(t + 1)$.

8 $y = \tan^{-1}(\sin x)$.

9 $v = 4(e^{4x} - 1)^{\frac{1}{2}}$.

10 $y = x$.

11 $y = -\dfrac{1}{2}\ln|2(2 - e^x)|$.

12 $y = \dfrac{3(x+1)}{x+2}$.

13 $y = \dfrac{2e^x}{x+1}$.

14 $y = 5e^x - 2$.

15 $y = \cos^{-1}(\operatorname{cosec} x)$.

16 $y = 1 - e^{\frac{1}{x}}$.

17 $y = \dfrac{x}{x+2}$.

18 $y = \dfrac{2(5 + x^4)}{5 - x^4}$.

MIXED EXERCISE

1 **(a)** $\dfrac{dy}{dx} = \dfrac{1}{4}(x + 2y)$; **(c)** $z = 2\left(e^{\frac{1}{2}x} - 1\right)$; **(d)** $y = e^{\frac{1}{2}x} - 1 - \dfrac{1}{2}x$.

2 **(a)** $xe^x - e^x + c$; **(b)** $\ln y = xe^x - e^x + 1$.

3 **(a)** $-\dfrac{1}{3}xe^{-3x} - \dfrac{1}{9}e^{-3x} + c$; **(b)** $\tan y = -\dfrac{1}{3}xe^{-3x} - \dfrac{1}{9}e^{-3x} + \dfrac{1}{9}$;

4 **(a)** $y = \sqrt[3]{3(x + c)}$; **(b)** $y = \sqrt[3]{3x - 4}$.

5 $\dfrac{2}{3}x^{\frac{3}{2}}\ln x - \dfrac{4}{9}x^{\frac{3}{2}} + c$; **(b)** $y^{\frac{1}{2}} = \dfrac{1}{3}x^{\frac{3}{2}}\ln x - \dfrac{2}{9}x^{\frac{3}{2}} + \dfrac{11}{9}$.

6 **(a)** **(i)** $t = 5\ln\left|\dfrac{9}{10 - x}\right|$, **(ii)** $x = 10 - 9e^{-\frac{t}{5}}$;

 (b) 0.589.

8 **(a)** $\dfrac{1}{3 + \cos x} + c$; **(b)** $\ln y = \dfrac{1}{3 + \cos x} + \dfrac{3}{4}$.

9 **(b)** $t = \sqrt{A} - \dfrac{1}{2}$; **(c)** 17 days.

10 **(a)** $\dfrac{1}{3}xe^{3x} - \dfrac{1}{9}e^{3x} + c$; **(b)** $y^2 = xe^{3x} - \dfrac{1}{3}e^{3x} + \dfrac{4}{3}$.

11 **(a)** $k = \dfrac{1}{5}$ and $c = 1200$; **(b)** $x = 1200\left(1 - e^{-\frac{1}{5}t}\right)$;

 (c) 1037 (1038 and 1040 to 3 sf accepted).

12 **(a)** $x = \dfrac{a}{b}\left(1 - \dfrac{3}{4}e^{-bt}\right)$; **(b)** $\dfrac{1}{b}\ln 1.5$;

 (c) $\dfrac{a}{b}$.

13 **(a)** **(i)** $\dfrac{dP}{dt} = kP$, where k is a positive constant,

 (iii) 0.009 95, **(iv)** 55 200;

 (b) To 3 sf accuracy the model is valid.

9 Vector equations of lines

EXERCISE 9A

1 (a) $\begin{bmatrix} 8 \\ 8 \end{bmatrix}$;   (b) $\begin{bmatrix} 14 \\ 0 \end{bmatrix}$;   (c) $\begin{bmatrix} 0 \\ 56 \end{bmatrix}$.

2 (a) $\begin{bmatrix} 6 \\ -1 \end{bmatrix}$;   (b) $\begin{bmatrix} 0 \\ -19 \end{bmatrix}$;   (c) $\begin{bmatrix} -19 \\ 0 \end{bmatrix}$.

3 (a) $\begin{bmatrix} 0 \\ -11 \\ 8 \end{bmatrix}$;   (b) $\begin{bmatrix} 11 \\ 0 \\ -6 \end{bmatrix}$;   (c) $\begin{bmatrix} -21 \\ 2 \\ 10 \end{bmatrix}$.

4 (a) $\begin{bmatrix} 10 \\ 1 \\ 0 \end{bmatrix}$;   (b) $\begin{bmatrix} 0 \\ -17 \\ 60 \end{bmatrix}$;   (c) $\begin{bmatrix} 17 \\ 0 \\ 6 \end{bmatrix}$.

5 (a) $\mathbf{e} = \begin{bmatrix} -1 \\ 4 \end{bmatrix}, \mathbf{f} = \begin{bmatrix} 5 \\ -3 \end{bmatrix}$;   (b) $\begin{bmatrix} 6 \\ -7 \end{bmatrix}$;   (c) $\sqrt{85}$.

6 (a) $\mathbf{r} = \begin{bmatrix} 1 \\ -2 \\ 4 \end{bmatrix}, \mathbf{s} = \begin{bmatrix} -1 \\ 3 \\ 1 \end{bmatrix}$;   (b) $\begin{bmatrix} -2 \\ 5 \\ -3 \end{bmatrix}$;   (c) $\sqrt{38}$.

7 (a) 5;   (b) 13;   (c) $\sqrt{128} = 8\sqrt{2}$.

8 (a) $\sqrt{41}$;   (b) $\sqrt{10}$;   (c) 19.

9 (a) 5;   (b) 3;   (c) $\sqrt{38}$.

10 (a) 13;   (b) 7;   (c) $\sqrt{86}$.

11 (a) 5;   (b) $\sqrt{14}$;   (c) $\sqrt{50} = 5\sqrt{2}$;   (d) $\sqrt{51}$.

EXERCISE 9B

2 (a) $\begin{bmatrix} -\dfrac{5}{13} \\ \dfrac{12}{13} \end{bmatrix}$; (b) $\begin{bmatrix} \dfrac{7}{25} \\ -\dfrac{24}{25} \end{bmatrix}$; (c) $\begin{bmatrix} \dfrac{4}{5} \\ \dfrac{3}{5} \\ 0 \end{bmatrix}$; (d) $\begin{bmatrix} \dfrac{2}{3} \\ -\dfrac{1}{3} \\ \dfrac{2}{3} \end{bmatrix}$; (e) $\begin{bmatrix} \dfrac{6}{7} \\ \dfrac{3}{7} \\ -\dfrac{2}{7} \end{bmatrix}$.

3 (a) $\begin{bmatrix} -\dfrac{4}{\sqrt{17}} \\ \dfrac{1}{\sqrt{17}} \end{bmatrix}$; (b) $\begin{bmatrix} -\dfrac{3}{\sqrt{13}} \\ -\dfrac{2}{\sqrt{13}} \end{bmatrix}$; (c) $\begin{bmatrix} \dfrac{1}{\sqrt{11}} \\ \dfrac{3}{\sqrt{11}} \\ -\dfrac{1}{\sqrt{11}} \end{bmatrix}$; (d) $\begin{bmatrix} \dfrac{2}{\sqrt{29}} \\ -\dfrac{4}{\sqrt{29}} \\ \dfrac{3}{\sqrt{29}} \end{bmatrix}$; (e) $\begin{bmatrix} \dfrac{5}{\sqrt{38}} \\ -\dfrac{3}{\sqrt{38}} \\ \dfrac{2}{\sqrt{38}} \end{bmatrix}$.

4 (a) (i) $13\mathbf{i} - 10\mathbf{j} + 16\mathbf{k}$, (ii) $6\mathbf{i} - 2\mathbf{j} + 27\mathbf{k}$;
(b) (i) 7, (ii) 3, (iii) $\sqrt{66}$; (c) $\frac{1}{7}(3\mathbf{i} - 2\mathbf{j} + 6\mathbf{k})$.

5 (a) (i) $18\mathbf{i} - 14\mathbf{j} - 17\mathbf{k}$, (ii) $-11\mathbf{i} + 6\mathbf{j} + 27\mathbf{k}$;
(b) (i) 3, (ii) $\sqrt{38}$, (iii) $\sqrt{37}$; (c) $\frac{1}{3}(2\mathbf{i} - 2\mathbf{j} + \mathbf{k})$.

6 $\begin{bmatrix} -\dfrac{5}{\sqrt{34}} \\ \dfrac{3}{\sqrt{34}} \end{bmatrix}$.

7 $\begin{bmatrix} \dfrac{1}{\sqrt{11}} \\ -\dfrac{1}{\sqrt{11}} \\ \dfrac{3}{\sqrt{11}} \end{bmatrix}$.

8 **(a)** -8; **(b)** -22; **(c)** -12; **(d)** 38; **(e)** 16;
  (f) 4; **(g)** -5; **(h)** 0.

9 **(a)** 18; **(b)** 5; **(c)** -3; **(d)** -58.

10 **(a)** yes; **(b)** yes; **(c)** no; **(d)** yes.

11 **(a)** yes; **(b)** yes; **(c)** yes; **(d)** no.

12 **(a)** $86.6°$; **(b)** $102.5°$.

13 **(a)** $\frac{4}{9}$; **(b)** $\frac{13}{21}$; **(c)** $\frac{22}{39}$; **(d)** $-\frac{6}{7}$; **(e)** $\frac{1}{14}$.

14 **(a)** $63.6°$; **(b)** $113.1°$; **(c)** $79.7°$; **(d)** $91.5°$.

EXERCISE 9C

1 **(a)** $(0, 1)$; **(b)** $\cos^{-1}\left(\dfrac{2}{\sqrt{5}}\right)$.

2 **(b)** $(-1, 2)$.

3 **(a)** $(3, 3, 1)$; **(b)** $\dfrac{1}{\sqrt{15}}$.

5 **(a)** $\cos^{-1}\left(\dfrac{1}{2\sqrt{7}}\right) \approx 79.1°$; **(b)** $(23, 28, 34)$.

6 **(a)** $(7, 2, 4)$.

7 **(a)** $(3, 8, 4)$; **(b)** $77°$; **(c)** they are parallel.

EXERCISE 9D

1 **(b)** $(-3, -2)$.   2 $(4, 0)$, $\sqrt{5}$.   3 $(2, 11)$

4 **(b)** $(0, 5, 4)$; **(c)** $\sqrt{21}$.

5 **(b)** $(1, 5, 1)$; **(c)** $\sqrt{66}$.

6 **(a)** $\mathbf{r} = \begin{bmatrix} 3 \\ -1 \\ 2 \end{bmatrix} + \lambda \begin{bmatrix} -1 \\ 1 \\ 0 \end{bmatrix}$; **(b)** $(1, 1, 2)$; **(d)** $(5, -3, 2)$.

7 **(a)** $(1, 2, 1)$; **(b)** $\mathbf{r} = \begin{bmatrix} 5 \\ 1 \\ 2 \end{bmatrix} + \lambda \begin{bmatrix} 1 \\ -1 \\ -1 \end{bmatrix}$.

8 **(b) (i)** -1.

C4 Exam style practice paper

1 **(b)** $\dfrac{3 + 2x - 2x^2}{2x + 1}$; **(c)** 3.

2 **(a)** $-14 + \dfrac{2}{2x + 1} + \dfrac{36}{3 - x}$; **(b)** $37 \ln 3 - 36 \ln 2 - 14$.

3 **(a)** $2\cos(\theta - 30°)$; **(b)** $150°, 270°$.

4 **(a) (i)** $t = \frac{1}{3}(\ln N - \ln 400)$; **(b) (i)** $161\,000$, **(ii)** 11 minutes.

5 **(a)** $(5, 3)$; **(b)** $\frac{9}{7}$; **(c)** $9y + 7x = 62$; **(d)** $k = 32$.

6 **(a) (i)** $1 + 2x + 3x^2 + 4x^3$, **(ii)** $\dfrac{1}{3} - \dfrac{x}{9} + \dfrac{x^2}{27} - \dfrac{x^3}{81}$;

   (c) $\dfrac{7}{3} + \dfrac{35}{9}x + \dfrac{163}{27}x^2 + \dfrac{647}{81}x^3$.

7 **(b)** $0.253, 1.57, 2.89, 4.71$.

8 **(a)** 7; **(b)** $\mathbf{r} = \begin{bmatrix} 3 \\ -1 \\ 4 \end{bmatrix} + \mu \begin{bmatrix} 6 \\ -3 \\ 2 \end{bmatrix}$; **(c)** $\cos^{-1}\left(\dfrac{17}{7\sqrt{14}}\right) \approx 49.5°$; **(d)** $(4\frac{1}{2}, 3\frac{1}{2}, 4)$.

Index

S 666 ADM A

Your best preparation for the new 2004 specifications...

Advancing Maths for AQA, 2nd editions.

- Written by a team of practising Senior Examiners to provide everything students need to succeed.

- The student-friendly, accessible approach eases transition from GCSE to AS, then up to A Level, in line with the new specification.

- Flexible support includes clear explanations, detailed worked examples and self-assessment tests to prepare students thoroughly.

- Includes favourite features such as test-yourself sections and key point summaries at the end of each chapter.

- Greater coverage for the combined AQA A and AQA B 2004 specification.

Statistics 1 (S1)
0 435 51338 9

Mechanics 1 (M1)
0 435 51336 2

Further Pure Mathematics 1 (FP1)
0 435 51334 6

Simply contact our Customer Services Department for more details:

(t) 01865 888068 (f) 01865 314029 (e) orders@heinemann.co.uk (w) www.heinemann.co.uk

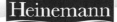

Heinemann
Inspiring generations